Testicular Cancer and Other Tumors of the Genitourinary Tract

Ettore Majorana International Science Series
Series Editor
Antonino Zichichi
European Physical Society,
Geneva, Switzerland

(LIFE SCIENCES)

Recent volumes in the series

A Continuation Order Plan is available for this series. A continuation order will bring delivery of
each new volume immediately upon publication. Volumes are billed only upon actual shipment.
For further information please contact the publisher.

Testicular Cancer and Other Tumors of the Genitourinary Tract

Edited by

M. Pavone-Macaluso
University Polyclinic Hospital
Palermo, Italy

P. H. Smith
St. James's University Hospital
Leeds, England

and

M. A. Bagshaw
Stanford University
Stanford, California

Plenum Press • New York and London

Library of Congress Cataloging in Publication Data

International School of Urology and Nephrology. Course (5th: 1983: Erice, Italy)
  Testicular cancer and other tumors of the genitourinary tract.

  (Ettore Majorana international science series. Life sciences; 18)
  "Proceedings of the Fifth Course of the International School of Urology and Nephrology, held July 16–24, 1983, at the Ettore Majorana Center for Scientific Culture, Erice, Sicily, Italy"—T.p. verso.
  Includes bibliographies and index.
  1. Testis—Cancer—Congresses. 2. Urinary organs—Cancer—Congresses. I. Pavone-Macaluso, Michele. II. Smith, P. H. (Philip Henry) III. Bagshaw, Malcolm A., 1925–   . IV. Title. V. Series. [DNLM: 1. Testicular Neoplasms—congresses. 2. Urogenital Neoplasms—congresses. W1 ET712M v.18 / WJ 160 I61 1983t]
  RC280.T4I58 1983   616.99′463   84-26493

ISBN-13: 978-1-4612-9490-0   e-ISBN-13: 978-1-4613-2453-9
DOI: 10.1007/978-1-4613-2453-9

Proceedings of the Fifth Course of the International School of Urology and Nephrology, held July 16–24, 1983, at the Ettore Majorana Center for Scientific Culture, Erice, Sicily, Italy

©1985 Plenum Press, New York
Softcover reprint of the hardcover 1st edition 1985

A Division of Plenum Publishing Corporation
233 Spring Street, New York, N.Y. 10013

PREFACE

The proceedings of this volume represent a record of a meeting
in the Ettore Majorana center in Erice, Sicily from 16-24 July, 1983.
This was the fifth course of the International School of Urology and
Nephrology, the fourth in a series of meetings devoted to different
aspects of Urological Oncology. Speakers and delegates came from
nearly every part of the world and contributed to a detailed analysis
of the current state of knowledge of tumors of the testicle and to
the development of knowledge of cancer of the bladder, prostate and
kidney.

Previous volumes resulting from this series of meetings have
been well received and the Editors hope that the reader may find the
record of this meeting equally attractive.

We are indebted to the Department of Medical Illustration, St.
James's University Hospital, for their kindness in producing several
of the illustrations in this book.

The editors are very grateful to Dr. Cavallo of Palermo for the
major part which he has played in preparing the index.

                              M. Pavone-Macaluso
                              P.H. Smith
                              M.A. Bagshaw

CONTENTS

TREATMENT

GENERAL CONSIDERATIONS

SURGERY

RADIOTHERAPY

CHEMOTHERAPY AND IMMUNOTHERAPY

FERTILITY IN PATIENTS WITH TESTIS TUMORS

HOW TO TREAT TESTIS TUMORS

II ADVANCES IN RENAL CANCER

IV   ADVANCES IN PROSTATE CANCER

V   TUMORS OF OTHER ORGANS OF THE GENITO-URINARY APPARATUS

CONTENTS

PRE-TREATMENT CLASSIFICATION OF TUMORS OF THE TESTIS

G. Pizzocaro, A. Milani, and M. Pasi

National Tuma Instituti
Milan
Italy

ABSTRACT

Pre-treatment classification of tumors (clinical staging) primarily depends on the natural history of the disease, and should be a guide to the proper application of actual therapeutic modalities. Today there is a need to reclassify testicular tumors as follows: stage I (no evidence of metastases); small volume stage II (subdiaphragmatic lymph node metastases not larger than 5 cm); and advanced disease. Advanced disease can usually be subdivided into small volume advanced disease (lymph node metastases no larger than 5 cm and-or pulmonary metastases no larger than 2 cm and low levels of serum tumor markers) and bulky advanced disease (larger metastases, or AFP > 1,000 ng/ml and beta-HCG > 10,000 mUI/ml).

The special circumstances in the subdivision of advanced disease are: persistent elevation of serum tumor markers only with no clinical evidence of metastases (occult disease); very bulky disease (lymph node metastases > 10 cm or lung metastases > 5 cm); or extrapulmonary metastases (liver, brain, bone, skin, etc.)

INTRODUCTION

Pre-treatment classification of cancer subdivides patients into homogeneous groups according to the extent of the disease in order to plan appropriate treatment. The natural history of the tumor and actual treatment possibilities usually are the major guidelines for this classification.

Testicular tumors usually metastatize to the paraaortic nodes and to the lung. Metastases in other organ sites (e.g. liver, brain, bones, and skin) are rare, and usually occur late in the course of the disease[1]. Inguinal and pelvic nodes may be involved primarily in patients with cryptorchidism or following inguino-scrotal surgery. From a practical point of view, both retroperitoneal and inguinal nodes can be considered together in the management of regional nodes in these tumors.

Mediastinal and supraclavicular lymph node metastases are not very common and they usually occur in patients with bulky retro-peritoneal disease. Lung metastases are usually rare and occur late in pure seminoma, but are very common in non seminomatous tumors. Approximately 10% of these patients with negative retroperitoneal lymph nodes and 50% of those with resected positive lymph nodes develop lung metastases and they are present in the great majority of patients with advanced disease.

In recent years great progress has been made in the management of germ cell testicular tumors which allows the following statements to be made: 1) orchidectomy alone is curative in approximately 80% of patients without clinical evidence of metastatic disease, 2) retro-peritoneal metastases from pure seminoma which are not larger than 5 cm can be cured with radiotherapy in almost all cases, and 3) in approximately two thirds of patients with non-seminomatous tumors cis-platin combination chemotherapy can cure approximately 60-70% of patients with advanced disease, with much better results in patients with small volume metastases than in those with bulky disease[4-6]. Furthermore, very high titers of serum beta-HcG and AFP signal the same prognosis as bulky disease[7].

CURRENT PRE-TREATMENT STAGING SYSTEMS

Usually, testicular tumors are divided into 3 stages:

Stage I   -   no evidence of metastases
Stage II   -   retroperitoneal lymph node metastases only
Stage III -   mediastinal, supraclavicular, or distant
                metastases.

This staging system, while very simple, is inadequate, in actual clinical practice, essentially because of the great variability in stage II disease, which must be further subdivided into small volume and bulky metastatic disease.

The TNM classification of malignant tumors (Table 1) regarding only the paraaortic nodes as regional presents the difficulty of considering the pelvic nodes as well as mediastinal and supraclavicu-lar nodes as juxtaregional. Furthermore, the differences in the

Table 1.   TNM Pre-Treatment Clinical Classification

T   —   Primary Tumor

In the absence of orchidectomy the symbol TX must be used.
T0 — No evidence of primary tumor
T1 — Tumor limited to the body of the testis
T2 — Tumor invading beyond the tunica albuginea
T3 — Tumor invading the rete testis or epididymis
T4 — Tumor infiltrating the spermatic cord and/or the scrotal wall
T4a — infiltrating the spermatic cord
T4b — infiltrating the scrotal wall
TX — The maximum requirements to assess the primary tumor can not be met

N   —   Regional and Juxta-regional Lymph Nodes

N0 — No evidence of regional lymph node involvement
NI — Evidence of involvement of a single homalateral regional lymph node, which if inguinal is mobile
N2 — Evidence of involvement of contralateral or bilateral or multiple regional lymph nodes, which if inguinal are mobile
N3 — Palpable abdominal mass present or there is evidence of involvement of fixed inguinal lymph nodes
N4 — Evidence of involvement of juxta-regional lymph nodes
NX — The minimum requirements to assess the regional and/ or juxta-regional lymph nodes can not be met

M   —   Distant Metastases

M0 — No evidence of distant metastases
MI — Distant metastases present
MX — The minimum requirements to assess the presence of distant metastases can not be met

---

extent of distant metastatic disease is not taken account of in the single Ml category.

The need for classifying patients with advanced disease became urgent with improving chemotherapy, and Samuels proposed a very clear classification for these patients (Table 2). This system is widely used, even though it is somewhat out of date. However it is problematical in that mediastinal lymph node metastases, as well as "mediastinal mass" should be considered advanced pulmonary disease, and intra-venous pyelography is no longer the most reliable method for assessing the extent of retroperitoneal metastases. A CT-scan is essential today.

Table 2. Samuels' Classification For Advanced Disease[8]

---

1) <u>Minimal pulmonary disease</u>: no more than five metastases per each
   lung field, with maximum diameter less than 2.0 cm;
2) <u>Advanced pulmonary disease</u>: any mediastinal mass, hilar mass,
   pleural effusion, or intrapulmonary mass greater than 2.0 cm;
3) <u>Minimal abdominal disease</u>: minimal pulmonary disease plus
   positive lymphangiogram with no ureteral displacement;
4) <u>Advanced abdominal disease</u>: palpable mass, enlarged liver with
   metastases, ureteral displacement with or without obstructive
   uropathy, inferior vena cava distortion from metastatic nodes
   and, rarely, inguinal adenopathy.

More recently, Peckham[9] proposed a very detailed classification (Table 3), which allows classification of most clinical
situations which are usually found in testicular tumors. However, in
our experience it does not accommodate many patients with very large
metastases[8]. Therefore we suggest adding to Peckham's staging
classification, stage IID, for patients with retroperitoneal lymph
node metastases larger than 10 cm, and the suffix L4 for patients
with lung metastases larger than 5 cm.

From a practical point of view, it is essential to divide
patients with testicular tumors into three broad categories:

- <u>Stage I</u>: no evidence of metastases,
- <u>Small volume stage II</u>: retroperitoneal metastases $\leq$ 5 cm;
- <u>Advanced disease</u>: bulky stage II and stage III.

In stage I it is essential to assign T categories (Table 1).
Small volume stage II includes Peckham's clinical Substages IIA and
IIB. Following lymphadenectomy for non seminomatous tumors in these
patients, clinical substages IIA and IIB are converted into pathological stages IIA, II B and IIC (Table 4). It is important to point
out that approximately 20% of clinical stage IIA and IIB non seminomatous germ cell testicular tumors are converted into pathologic
stage I and approximately 30% are converted into pathologic stage
IIC[6]. Pathologic stage IIC is true advanced disease, and it is a
clear indication for post-operative adjuvant chemotherapy.

Patients with advanced disease are usually sub divided into two
broad categories:

- Small volume advanced disease: no more than 5 lung metastases
  with none > 2 cm; with or without lymph node metastases $\leq$ 5 cm.
- Bulky advanced disease: lymph node metastases > 5 cm, and/or
  more than 5 lung metastases or any lung deposit > 2 cm.
- Patients with beta-HCG > 50,000 mUI/ml and/or AFP > 1,000 ng/ml
  are allocated to the bulky group, regardless of disease extent.

Table 3. Peckham's Staging Classification[9]

  I. Lymphogram negative, no evidence of metastases.

 II. Lymphogram positive, metastases confined to abdominal nodes,
      3 subgroups recognised:

           A - maximum diameter of metastases > 2 cm;
           B - maximum diameter of metastases 2-5 cm;
           C - maximum diameter of metastases > 5 cm;

III. Involvement of supradiaphragmatic and infradiaphragmatic lymph
      nodes. No extralymphatic metastases:

      Status of abdominal nodes:

           A
           B   as for Stage II
           C

 IV. Extralymphatic metastases suffixes:

           O - lymphogram negative
           A
           B   as for Stage II
           C

      Lung Status:

           L1- ≤3 metastases present
           L2- multiple metastases ≤2 cm maximum diameter
           L3- multiple metastases

      Liver Status

           H+- liver involvement

     As in clinical practice we met patients with very different
clinical situations which were not easily included within the small
volume or the bulky disease group, we proposed a more detailed staging system for patients with disseminated disease (which includes, in
addition, categories for occult disease, very bulky disease and
extrapulmonary disease (Table 5)[10].

     In conclusion, pre-treatment staging of testicular tumor should
now describe clinical (or pathological) situations in sufficient
detail to provide clear guidelines for planning the optimal therapy
of these potentially curable cancers.

Table 4.   New Classification[6,10] of Pathologic Small Volume Stage
           II Subsets in NSGCT of Testis Following Radical RPLND
           (pN1, pN2 Categories)

IIA: Micro or macroscopic intranodal metastases (no more than
           5 retroperitoneal LN) none larger than 2 cm;
IIB: Metastases to more than 5 retroperitoneal LN; or any
           metastasis between 2 and 5 cm;
           or microscopic capsular invasion.
IIC: Tumor invasion into retroperitoneal veins; or any
           macroscopic extranodal extension;
           or at least one retroperitoneal metastasis larger than
           5 cm.

NSGCT = Non seminomatous germ cell tumors
RPLND = Retroperitoneal lymph node dissection
LN   = Lymph nodes

Table 5.   New Subdivisions of Tumor Bulk in Advanced Disease (GCT of
           Testis)[10]. (N3, N4, M1 Categories)

Tumor bulk	Definition
Occult disease	Tumor markers AFP and/or b-HCG persistently elevated, with no additional demonstrable disease.
Small volume advanced disease	No more than 5 pulmonary metastases in each lung, none larger than 2 cm, with or without LN metastases smaller than 5 cm.
Bulky advanced disease	LN metastases $\geq$ 5 cm and/or more than 5 pulmonary metastases in each lung, or at least one pulmonary metastasis between 2 and 5 cm.
Very bulky disease	LN metastases $\geq$ 10 cm and/or pulmonary metastases occupying more than 50% of lung fields, or at least one pulmonary metastasis $\geq$ 5 cm.
Extrapulmonary disease	Any metastasis outside LNs and lungs (specify).

Acknowledgement

Supported in part by C.N.R. Rome, special project "control of Neoplastic Growth", grant no. 83.00915.96.

REFERENCES

1. F. K. Mostofi, and E. B. Price Jr., Tumor of the male genital system, in: "Atlas of Tumor Pathology," 2nd series, fascicle 8 p1-175, AFIP, Washington DC (1973).
2. R. T. D. Oliver, G. Read, W. G. Jones, C. J. H. Williams, and M. J. Peucham, Justification for a policy of surveillance in the management of Stage I Testicular Teratoma, in: "Controlled Clinical Trials in Urologic Oncology," p. 73-78, L. Denis, G.P. Murphy, G.R. Prout and F. Schroder, eds., Ravon Press, New York (1984).
3. D. Ball, A. Barrett, and M. J. Peckham, The Management of metastatic seminoma testis, Cancer, 50:2289-2294 (1982).
4. L. H. Einhorn, and S. D. Williams, Is adjuvant chemotherapy necessary to ensure high cure rates in non seminomatous cancer? Internat.Cong.Adjuv.Therapy of Cancer, 17 (1981) (Abst).
5. G. Pizzocaro, The case for radical surgery and combined therapy in testicular non seminoma, in: "Germ Cell Tumors" p.315-328, C. K. Anderson, W. H. Jones, and A. Milford Ward, eds., Taylor & Francis, London (1981).
6. G. Pizzocaro, F. Zanoni, and A. Milani, Retroperitoneal lymphadenectomy and aggressive chemotherapy in non bulky stage II non seminomatous germ cell tumors of testis, Cancer, in press.
7. A. Milford Ward, Markers in germ cell tumors: the current state of the art, AFP, b-HCG and apparent half like kinetics, in: "Germ Cell Tumors", p. 207-215, C. K. Anderson, W. G. Jones, and A. Milford Ward, eds., Francis & Taylor, London (1981).
8. M. L. Samuels, D. F. Johnson, and P. Y. Holoye, Continuous intravenous bleomycin therapy with vinblastine in stage III testicular neoplasia, Cancer Chemother.Rep., 59:563-570 (1975).
9. M. J. Peckham, A. Barrett, T. J. McElwain, and W. F. Hendry, Combined management of malignant teratoma of the testis, Lancet, 2:267-270 (1979).
10. G. Pizzocaro, F. Zanoni, and L. Piva, Bulk of the disease and management of metastatic non seminomatous germ cell tumors of testis, in: "Controlled Clinical Trials in Urologic Oncology," L. Denis, G.P. Murphy, G.R. Prout and F. Schroder, eds., Raven Press, New York (1984).

HISTOPATHOLOGICAL CLASSIFICATION OF TESTICULAR TUMORS

F. K. Mostofi and Isabell A. Sesterhenn

Department of Genitourinary Pathology
Armed Forces Institute of Pathology
Washington, DC 20306, USA

In this presentation we plan to discuss the contributions of pathology to the management of patients with germ cell tumors of testis. However, before doing this let me briefly mention tumor markers, since we plan to discuss their presence in tumor cells.

Tumor Markers

The following markers have been studied: Alpha Fetoprotein (AFP), Human Chorionic Gonadotropin (HCG), Placental Lactogen (HPL), Pregnancy Specific Antigen (SP1), Placental Alkaline Phosphatase (PLAP), Lactic Dehydrogenase (LDH), Carcinoembryonic Antigen (CEA), and Gamma Glutamyl Transpeptidase (GGT). Of these, the most widely utilized are HCG and AFP.

The clinical application of these markers will be discussed by others but let me say that after initial enthusiastic claims about the value of the two major markers (AFP and HCG) it has now been demonstrated that normal values may be found in the presence of a germ cell tumor either in the primary site or in the metastasis, even when these are bulky. May we also say that a number of confusing reports about histological localization and source of the markers have appeared and we hope to clarify these. We have demonstrated that although two tumors may be histologically identical, one may show many positive cells while in the other the reaction may be negative.

Our comments on histological demonstration of the markers in tissue are based on studies that have been conducted in our laboratory covering about 1,000 tumors.

Pathologic Classification

A decade ago there were one or more American, British, French, Scandinavian and Russian classifications. The reasons for multiplicity of the classification were mainly ignorance of early placentation and embryogenesis and failure to take into account the potentialities of the germ cell. The effect has been that it has been impossible to compare the results of therapy and survival and mortality rates.

In 1973, in a heroic attempt to standardize pathological classification of these tumors, the World Health Organisation (WHO) convened a panel of Investigators on the Histological Classification of Testes Tumors* bringing together the adversaries. The Panel considered the existing classification and adopted a slight modification of a classification that had been proposed earlier by Mostofi[1]. The classification was field tested and finally published in 1977[2]. It is of interest to note that although the classification was proposed long before the introduction of tumor markers it correlates well with markers both clinically and pathologically, and it provides the clinician with specific information about the structure of the tumors to guide him in the type of treatment best suited to that specific histological type. It is the only classification that provides adequate explanation for the unusual behavior of these neoplasms. Although Dr. Pugh was an active member of the Panel, he and his associates[3] have persisted in advocating their own classification which cannot be correlated with tumor markers. Per se this would seem to be unimportant but the Collins-Pugh[4] and Pugh-Cameron[3] categories have been inaccurately equated to the WHO categories so that comparison of results of therapy of testicular tumors in Great Britain with those of other nations is difficult.

The Mostofi-WHO classifications[4] recognized seven basic histological types which could occur either in pure form or in admixtures. Thus tumors are classified as those of single histologic type and those of more than one histologic type.

TUMOR OF ONE HISTOLOGICAL TYPE

Seminoma (S)

The tumor has been defined as a neoplasm of fairly uniform cells typically with clear cytoplasm and well defined cell borders (Figure 1).

---

*Drs. F. Cabanne, K. Ganina, Chr. Hedinger, V. McGovern, F. K. Mostofi, R. C. B. Pugh, M. Rapaport, R. E. Scully, S. F. Serov, G. Teilum, E. D. Williams, and L. H. Sobin.

The cytoplasm contains varying amounts of glycogen. The nuclei are round or oval and vesicular, usually with one distinct nucleolus. Mitotic figures and giant cells are rare.

The cells form a mosaic or cobblestone arrangement forming lobules, separated from each other by the delicate fibrovascular stroma containing varying amounts of lymphocytic infiltration and granulomatous reaction with Langerhans' cells. In about 20% these are marked. Occasionally, the stroma is scirrhous. The cells may also occur in columns or rows and rarely have an alveolar pattern. No papillary or acinar structures are seen. The adjacent seminiferous tubules may also contain tumor cells either identifiable as seminoma cells or as malignant undifferentiated germ cells. Extension of the tumor to the rete testis may simulate carcinoma of rete testis. While Follicle Stimulating Hormone (FSH) may be elevated in seminoma, neither HCG nor AFP are elevated in pure seminoma. In 200 seminomas not a single cell was positive for HCG, HPL, SP1, or AFP. Cells that were positive for HCG or AFP could be demonstrated to be syncytiotrophoblasts or yolk sac elements, respectively. Therefore, elevation of HCG or AFP in patients with seminoma is indicative of a neoplasm that is not a pure seminoma.

Seminoma tends to occur in patients in their 30's and 40's. Grossly, the cut surface shows a bulging grayish-white lobulated rather homogeneous glistening tissue. The consistency is usually soft but it may be hard. Areas of hemorrhage and necrosis are absent except in large tumors. The tumor is usually demarcated from testicular tissue.

Anaplastic seminoma (AS) has been variously defined as a tumor that shows marked variation in size and shape of cells and nuclei, one which shows high mitotic activity, one with elevated HCG or AFP, or one which has metastasized. Since higher death rates have been reported in tumors with increased mitotic activity[5,6] we would like to propose that the term of "with high mitotic index" be substituted for anaplastic seminoma. This category is defined as a seminoma with 30 mitotic figures in 10 consecutive high power fields throughout the tumor. In other words, taking any field the subsequent nine fields should show an average of 3 mitoses per high power field.

Spermatocytic Seminoma (SS)

The tumor is defined as one composed of cells which vary in size from lymphocyte-like to giant cells of about 100 μm. Most of the cells are intermediate sized with eosinophilic cytoplasm and round or oval nuclei (Figure 2). These and the larger giant cells have a distinct nuclear chromatin distribution, spireme or filamentous, similar to that seen in spermatogonia. The small cells simulate lymphocytes except that they have considerable amounts of cytoplasm.

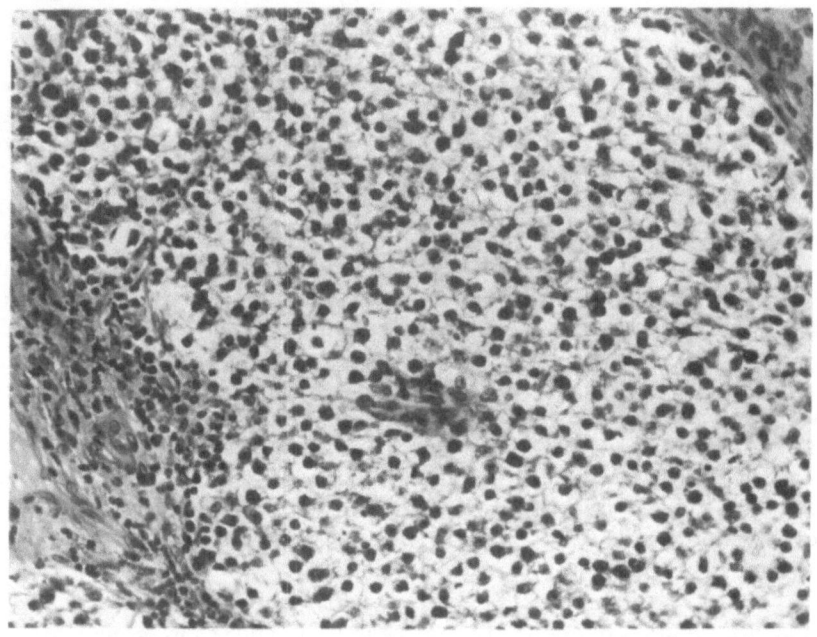

Fig. 1.   Seminoma.   AFIP Neg. 83-8645, H&E X100.

The stroma is scant and edematous, lymphocytic infiltration and
granulomatous reaction are absent.   Both AFP and HCG are absent.   SS
usually occurs in pure form but we have seen several instances of
association with sarcomas.   SS is seen only in the testis usually in
older patients.   Grossly, the tumor tends to be large, yellowish and
soft-brain-like and more mucoid than seminoma.   Spongy and cystic
areas with ragged edges are present, filled with blood tinged fluid.

Embryonal Carcinoma (ECa)

The tumor is defined as one composed of cells of primitive
epithelial appearance often with clear cytoplasm and indistinct cell
borders growing in a variety of patterns: acinar, tubular, papillary
and solid (Figure 3).   The cells are larger and more epithelial with
more abundant and denser cytoplasm.

The nuclei are vesicular and mitotic figures are frequent.   The
solid pattern may simulate S but the cells lack the distinct cell
borders and the lobular pattern.   Lymphocytic and granulomatous
reactions are usually absent.   The stroma may be primitive.

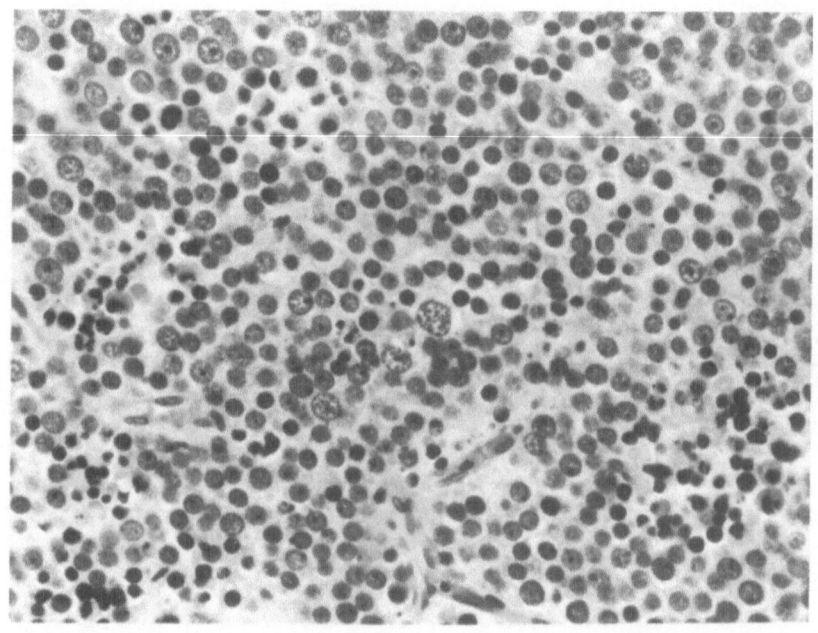

Fig. 2.   Spermatocytic seminoma.   AFIP Neg. 83-8745, H&E X100.

Elevated levels of AFP in sera and urine of patients with embryonal carcinoma and positive reaction for AFP in tumor cells have been
reported.   In 356 embryonal carcinomas, either pure or in tumors of
more than one cell type, only 47 were AFP positive (13.3%).   AFP was
positive in only 2 of 24 pure embryonal carcinomas and only in scattered cells.   When embryonal carcinoma and yolk sac tumor occurred
together small clusters of AFP positive ECa cells were seen[7].   HPL
was present in a few tumors both in the primary and in the metastasis
and this was usually in many cells but HCG and SP1 were negative in
pure ECa.

Grossly, ECa is usually the smallest of germ cell tumors.   Cut
surfaces show a variegated appearance with grayish-white, granular or
smooth bulging soft tissue containing extensive areas of hemorrhage
and necrosis.   The tumor is not encapsulated.

In their various publications Pugh et al.[3,4] have equated
embryonal carcinoma with malignant teratoma undifferentiated (MTU).
This tumor which includes both the old MTA and MTIB contains not only
ECa but some ECa and teratoma, yolk sac tumor in adults, ECa and yolk
sac tumors (since these are not recognized in mixtures), immature
teratomas and teratomas with malignant transformation.

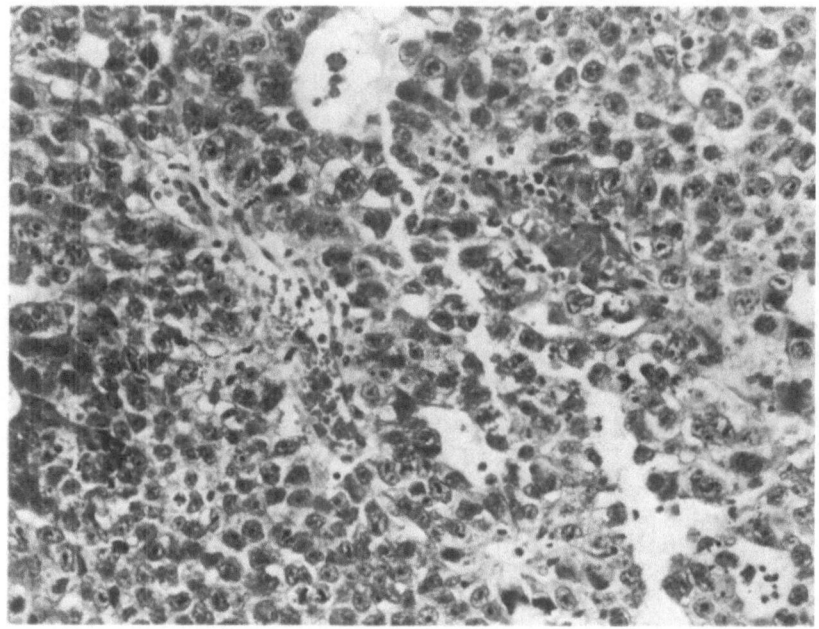

Fig. 3.   Embryonal carcinoma.   AFIP Neg. 83-8643, H&E X100.

Yolk Sac Tumor (YST) (Embryonal Carcinoma, Infantile Type, Endodermal Sinus Tumor

This neoplasm is defined as a tumor characterized by small cells of primitive appearance growing typically in a loose vacuolated network (Figure 4). The cells vary from flat endothelial cells to cuboidal and low columnar. They are small, the cytoplasm is often vacuolated but may be eosinophilic, the nuclei are small, often round and vacuolated. The cells grow in a reticular pattern, and form anastomosing tubular, acinar and papillary structures. At times the cells form solid sheets and this has been confused with seminoma and embryonal carcinoma. The cells show varying amounts of PAS and lipid positive material and there are often hyaline droplets.

Grossly, the testis is invariably enlarged. The cut surface is homogeneous, grayish-pink and lobulated. It may exude mucoid and lipid substance. Small areas of hemorrhage and necrosis may be encountered.

YST is the most common germ cell tumor of testis of infancy and it occurs rarely in pure form in adults, but its frequency in combination with other germ cell tumors of adults has escaped attention.

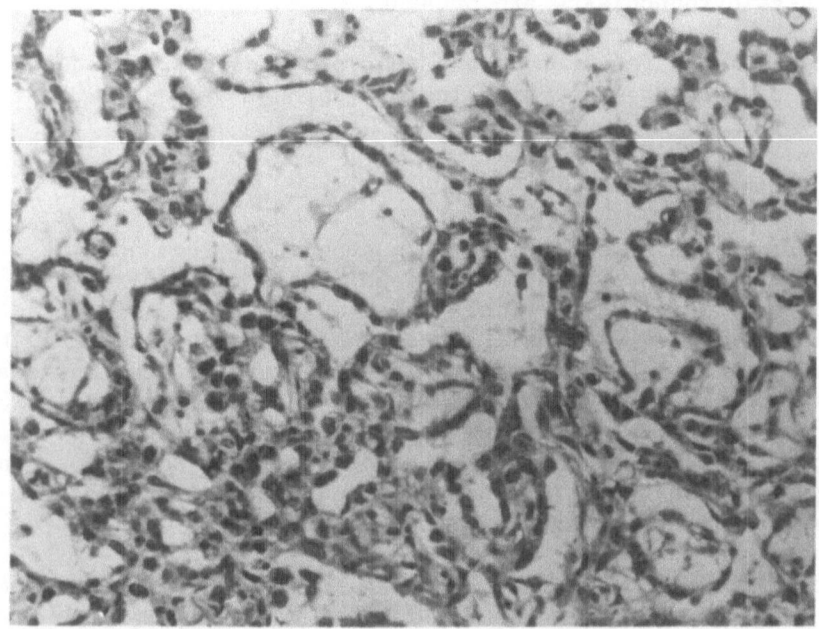

Fig. 4.   Yolk sac tumor.   AFIP Neg. 83-8755, H&E X100.

Talerman reported that 44.4% of germ cell tumors he had studied contained YST elements[8]. We observed these elements in 41% of about 1,000 testicular tumors.

YST is usually associated with elevated AFP and AFP can be demonstrated in all cell types seen in this tumor. In 19 of 20 pure YST in infants AFP was present in the tumor cells. AFP was positive in 92% of our testicular tumors where YST was seen in association with other tumor types. The reaction was focal.

Initially designated as orchioblastoma by Collins and Pugh[4] YST was accepted by Pugh and Cameron[3]. However, the term is limited to tumors seen in infants and children. When the neoplasm occurs in adults in pure form it is classified as malignant teratoma, undifferentiated and it is not recognized when it occurs in association with other types.

Polyembryoma

This is defined as a tumor which is composed predominantly of embryoid bodies (Figure 5). These structures present a variety of appearances, but in their most recognizable form they consist of a disc and one or two cavities surrounded by loose mesenchyme. Tubular structures suggesting endoderm, cysts reminiscent of amnion and yolk sac and syncytiotrophoblasts are present.

The tumor is extremely rare but embryoid bodies are frequent in many tumors of more than one histologic type.

Choriocarcinoma (CCa)

This is defined as a highly malignant neoplasm composed of cytotrophoblasts and syncytiotrophoblasts. This term is reserved for a rare category of germ cell tumors whose existence was first reported by Mostofi[1]. Clinically these tumors are characterized by metastatic symptoms: haemoptysis, hematemesis and central nervous symptoms without any complaints relative to the testis. The patients had gynecomastia and, where known, very high levels of chorionic gonadotropins. Microscopically, the tumor consisted solely of syncytiotrophoblasts and cytotrophoblasts (Figure 6). Grossly, the tumors were usually small and hemorrhagic.

The pure form is quite rare. We saw it in 18 of 6,000 testicular germ cell tumors and it has the highest mortality. Since then

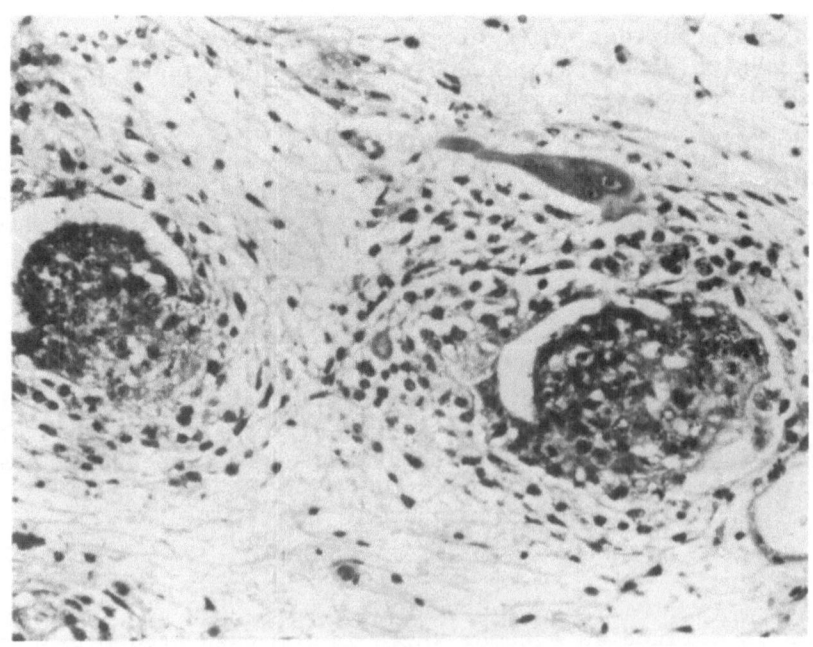

Fig. 5.   Embryoid bodies and syncytiotrophoblast. AFIP Neg. 83-9716, H&E X100.

we have seen a patient in whom the initial symptoms were pain and tenderness in the testis.

In discussing this group mention should be made that while choriocarcinoma is rare, syncytiotrophoblasts alone are frequently encountered in association with other germ cell tumors. The cells may be intermingled with other histological types or in the fibrous stromal tissue. Syncytiotrophoblasts are seen in 42.5% of testicular tumors. A third group are those in which foci of CCa are seen in association with other germ cell tumors. This will be discussed later.

Teratoma

The Mostofi and WHO classifications have defined teratoma as a tumor typically composed of several types of tissue representing different germinal layers (endoderm, mesoderm, ectoderm). This is in sharp contrast to teratoma as defined by Collins and Pugh[4], and Pugh and Cameron[3] who include all non seminomatous tumors in teratoma. WHO Teratoma is subdivided into mature, immature and teratoma with malignant transformation. It includes multiple tissues repre-

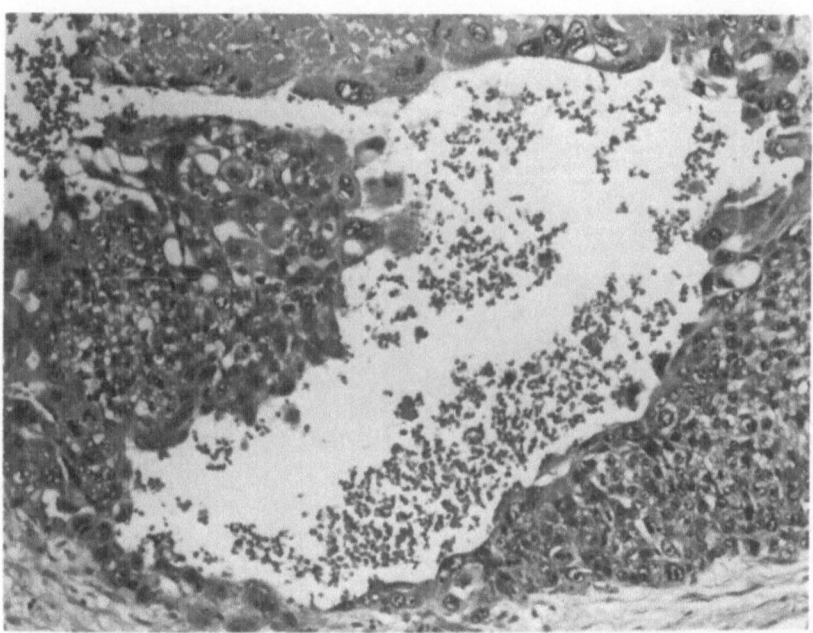

Fig. 6.   Choriocarcinoma. AFIP Neg. 83-8752, H&E X70.

senting one germinal layer, e.g., skin and brain in a testicular
tumor.   If a single type of differentiated tissue, e.g., cartilage,
mucous gland, squamous cyst is associated with seminoma, ECa or YST
that tissue is considered to be teratomatous.   Dermoid cysts, although regarded as teratoma, would be classified as dermoids while
epidermal cysts are classified as tumor-like lesions and not as
teratoma.

     Grossly, teratomas are characterized by the presence of solid
and cystic areas.

     a) Mature teratoma is defined as a teratoma consisting exclus-
ively of well differentiated tissue which may be arranged in an
organoid fashion.   In its simplest form mature teratoma contains
cysts lined by squamous epithelium, mucous glands, cartilage, bone
and smooth muscle.   More complex tumors contain abortive gastrointestinal tract, brain, eye, pancreas, prostate, salivary gland, etc.
(Figure 7).

     We have demonstrated that mature teratoma can invade vascular
spaces.   The clinical course in the adult is unpredictable while in
infants the prognosis is very favorable.

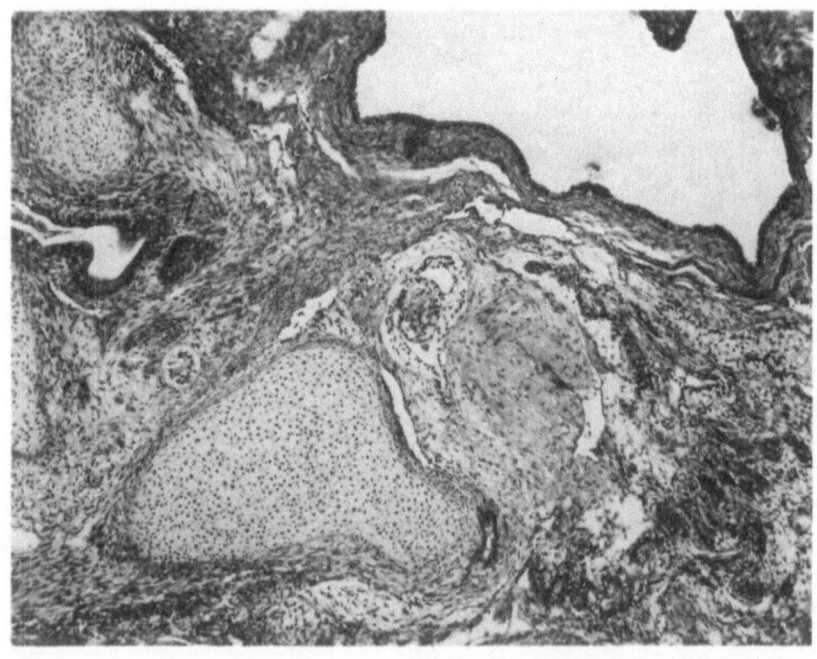

Fig. 7.   Mature teratoma. AFIP Neg. 83-9717, H&E X100.

The WHO mature teratoma has been equated with differentiated teratoma[3]. However, since several of the 24 differentiated teratomas of Pugh-Cameron were composed almost entirely of cysts lined by squamous epithelium[3], mature teratoma and differentiated teratoma cannot be equated with each other.

b) Immature teratoma is defined as a teratoma in which there are incompletely differentiated tissues (Figure 8). Neuroectodermal tissues, and primitive gut are the most common. We have seen vascular invasion by immature teratoma. There are almost always some mature elements in association with immature teratoma but since very little is known about the natural history of immature teratoma in testis, separating them from mature teratoma seemed appropriate. It is of interest to note that immature teratomas are rare in infants.

The mucous glands of mature and immature teratoma and the clusters of eosinophilic cells resembling liver may give a positive reaction for AFP. We have seen positive AFP reactions in 19.2% of our teratomas.

c) Teratoma with malignant transformation is defined as a rare form of teratoma containing a malignant component of a type typically encountered in other organs and tissues. Sarcoma, squamous cell

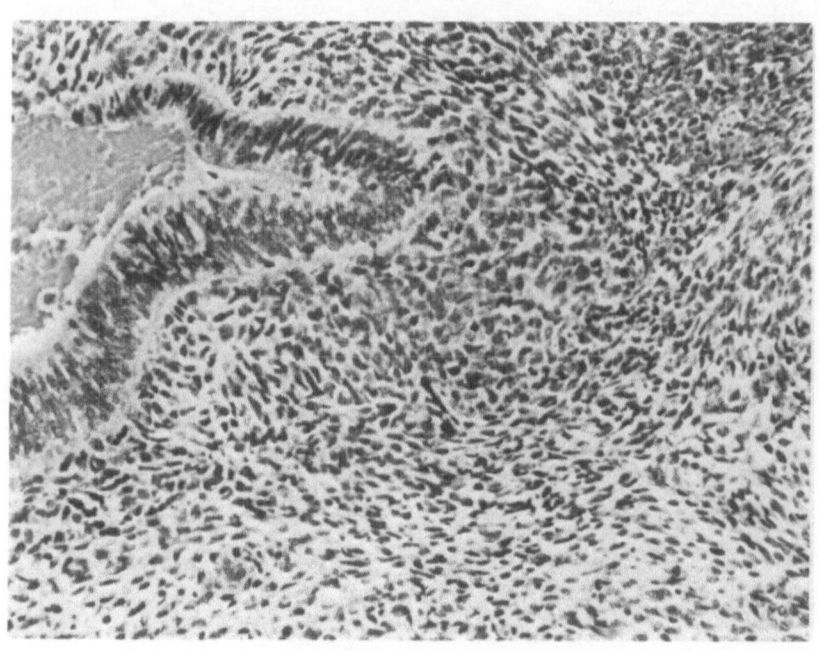

Fig. 8.   Immature teratoma.   AFIP Neg. 83-9718, H&E X100.

carcinoma and adenocarcinoma are its most common forms. Carcinoid in
a teratoma is included here but if pure it is diagnosed as carcinoid.

TUMORS OF MORE THAN ONE HISTOLOGIC TYPE

     This category is defined as those in which two or more of the
seven histological types are found. Application of tumor markers has
led to the realization that over 60% of testicular germ cell tumors
fall in this category. The most frequent combinations are ECa, YST,
T and CCa and/or syncytiotrophoblast constituting 14.3% of all germ
cell tumors (Figure 9); seminoma and syncytiotophoblasts are seen in
8.1% (Figure 10): ECa, YST, T and S, and syncytiotrophoblasts and/or
choriocarcinoma in 7.4%; and ECa and T constitute 1.4%; this last
category is the one to which teratocarcinoma may be applied.

     Obviously the classification requires that each histological
type should be listed and the relative proportions of each mentioned.
Included in this category is one group that has been a source of much
confusion.

     HCG is elevated in a number of patients with seminoma. Such
cases have been reported as HCG producing pure S.

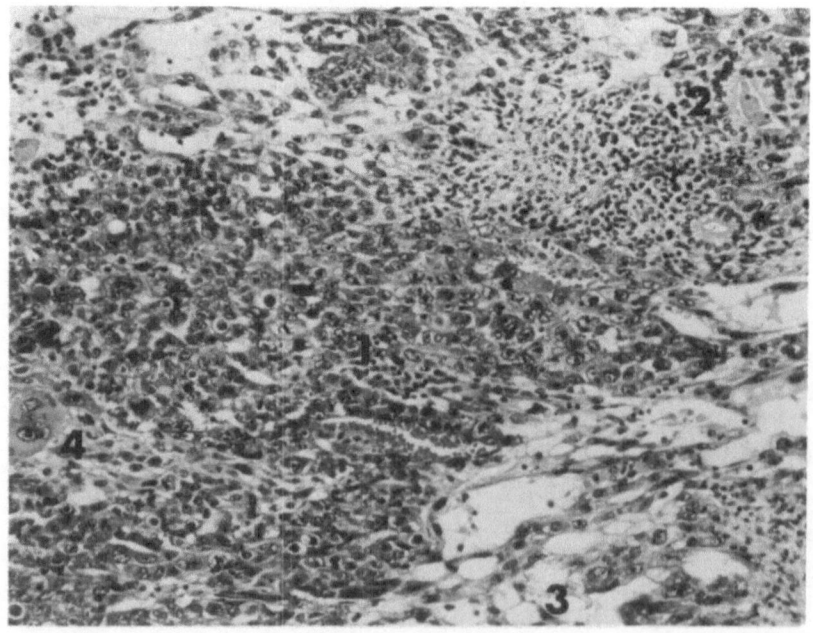

Fig. 9.   Embryonal carcinoma (1) teratoma (2) yolk sac tumor (3)
          syncytiotrophoblasts (4). AFIP Neg. 83-8640, H&E X70.

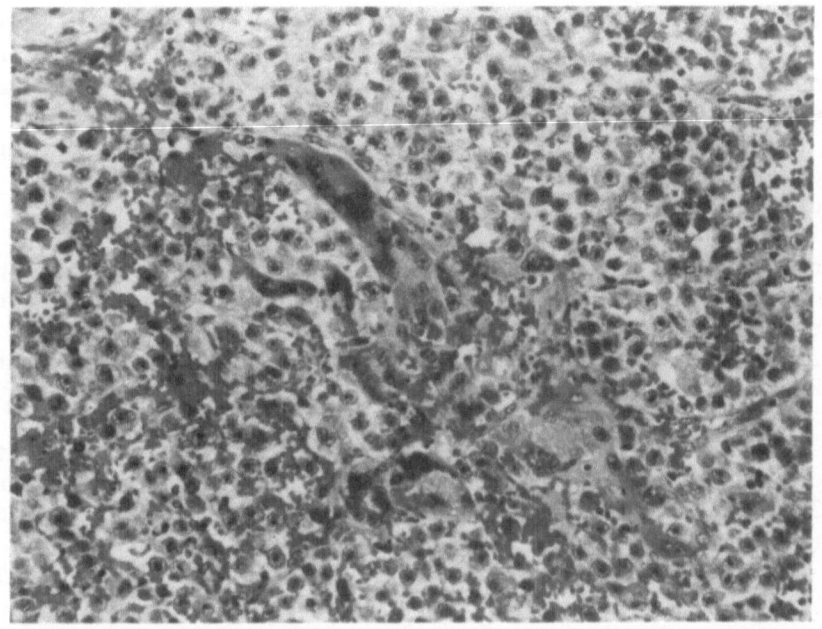

Fig. 10. Seminoma and syncytiotrophoblast. AFIP Neg. 83-8650,
          H&E X100.

        We believe this is erroneous. Failure to find syncytiotropho-
blasts is due to four reasons:

        Inadequate sampling. We have seen the 11th section loaded with
syncytiotrophoblasts whereas the first ten did not have any.

        The second reason is failure to use immunopathologic technics.
We found that if study was limited to H&E stains the cells could be
identified in only 7%, while with HCG stains the incidence was 24%.

        The third source of misconception has been that the diagnosis of
syncytiotrophoblasts has been restricted to multinucleated giant
cells. It must be remembered that in its evolution syncytiotrophoblast begins as mononuclear cell, it develops into cuboidal cells
with eosinophilic cytoplasm, multinucleated cells with eosinophilic
cytoplasm, to vacuolated cells, to cells with large lacunae and
spindle shaped and endothelial-like cells.

        Fourthly, there may be metastases which contain syncytiotropho-
blasts or CCa.

Comments relative to S are equally applicable to ECa with elevated HCG.

Elevation of AFP in S has also resulted in serious misinterpretation. Sometimes the tumor is, in fact, solid YST which has been misdiagnosed as S. Sometimes YST elements have been overlooked in S, either because of inadequate sampling, failure to use immunocytochemical technics or ignore the variable forms of yolk sac elements. In a study of many pure S or seminoma occurring with other cell types, not a single S cell has contained AFP.

In these tumors of more than one histological type either one or both tumor markers may be elevated depending on whether there are syncytiotrophoblasts and/or choriocarcinoma or YST elements. If neither is present the markers are negative. If both are present, both markers will be present.

These tumors of more than one histological type can be equated with almost all malignant teratomas and combined tumors of Pugh and Cameron[3]. They could be malignant teratoma undifferentiated (if they contain YST), malignant teratoma intermediate (if they had teratoma and YST), malignant teratoma trophoblastic if they had T, ECa, YST, SCT and CCa.

COMPARISON OF THE WHO, AND PUGH AND CAMERON CLASSIFICATIONS

Throughout the presentation we have demonstrated that except for S, SS, and YST in infants and children, the categories cannot be equated with each other simply because the former classification is based on histology and the latter on a now abandoned theory of origin of non seminomatous tumors.

Attempts by Pugh and Cameron[3] to equate the 2 classifications are regrettably not acceptable. Malignant teratoma undifferentiated is not identical to ECa because the former contains not only ECa but ECa and T, immature teratoma and YST in adults and mixtures of these: malignant teratoma trophoblastic is not identical to CCa as it includes both pure choriocarcinoma and those with mixtures. Differentiated teratoma is not comparable to mature teratoma as the former includes epidermal cysts.

TISSUE DEMONSTRATION OF TUMOR MARKERS

In each category of germ cell tumors of testis we have called attention to the presence or absence and type of marker that may be demonstrated in the tissue. AFP is present predominantly in YST and rarely in ECa and T. We have not seen elevated AFP levels in pure ECa and/or T. HCG is seen only in syncytiotrophoblasts whether alone

or in CCa. HPL and SP1 are found only in HCG producing syncytiotrophoblasts. Some cells that are histologically typical syncytiotrophoblasts do not give positive reaction with any of the three
markers. HPL is occasionally seen in ECa. Thus a positive AFP
indicates either pure YST or YST with other types. A positive HCG
indicates syncytiotrophoblasts either alone or as CCa.

     In 36.6% both markers are positive; in 32.6% both are negative.
One or the other is present in 30.8%.

     In Collins and Pugh[4] and Pugh and Cameron classifications[3]
HCG may be present or absent in seminoma, differentiated teratoma,
malignant teratoma intermediate, malignant teratoma undifferentiated
and combined tumors. In malignant teratoma, trophoblastic, both HCG
and AFP may be positive or only HCG may be positive. The reason for
this lack of correlation is that the classification does not recognize syncytiotrophoblasts when they occur alone and YST elements when
they occur in adults. In contrast, as we have demonstrated, the WHO
classification correlates remarkably well with markers.

HISTOGENESIS OF THE TUMORS

     Friedman and Moore[10], Dixon and Moore[11], and Pierce and
Abell[12] have postulated that germ cells may go into S or into
totipotent cells - ECa. Teratoma represents somatic differentiation;
YST, and CCa, trophoblastic differentiation. This is the most widely
accepted theory.

     Collins and Pugh favored the theory that S arose from germ cell
while non S tumors arose from embryonic totipotent cells that had
escaped the influence of organizers[3]. In a later publication, Pugh
and Cameron[4] have indicated that they reserved opinion on the
histogenesis of these tumors.

     In a number of reports Rhagavan and his associates[13,14,15]
have proposed that seminoma, being of germ cell origin, is the pre-
curser for a series of stages of further differentiation forming a
continuum that can result in various forms of teratomatous and extra
embryonic type of tumors.

     The authors have reported that transplantation and passage of
seminoma through mice yielded YST. It is not known precisely what
the authors transplanted. It is quite probable that whatever they
transplanted contained YST elements. The authors have also claimed
that ultrastructurally, S-like YST was identical to YST. This is to
be expected because it is the solid YST which has been misdiagnosed
as S and not a true S. It is regrettable that the authors have not
recognized the solid type of YST. The cells of this tumor are smaller than S cells, they lack the lobular arrangement and lymphogranul-

omatous reactions of S, they invariably have tubular structures and
the cells contain varying amounts of AFP. Their illustrations in
various publications fulfill these criteria for YST.

The authors have also claimed that since a number of S contained
syncytiotrophoblasts, CCa can also arise from S. The coexistence of
2 or 3 different cell types in a testicular tumor does not necessarily prove that one cell type arose from the other. Also, regrettably, the authors did not know that syncytiotrophoblasts can be
mononuclear.

Certain observations have led the senior author to suggest a
different relationship. In infants we see YST and teratoma but no S
or ECa. Most YST in adults are associated with T and ECa. Syncytiotrophoblasts and YST elements are a frequent accompaniment of other
testicular tumors. We have seen S and CCa in the same tumor, and in
addition to malignant intratubular germ cells we have demonstrated
intratubular S, SS, ECa, YST and syncytiotrophoblasts[16]. These
facts have led us to conclude that each of these cell types is capable of developing directly from malignant transformation of intratubular germ cells (Figure 11).

Azzopardi et al.[17] reported undifferentiated malignant intratubular germ cells in testis of patients with burned out testicular
tumors. Skakkebaek[18] has written extensively about these cells and
referred to them as carcinoma in situ, a term that we believe should
be reserved for intratubular ECa and YST. We have also demonstrated
that these undifferentiated malignant cells can invade the testicular
parenchyma precisely as the S, SS, ECa, YST, and CCa do. They can
also incite immunological response in testicular parenchyma.

These observations provide strong evidence for the hypothesis of
origin of each cell type directly from malignant transformation of
the intratubular germ cell.

METASTASES IN GERM CELL TUMORS

We have discussed these in detail[9,16]. Suffice it to say that
whether of single histologic type or of more than one histologic type
metastases usually reflect the histology of the primary. There is,
however, frequent discrepancy between the reported histology of the
primary and that of the metastases. We believe that this is entirely
because the various histological types have not been recognized in
tumors of more than one histologic type.

Dixon and Moore[11] reported that 50% of their tumors had T with
or without ECa, and 8% of ECa showed either pure T or T with ECa.
Other reports have confirmed these observations. More recently
several reports have claimed that chemo and/or radiation therapy have
resulted in transformation of ECa to T.

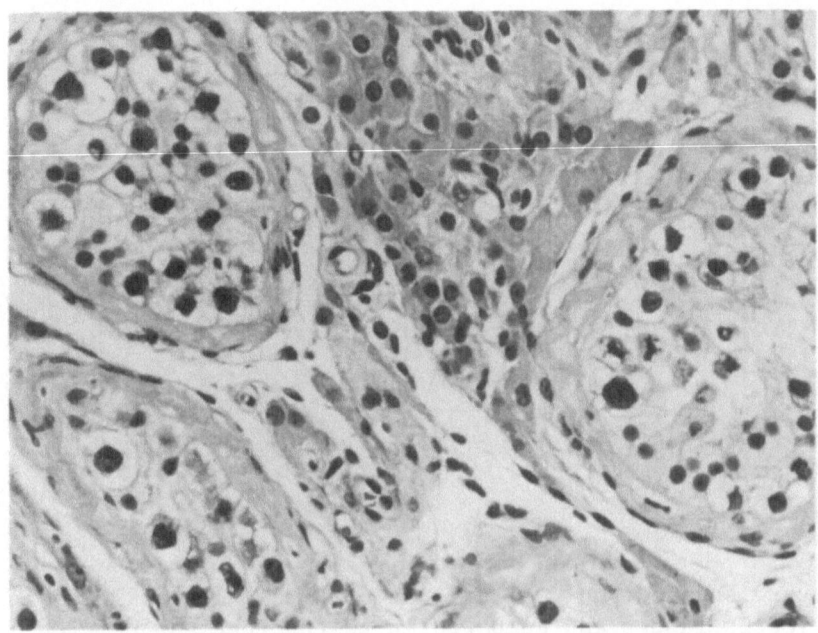

Fig. 11.   Intratubular malignant germ cell.   AFIP Neg. 83-8748,
           H&E X100.

We have demonstrated vascular invasion by mature teratoma[2] and
have observed several instances of invasion by immature teratoma.
Since malignant transformation of germ cells can lead directly to all
cell types but T, and since in mice Stevens[19] has shown origin of
teratoma from germ cell, we believe that the malignant germ cell may
develop into one or more histological types in the testis and the
same or different cell type in the metastases (Table 1).

Whether chemo or radiation therapy serve as organizers and
convert embryonal carcinoma in the metastasis to teratoma has not
been resolved. We believe that it is more likely that embryonal
carcinoma is destroyed by therapy while the teratoma, untouched by
radiation and/or chemotherapy, continues to grow. Yet, as mentioned
earlier, the fact that teratomas in infants often are organoid consisting of brain, gastrointestinal tissue, etc., and immature teratomas are rare in this age group, makes one wonder about the possibility that organizers could affect embryonal carcinoma causing a
transformation to teratoma.

Table 1.   Schematic Relationshio of Testicular Germ Cell Tumors
           (Mostofi)

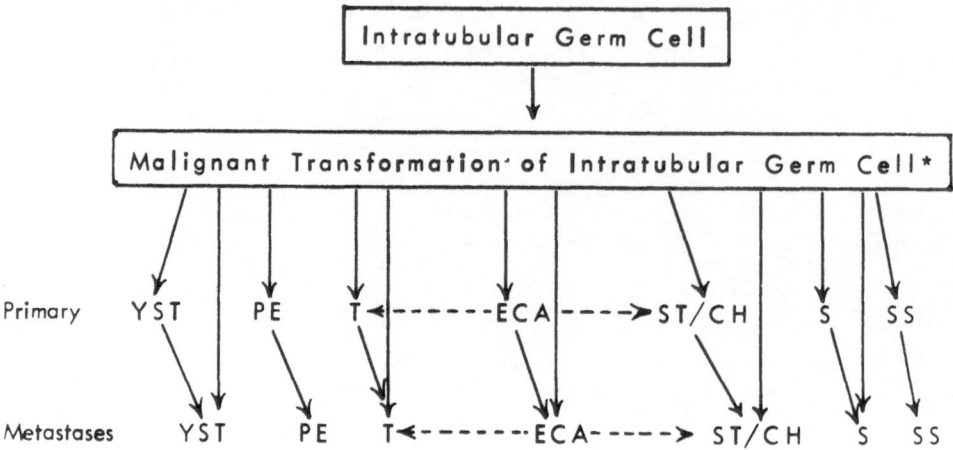

SUMMARY

We have discussed the WHO International Histological Classification of Testis Tumors and compared it to the Pugh and Cameron
classification. The WHO classification recognizes 7 histological
types which occur as a single histologic type in 38% and as more than
one histologic type in 62%. Seminoma, spermatocytic seminoma, embryonal carcinoma, yolk sac tumor, choriocarcinoma and teratoma are the
main tumor types. Polyembryoma rarely occurs in pure form – it is
usually in association with embryonal carcinoma.

Comparison with the Collins and Pugh, and Pugh and Cameron
classifications shows that except for seminoma, spermatocytic seminoma and yolk sac tumor in infants and children, the two classifications cannot be equated.

The WHO classification correlates much better with tumor markers
and with clinical behavior.

STATEMENT

The opinions or assertions contained herein are the private
views of the author and are not to be construed as official or as
reflecting the views of the Department of Defense.

REFERENCES

1. F. K. Mostofi, Testicular tumors: epidemiologic, etiologic and
        pathologic features, Cancer, 32:1186 (1973).

2. F. K. Mostofi and L. H. Sobin, "International Histological
    Classification of Tumors of Testis" (No.16) WHO, Geneva
    (1977).
3. R. C. B. Pugh and K. M. Cameron, Teratoma, in: "Pathology of the
    Testis", R. C. B. Pugh, ed., Melbourne, Blackwell Scientific
    Publ., Oxford, London, Edinburgh, p.199 (1976).
4. D. J. Collins and R. C. B. Pugh, "Pathology of Testicular
    Tumors", E. S. Livingstone, 2d., Edinburgh and London,
    (1964). F. J. Dixon and R. A. Moore, Tumors of Male Sex
    Organs (Fasc. 31b, 32), Atlas of Tumor Pathology, Washington
    D.C., Armed Forces Institute of Pathology (1952).
5. J. G. Maier and M. H. Sulak, Radiation therapy in malignant
    testis tumors, Part 1, Seminoma, Cancer, 32:1212 (1973).
6. A. C. Thackray and W. A. J. Crane, Seminoma, in: "Pathology of
    the Testis", R. C. B. Pugh, ed., Blackwell Scientific,
    Oxford, London, Edinburgh and Melbourne, p.164 (1976).
7. I. Sesterhen, "Demonstration of tumor markers in testicular germ
    cell tumors", (unpublished).
8. A. Talerman, Endodermal sinus (yolk sac tumor) elements in
    testicular germ cell tumors in adults: Comparison of prospec-
    tive and retrospective studies, Cancer, 46:1213 (1980).
9. F. K. Mostofi and E. B. Price, Tumors of the Male Genital System
    (Fasc 8), Atlas of Tumor Pathology, 2d Series, Washington
    D.C., Armed Forces Institute of Pathology.
10. N. B. Friedman and R. H. Moore, Tumors testis: A report of 922
    cases, Milit.Med., 99:573 (1957).
11. F. J. Dixon and R. H. Moore, Tumors of Male Sex Organs, (Fasc.
    31, 32), Atlas of Tumor Pathology, Armed Forces Institute of
    Pathology, Washington D.C. (1952).
12. G. B. J. Pierce and M. A. Abell, "Embryonal carcinoma of the
    Testis", Pathologic Annual, C. Sheldon, ed., Sommers Series,
    New York: Appleton-Century-Crofts (1970).
13. D. Raghavan, E. Heyderman, P. Monaghan, and J. Gibbs,
    Hypothesis: When is seminoma not a seminoma? J.Clin.Pathol.,
    v.34 (1981).
14. D. Raghavan and M. Neville, "Biology of Testicular Tumors",
    Scientific Foundation of Urology (2d ed), G. D. Chisholm and
    D. I. Williams, eds., Yearbook Med. Publisher, Chicago,
    (1982).
15. D. Raghavan, A. L. Sullivan, N. J. Peckham, and M. Neville,
    Elevated serum Alphafetoprotein and seminoma: Clinical evi-
    dence for a histologic continuum, Cancer, 50:982 (1982).
16. F. K. Mostofi, Pathology of germ cell tumors of testis. A
    Progress Report, Cancer, 45:1735 (1980).
17. J. G. Azzopardi, F. K. Mostofi, and E. A. Theiss, Lesions of
    testis observed in certain patients with widespread chorio-
    carcinoma and related tumors, Am.J.Pathol., 38:207 (1961).
18. N. E. Skakkebaek, Possible carcinoma in situ of the testis,
    Lancet, 2:516 (1972).
19. L. C. Stevens, Origin of testicular teratomas from primordial
    germ cells in mice, J.Nat.Cancer Inst., 38:549 (1967).

RARE TUMORS OF TESTES AND ADNEXAL TUMORS

F. K. Mostofi and Charles J. Davis Jr.

Department of Genitourinary Pathology
Armed Forces Institute of Pathology
Washington, DC 20306

In this presentation the following entities will be discussed.
These have been discussed in detail elsewhere[1,2,3].

    I   Sex cord/gonadal stromal tumors
    II   Tumors of collecting system
    III Malignant lymphoma
    IV   Secondary tumors
    V   Adnexal tumors
    VI   Tumor-like lesions

I SEX CORD/GONADAL TUMORS

These constitute about 6% of testicular tumors. They have been
classified by WHO as follows:

a)   Fully differentiated forms: Leydig cell, Sertoli cell and
      granulosa tumors
b)   Incompletely differentiated forms

The WHO classification of ovarian tumors[4] categorized the
group as sex cord/stromal tumors. The same designation was carried
over to testicular tumors even though there is no evidence of sex
cord in the development of testes.

Leydig Cell Tumors

These tumors occur in all ages and races. In children they are
invariably associated with macrogenitosomia, in adults 40% show
gynecomastia and other feminizing features.

Grossly, the tumors are uniformly mahogony brown, lobulated, more or less encapsulated. Histologically most frequently, the tumors are made up of hexagonal cells with ground glass eosinophilic or vacuolated cytoplasm and a vesicular nucleus with a delicate nucleolus (Fig. 1). Reinke's crystals are found in about 40%.

This is the most common histology, but the tumors may reflect the histologic spectrum of the evolution of Leydig cells from spindle shaped cells to large cells with or without vacuolated cytoplasm. Mitoses and giant cells and areas of necrosis are absent. Binu-cleated cells are fairly common. The supporting stroma is usually highly vascular presenting an endocrine type of vascularity, but it may be hyalinized or even calcified.

About 10% of Leydig cell tumors are malignant. The criteria for malignancy are: hemorrhage and necrosis, anaplasia, increased mitotic activity (Fig. 2), vascular invasion and/or local extension. Metastases, however, are usually delayed occurring 3 to 4 years post orchiectomy.

Bilateral Leydig cell tumors raise the suspicion of adreno-genital syndrome. Testicular involvement in this syndrome is mani-fested in early stages as hyperplasia and the distinction between hyperplasia and tumor may be quite difficult.

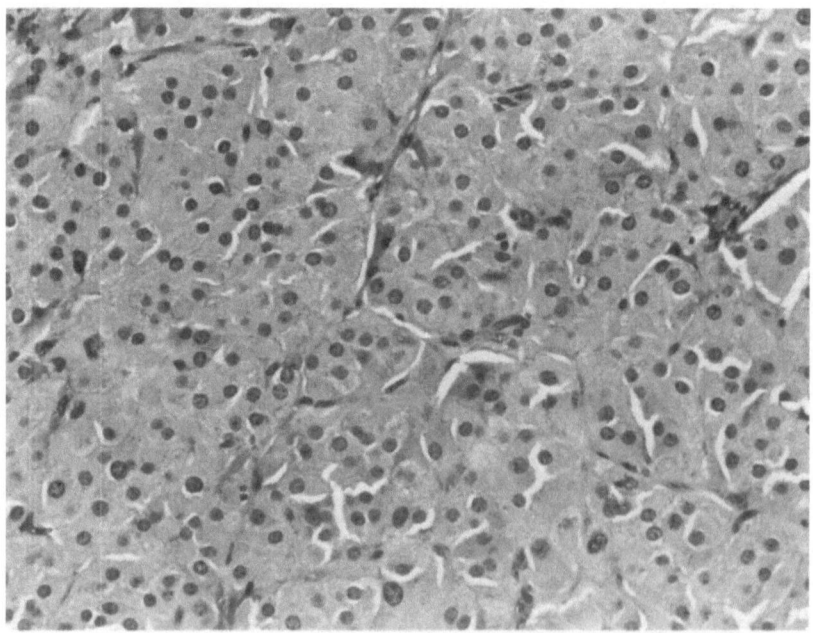

Fig. 1. Leydig cell tumor. AFIP Neg. 83-9719, H&E X100.

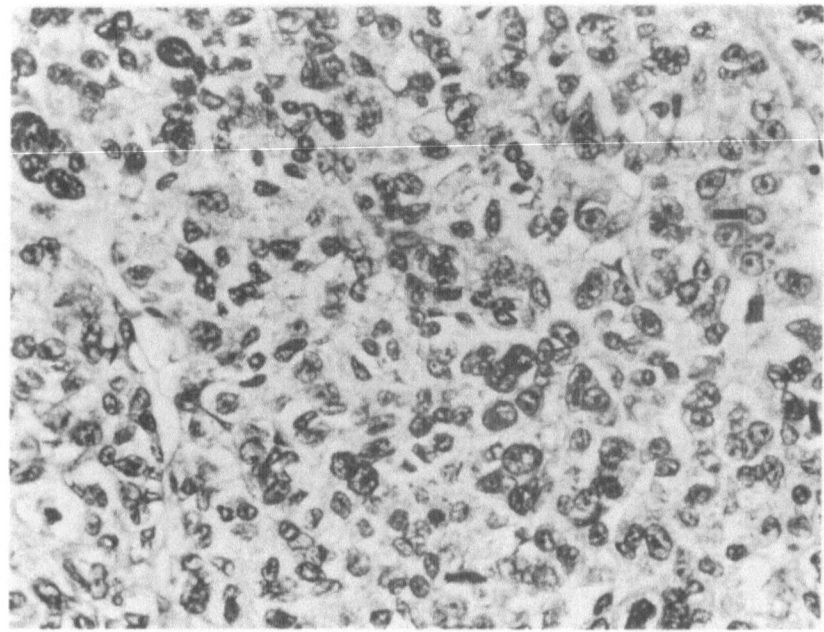

Fig. 2. Malignant Leydig showing cell pleomorphism and mitotic
figures (arrows). AFIP Neg. 83-9719, H&E X100.

Sertoli, Granulosa Cell, and Incompletely Differentiated Forms

These will be discussed together. These tumors occur in all
ages but predominantly in infants and children. In adults, about 40%
are associated with gynecomastia and other feminizing features. The
tumors present a homogeneously yellowish lobulated, usually well
circumscribed cut surface. Microscopically, the histology may be
that of Sertoli cells (Fig. 3), granulosa cell, theca cell or admixture of these with or without Leydig cells. The criteria for malignancy and the frequency of malignant behavior are similar to Leydig
cell tumors except that if the lesion is malignant, metastases appear
usually within a year.

Because these tumors often show more than one histological
pattern and one may see Leydig cell elements in these tumors, Mostofi
et al.,[3] postulated that these tumors were derived from primitive
specialized gonadal stromal cells which normally have the capability
of differentiating in the male to Leydig and Sertoli cells and in the
female to granulosa, theca and lutein cells. We may see many combinations of these elements in a single tumor; these could best be
designated as tumors of specialized gonadal stroma.

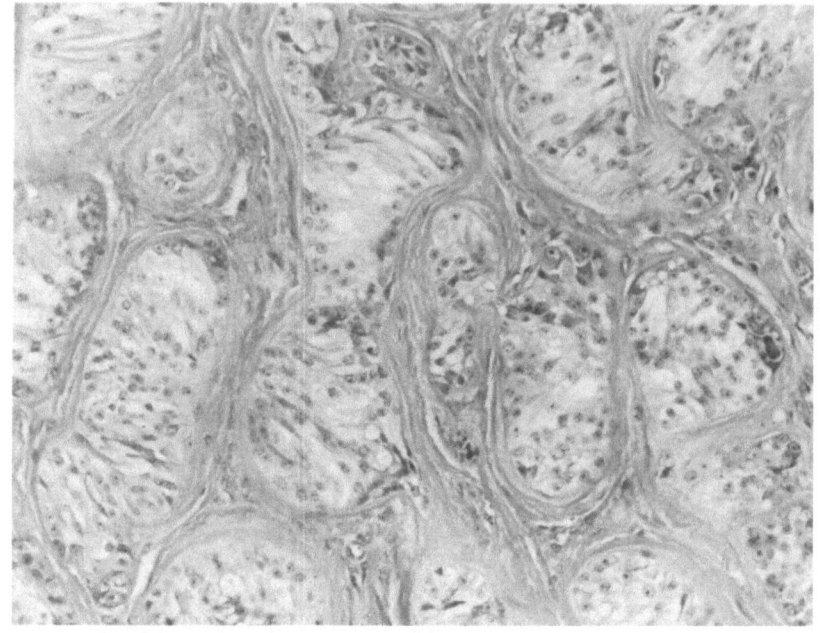

Fig. 3.   Sertoli cell tumor.   AFIP Neg. 83-9720, H&E X100.

II   TUMORS OF COLLECTING SYSTEM

Tubuli recti, the rete testes and the efferent tubules are lined by epithelium which varies from cuboidal to flattened cells to low columnar. Hyperplasia, adenoma and adenocarcinoma of the system are rare and difficult to distinguish from each other except in early stages.

III   LYMPHOID/HEMATOPOIETIC TUMORS

Enlargement of the testes may be the initial manifestation of malignant lymphoma or plasmacytoma. Characteristically the tumors are more frequently seen in patients in their 50's and 60's. The testicle is enlarged. It presents a grayish-pink cut surface. Areas of necrosis and hemorrhage are frequent. Histologically there is intense intertubular infiltration (Fig. 4). Various histologic types are encountered, the description of which is beyond the scope of this paper. In the majority of cases, systemic lymphoma is encountered within 2 years. Bilaterality is seen in almost 20%; cutaneous and central nervous system involvement are not infrequent. It may be mentioned that children whose systemic diseases have been controlled by chemotherapy may develop malignant lymphoma of the testes.

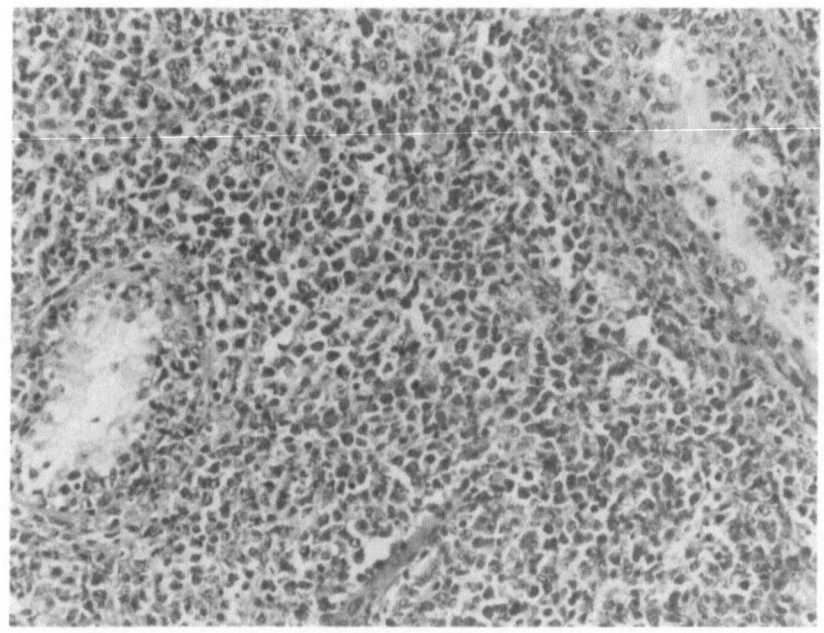

Fig. 4. Malignant lymphoma. AFIP Neg. 83-9721, H&E X100.

IV SECONDARY TUMORS

     Testicular enlargement may rarely be the initial or incidental
manifestation of secondary carcinoma. The most frequent primary
sites are lung, prostate, stomach, kidney, colon, pancreas, bladder
or rectum. The characteristic histologic picture is interstitial
infiltration with neoplastic cells that are not recognizable as
typical germ cell or gonadal stromal tumor (Fig. 5). There is extensive vascular and lymphatic involvement.

V TUMORS OF ADNEXAE

Benign

     Adenomatoid Tumors. These are the most common tumors of male
adnexae constituting 32% of paratesticular structures. They are
localized to epididymis, tunica and rarely spermatic cord. They are
usually small, less than 5 cm in diameter, oval or flattened disc-
like, firm to hard and well circumscribed. Although most are super-
ficially located, occasionally they may be embedded deeply in the
tissue. They cut with fibrous consistency. The cut surface has a

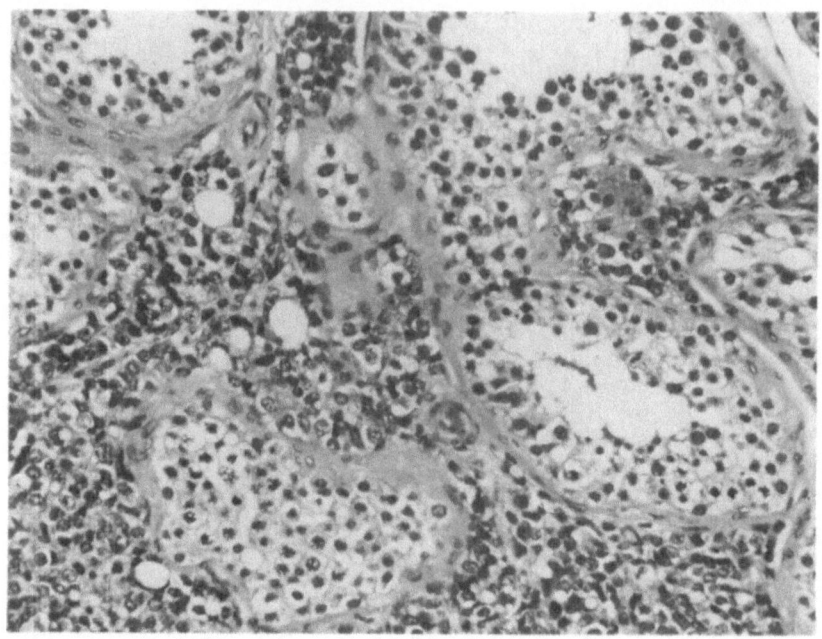

Fig. 5.   Secondary carcinoma of testis.   AFIP Neg. 83-9826, H&E X100.

grayish or white fibrous appearance. Histologically the tumor is
made up of 2 elements: epithelial and fibrous tissue. The epithelial elements may be arranged in cords or solid strands producing
a plexiform pattern, as tubuloglandular structures or as dilated
spaces resembling lymphangioma (Fig. 6). The cells may vary from
flat endothelial like to cuboidal or low columnar. Many of the cells
are vacuolated. The nuclei are round or oval, centrally placed and
mostly vesicular. Mitoses and giant cells are absent.

The stroma ranges from loose connective tissue to dense collagenized and even hyalinized tissue. Bands of smooth muscle cells
may be present. Almost invariably there are scattered or focal
lymphocytic infiltrates.

There has been considerable debate about the origin of the
cells, but it is now generally agreed that they are mesothelial in
origin; however, the lesion is not labeled mesothelioma as that
connotation is reserved for another entity.

Nodular Fibrous Periorchitis. As the name indicates, this
lesion is located on the tunica and may consist, at one extreme, of
localized thickening of the tunica and at the other end of single or

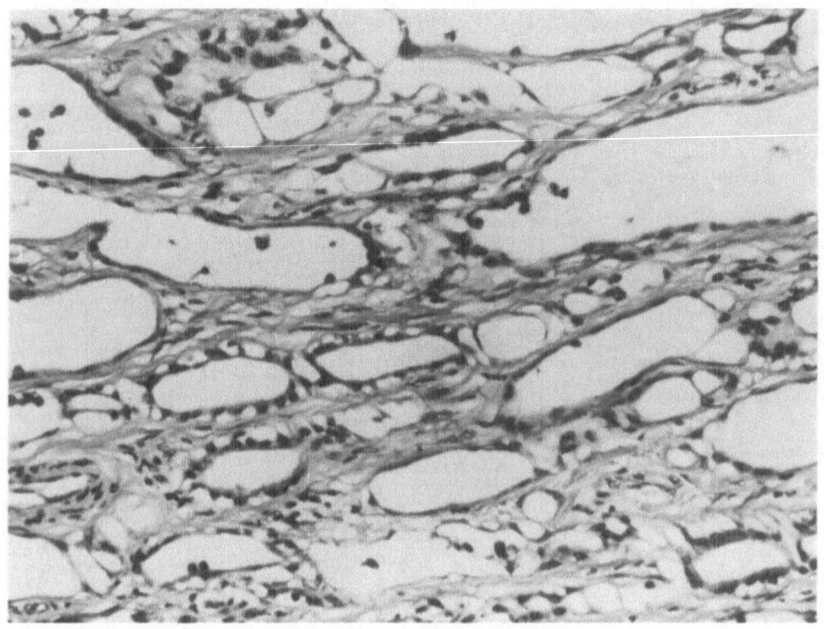

Fig. 6.   Adenomatoid tumor.   AFIP Neg. 83-9722, H&E X100.

multiple large or small nodules on the tunica.   The cut surfaces
reveal grayish-white fibrillar tissue.   Histologically the lesion
consists of bands of hyalinized fibrous tissue with scattered lymphocytic infiltration.

Papillary Cystadenoma.   This lesion which is often present as
the epididymal manifestation of von Hippel-Lindau's syndrome consists
of cuboidal epithelial cells with clear cytoplasm and vesicular
nuclei forming glandular and papillary structures (Fig. 7).   When
first seen, the lesion may be mistaken for metastatic renal cell
carcinoma.

Vasitis Nodosa.   Following surgical ligation or trauma to the
cord, there may be leakage of sperm with formation of sperm granuloma
and reactive proliferation of epithelial cells.   Small tubular,
sometimes dilated structures, are seen simulating metastatic adenocarcinoma.   The true nature of the structures can be detected by the
presence of spermatozoa in some of the tubules.

Malignant Tumors

Rhabdomyosarcoma.   This is the most frequent malignant tumor of
spermatic cord in the pediatric age group, most often in infants and

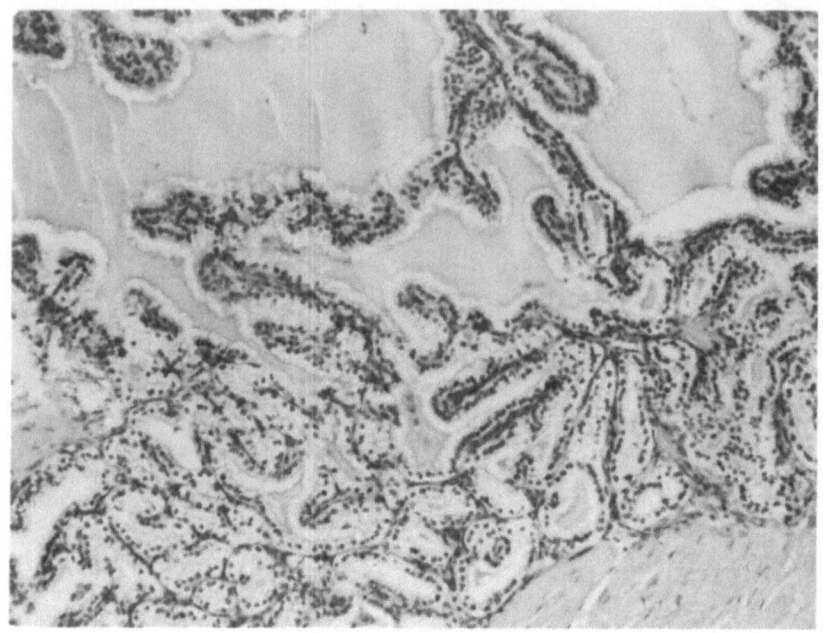

Fig. 7. Papillary cystadenoma. AFIP Neg. 83-9723, H&E X70.

young children. A second peak is observed at the age of 19. It may
occur in blacks. The tumors may range from 1 to 20 cm in diameter.
They are circumscribed but not encapsulated. They are firm with a
grayish-tan cut surface.

Microscopically, the whole range of rhabdomyosarcomas described
in other locations is encountered here. The histology has been
described in detail elsewhere[1]. It may range from undifferentiated
cells to myxomatous elements, spindle cells, pleomorphic cells with
bizarre tadpole shaped cells, uni or bipolar cells and multinucleated
giant cells (Fig. 8). Cross striations may rarely be found.

Liposarcoma, malignant fibrous histiocytoma, leiomyosarcoma,
fibrosarcoma, and undifferentiated sarcomas are rarely encountered,
usually in older patients.

Mesothelial lesions. Reactive proliferation of mesothelial
cells covering the tunica is often seen in association with hydrocele
or hematocele. Occasionally, a free floating nodule may be found in
the scrotal sac representing tissue culture like growth of mesothelial cells. The mesothelioma has a complex structure. The cellular elements may be epithelial-like spindled or mixed (Fig. 9). Many

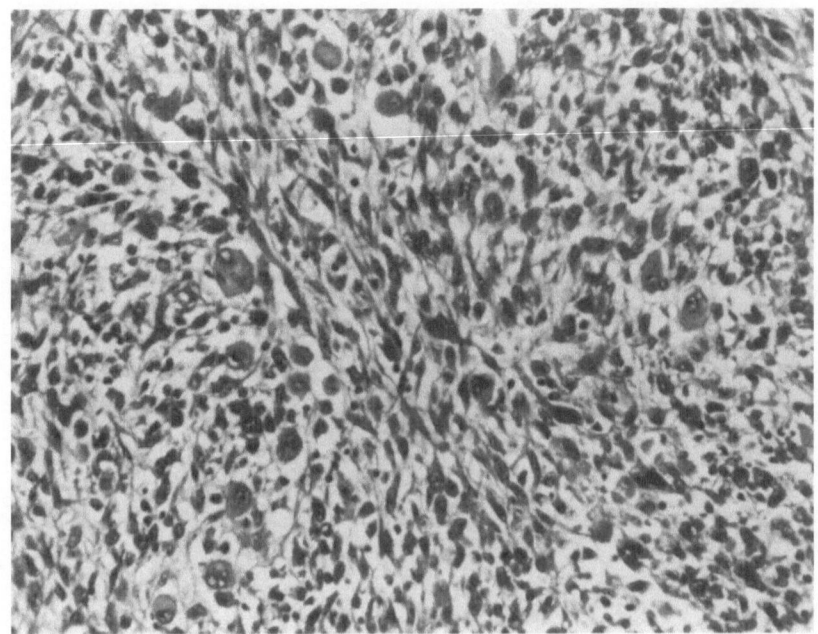

Fig. 8.   Rhabdomyosarcoma.   AFIP Neg. 83-9724, H&E X100.

malignant mesotheliomas are biphasic. The predominantly epithelial
type tends to be diffuse and spread over the tunica. The fibrous
variant resembles fibrosarcoma but is more variegated. It shows
fusiform cells generally larger than fibroblasts and arranged in an
irregular fashion. Highly cellular fields often alternate with more
collagenized areas. The fibrous type tends to grow as solid masses.

     Demonstration of origin from tunical mesothelial cells is essen-
tial for a diagnosis of mesothelioma and pleomorphism of cell population and invasion of tunica are indicative of malignancy. However,
before this diagnosis is made, a secondary tumor must be ruled out.

VI TUMOR-LIKE LESIONS

     Two categories may be mentioned:

Epidermal Cysts

     Epidermal cysts consist of cysts lined by keratinizing strati-
fied epithelium without skin appendages. The cyst is filled with
keratohyaline matter. Rupture of the cyst may occur with extrusion
of contents and secondary inflammation and giant cell reaction.

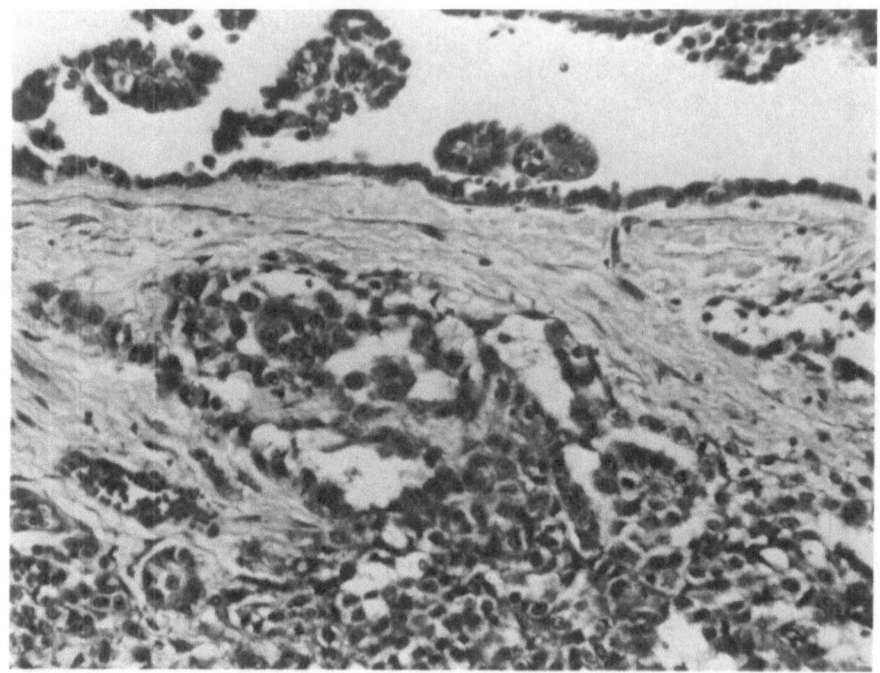

Fig. 9. Malignant mesothelioma. AFIP Neg. 83-9725, H&E X100.

Granulomatous Orchitis

Chronic inflammation of known or unknown etiology may result in enlargement and hardness of the testis and simulate neoplasia. Two categories are generally recognized: those in which the initial and primary manifestation is in the seminiferous tubules with secondary involvement of the interstitium and those in which the primary involvement is in the interstitium with secondary involvement of seminiferous tubules. It is the first type which may be confused with a neoplasm. There is destruction of germ cell elements and replacement by an admixture of cells - large round cells with eosinophilic or vacuolated cytoplasm and vesicular nuclei. Multinucleated giant cells and epithelioid cells are frequent.

STATEMENT

The opinion or assertions contained herein are the private views of the author and are not to be construed as official or as reflecting the views of the Department of the Army or the Department of Defence.

REFERENCES

1. F. K. Mostofi and E. B. Price, Jr., Tumors of male genital system (Fasc 8), in: "Atlas of Tumor Pathology," 2nd Series, Washington, DC., Armed Forces Institute of Pathology (1973).
2. F. K. Mostofi and L. H. Sobin, "International Histological Typing of Tumors of Testis," Geneva, World Health Organization (1977).
3. F. K. Mostofi, E. A. Theiss, and D. J. B. Ashley, Tumors of specialized gonadal stroma in human male patients: Androblastoma, Sertoli cell tumor, granulosa-theca cell tumor of testis, Cancer, 12:944-957 (1959).
4. S. F. Serov and R. E. Scully, Histological typing of ovarian tumors, Geneva, World Health Organization (1973).

TUMORS OF THE TESTIS -

AETIOLOGY, EPIDEMIOLOGY AND ANIMAL MODELS

William G. Jones

University Department of Radiotherapy
Leeds
Great Britain

Testicular tumors are rare, accounting for approximately 1% of all cancer in males in those countries with the highest incidence rates. The majority, approximately 90% are of germ cell origin[1,2], while the remainder are lymphomas or 'others'. The latter two categories will be ignored in this chapter because of their rarity.

EPIDEMIOLOGY

Epidemiology is the study of the behavior of disease in groups rather than in individuals. In the past it has had to rely on mortality statistics. The setting up of population-based cancer registries in many regions and countries over the last two decades has improved the situation. Information about incidence and mortality as well as the separation of data by both anatomical site and by histology are other significant advances[3]. Figure 1 illustrates the fact that testicular tumors are increasing in frequency in most of the Cancer Registries covered by Volumes II and III of "Cancer Incidence in Five Continents" [4,5], since there are more points above the line of equality than below. This increasing incidence, particularly in those under 45 years of age is now well documented in many different geographical areas, in those with low as well as high incidence rates, including the English county in which the author practices[6]. Some important studies are those of Davies in England and Wales[7], Clemmesen in Denmark[8,9], Shottenfield and Warshauer in the U.S.A.[10] and Lee, Hitosugi and Petersen in Japan[11]. The incidence has been increasing since at least the early 1940's and probably since the turn of the century[7]. According to Davies[7], one boy in every 500 in England and Wales can now expect to develop testicular cancer before the age of 50. Testicular cancer is now the commonest tumor in men 25 to 34 years of age, in England and Wales, whereas it was rare in 1911-15. These increases cannot be due to better diagnostic accuracy alone.

Figure 1 shows a considerable scatter of points in terms of incidence (rate/100,000), which vary not only with geographical situation but also within racial/ethnic groups in those areas[12]. Table 1 shows that testicular cancer is rare in Africa and Asia and that the rate for U.S. Blacks is much lower than for U.S. Whites. There is also considerable variation among the European countries, with a low rate for Finland but the highest rate of all for Denmark. There is some evidence of an excess incidence in urban over rural areas in some countries. Clemmesen[8] reported that Copenhagen had twice the rate of rural Denmark, a feature similar to that reported by Pedersen and Magnus for Norway in the years 1953-54 [13]. These facts suggest that environmental factors and/or possible genetic/ racial factors might be at play in the aetiology of these conditions. However Lipworth and Dayan, 1969 analyzing the figures for England and Wales for 1961-63 found a 30% excess of rural over urban incidence for seminoma, but an almost equal rural/urban distribution for the other tumors of the testis[14]. A higher rural incidence was reported by Graham and Gilson, for New York State[15], and for the Netherlands by Talerman, Kaalen and Fokkens[16].

Figure 2 shows the age incidence for seminoma and non-seminomatous germ cell tumors for countries included in the U.I.C.C. Gonadal Tumor Survey[3]. The age at diagnosis of seminomas (the late thirties) is approximately ten years later than that for non-seminoma, which has a peak incidences in the mid to late twenties. This observation is not new, but it is interesting to note that tumors with a histology of combined seminoma and non-seminoma have an intermediate peak age incidence[17], and exhibit a "mitigation" of malignancy by the seminomatous element, making for a better prognosis[6]. This observation may point to environmental factors affecting the way in which the malignant potential of germ cells is expressed (or possibly suppressed), since there is some evidence that these tumors may arise from a common malignant stem cell[18].

A predominance of right sided primary tumors has been recorded from both hospital and population based registries. Clemmesen[19] attributed this excess to chance, as did MacKay and Sellars[20]. Nevertheless it is surprising that this excess is reported with such uniformity. One possible explanation is that the right testis usually enters the scrotum later than the left so that a causative factor acting only on an intra-abdominal testis has a greater chance of doing so on that side.

The risk of the development of a second primary tumor of germ cell origin is small. Aristizabal et al., calculated a frequency of bilaterality of 1.5% from their review of the literature[21]. The

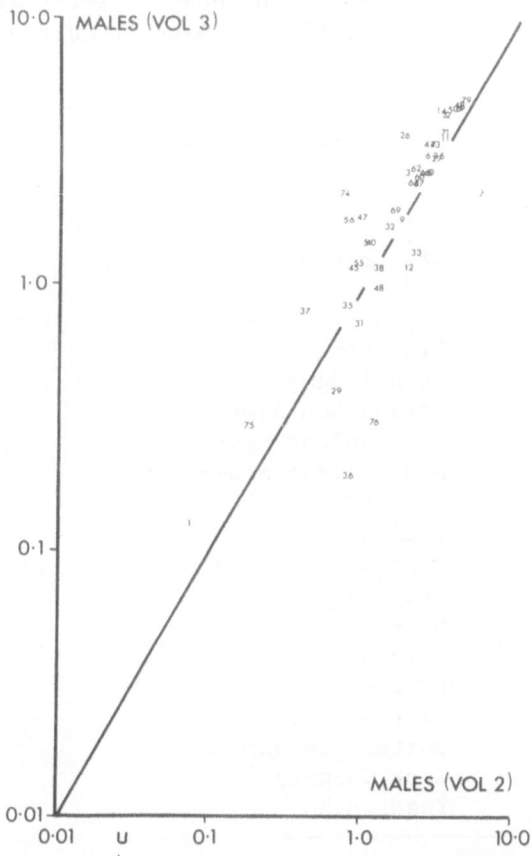

Fig. 1.   Increasing incidence of testicular tumors. Plot of
           incidence rates for registries included in Volumes 2 and 3
           of "Cancer Incidence in Five Continents" (from J. A. H.
           Waterhouse[3], reproduced by kind permission of the author
           and the publishers and editors of "Germ Cell Tumors").

incidence of bilateral tumors is considerably higher in patients with
a history of testicular maldescent, the risk increasing from an
estimated 0.7% for patients with descended testes, to 15% when both
testes are undescended, and as high as 30% in patients with intra-
abdominal testes[22]. Synchronous bilateral tumors are exceedingly
rare. Hoelsstra et al., found 24 cases in the literature[22].
However, Berthelson et al., found carcinoma-in-situ in the contralateral testis by fine needly biopsy in 8% of patients at the time of
diagnosis of the first lesion[23], although they may have been seeing
a particularly high risk group of men. In the series from the Royal
Marsden Hospital[24] nearly half of the second tumors occurred within
two years of the first but the range was wide (4-180 months). Of

Table 1. Testicular Cancer – Annual Incidence rates/100,000 – age
          adjusted to world population. (Data of Muir and Nectoux,
          1979[12])

Continent	Country/Ethnic Group	Rate
Africa	African	0.1
America	USA Blacks	0.5–1.0
	USA Whites	2.8–4.5
Asia	India	0.7
	Japan	0.7–1.2
	Singapore	0.3–0.9
(Middle East)	Israel Jews	1.6
	Israel Non-Jews	0.2
Australasia	New Zealand Maori	4.3
	New Zealand Non-Maori	3.7
Europe	Denmark	4.9
	Finland	1.1
	Hungary	1.0–1.8
	Iceland	2.1
	Norway	4.3–4.6
	Poland	0.6–2.3
	Sweden	2.5
	Switzerland	4.4
	United Kingdom	2.4–3.0
	West Germany	3.0–4.7
	Yugoslavia	1.9

those able to give information about maldescent, nearly half gave
such a history. There is a suggestion that there is an underlying
pathologic process affecting both testes in the unilateral cryptorchid[25,26].

Although testicular tumors have been reported in siblings and in
fathers and sons, Davies states that these cases are too few to
suggest a strong familial factor[7]. However, in view of different
incidence rates in different racial groups in the same community,
genetic factors may be playing a role in aetiology. Wobbes et al.,
report 17 cases of father/son cases in the literature and state that
it is impossible to decide whether this phenomenon is due to chance
or to genetics[27]. Abratt in his review of cases in non-twin
brothers in 1982 records only 11 instances in the literature[28], and
Wilbur et al., found only 8 cases of testicular tumors in twins[29].
Because of the rarity of germ cell tumors of the testis, no meaningful incidence figures concerning families currently exist, and the
last three groups of authors make the plea that all familial cases
should be reported to increase our knowledge. In my own personal
series there is one father/son case (histology unknown/seminoma) and

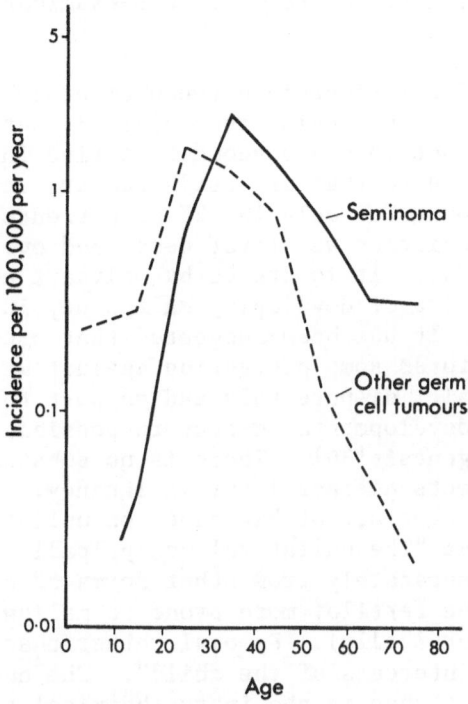

Fig. 2. Age incidence for seminoma and other germ cell tumors for
countries included in the UICC Gonadal Tumor Survey (from
J. A. H. Waterhouse[3], reproduced by kind permission of
the author and the publishers and editors of "Germ Cell
Tumors").

one instance of possible monozygotic twins, one with seminoma, the
other with a non-seminoma. One interesting observation about twins,
but stated in the reverse direction, is that there is a higher incidence of twins in families with sacrococcygeal teratomas[30] suggesting that an abortive attempt at twinning may be a possible aetiological mechanism. There is only one case of epidemiological space/
time clustering of cases with germ cell testicular tumors in the
literature, with four individuals living in close proximity for a
length of time. The only conclusion that can be drawn from this
observation is that it is possible environmental carcinogen was
aetiologically involved[31].

AETIOLOGY

     Despite the considerable and bewildering number of possible
aetiological factors suggested for testicular cancer, the aetiology
of this disease must be considered unknown. At present, according to
Davies, "there seems little immediate prospect of primary prevention,

and attention is turning to the role of self-examination and early diagnosis"[7].

There is a strong relationship between cryptorchism and the development of germ cell testicular tumors[32,33] but Davies[7] argues that there has not been a concommitant rise in the incidence of cryptorchism compared to that of testis cancer, and that patients with cryptorchism account for only 10-12% of patients with testicular malignancy. This association was first described over 200 years ago by Sir Percival Pott[34]. It is argued by Welvaart and Tijssen that the exact individual risk of developing malignancy in the cryptorchid is still unknown[35]. It has been suggested that orchidopexy before the age of 2 years affords some protection against malignancy[33]. However Van Cangh et al., dispute this and support the theory that there is an inherent developmental defect responsible for both cryptorchidism and tumorigenesis[36]. There is no substantive evidence that orchidopexy protects against later malignancy. Indeed, Hinman (1979) states, in the abstract of his paper on unilateral abdominal cryptorchidism[37] that "The unilateral non-palpable undescended testis is considered separately from other forms of cryptorchidism. It is less likely to be fertile, more prone to malignancy and more difficult to place (surgically). Removal rather than orchidopexy often is in the best interests of the child". The numerical risk factors for malignant change in the intra-abdominal testis are quoted above. Certainly pathological changes, particularly fibrosis, are evident in the cryptorchid testis within the first year of life[26] and it is suggested that these progressive degenerative changes may have an auto-immune basis. This conclusion comes from experimental studies in dogs, changes being present not only in the undescended testis, but also in the scrotal testis. This effect on the scrotal testis can be blocked by immunosuppressant therapy. In the very rare cases of polyorchidism (37 cases reported in the literature), one or both the duplicated testes may be undescended and thus at risk[38]. Intersex states with intra-abdominal testes are another high risk group[39].

The 1982 edition of the International Agency for Research on Cancer's Directory of On-Going Research in Cancer Epidemiology, records 18 studies in which a total of 44 possible aetiological factors in 7 countries were investigated[40].

Although the aetiology of the disease is unknown, it might be profitable to logically examine some of the suggested factors, certain of which are obviously interrelated. Some have already been discussed earlier in the chapter.

Preconception Factors

The differences in incidence rates for different races or ethnic groups within populations indicates a probable positive (or negative)

genetic influence[12,41]. However, familial occurrences are not
numerous enough to suggest a strong factor[7], and it is impossible
to decide whether it is the abnormal gene configuration in the inter-
sex/genetic mosaic states, the "cryptorchid" testis or some other
factor which causes the malignancy. Studies of HLA typing in
patients with testicular tumors have shown inconsistent
results[42,43]. Whatever the general overall genetic background,
another possible prenatal factor is the changing pattern of contra-
ceptive practice. It is possible that small doses of the oestrogen/
progestogen pill, taken in the presence of a fertilized ovum (because
of previous non-compliance in the same cycle) may have an effect on
incidence. Time alone will tell.

Prenatal Factors

     Recent studies in the U.K.[44,45] suggest that infections (par-
ticularly tuberculosis) during the index pregnancy, and maternal
epilepsy significantly raise the relative risk of testicular cancer.
Other factors which increase the risk, but not significantly, are a
maternal history of still-births, hyperemesis in the index pregnancy,
concomitant hernia, genito-urinary or other congenital defects in the
child[46,47], or still-born sibs. However, it is proposed that the
increasing incidence in young adults is more likely to be due to
postnatal factors[44].

     Another possible factor receiving much attention is the prenatal
exposure to exogenous hormones including diethylstilboestrol[48],
which has been implicated in causing cryptorchism, hypoplasia of the
testes, semen abnormalities and possible tumor formation[49,50,51].
However, Kinlen et al., present evidence to suggest that the use of
oestrogens during pregnancy is unlikely to be of importance since the
rise in incidence rates started before oestrogens were first introduced into obstetric practice in the 1940's[52]. Further case controlled studies are needed in this area, to determine whether the use
of drugs or other events, such as radiography in pregnancy, are of
importance.

Postnatal/Childhood Factors

     Maldescent or nondescent of the testis has been discussed above,
but the relationship between this and tumorigenesis is not clear. Is
it the maldescent that causes the tumor or is some other factor
causing both? Does the higher intra-abdominal temperature (and/or
other factors) have an effect, hence the slight predominance of right
sided lesions? The correlation with other congenital lesions, inguinal hernias, and maldescent (and hence scrotal/inguinal surgery) have
been discussed above. Environmental and other factors are discussed
below.

Possible Factors in Adulthood

Perhaps the most striking factor identified to date is that testicular tumors tend to affect more individuals in higher social classes than those in manual work[7,41]. A multiplicity of features of modern day life could be contributing to the increasing incidence. Such factors as increased testicular temperature (central heating, tight fitting clothes, especially "jockey" shorts[51], hot baths, and showers), a sedentary way of life, sedentary entertainment, car driving, changes in diet, earlier sexual maturation (hormone effects?), and urban existence may be operative here. Greater risk of exposure to environmental hazards such as exposure to certain metals, to nitrates in water supplies and to other chemicals exists in urban environments. Muir and Nectoux (1979) reported an excess of testicular cancer in chemists in the data for England and Wales[12].

Individuals with atrophic testes are at risk, and correlations with maldescent were mentioned above. Mumps orchitis has been suggested as an aetiological factor[51], but the work of Ehrengut and Schwartau would suggest that this is unlikely[53]. Similarly, recurrent genito-urinary tract infections giving rise to atrophy have been suspected without substantiating evidence.

Whether trauma is the initiator of the tumor, or the enlargement of the testis by tumor predisposes to trauma is a circular argument. It is not uncommon for patients with cancer at any site to become aware of a swelling after an accidental blow, or to use that event to acknowledge the presence of an abnormality worthy of medical attention. A recent study[54] suggests that chronic trauma from cycling and horse-riding in adolescence are definite risk factors. Further epidemiological studies are essential for these diseases since the causative factors are most unclear at present.

ANIMAL MODELS

The development of three types of animal model systems has furthered our knowledge of the biology and characterization of germ cell tumors and possible mechanisms of oncogenesis. They are also increasing our knowledge about embryogenesis[55,56,57,58]. However Graham (1981) has expressed disappointment about the slow progress being made with human tumors compared to mouse teratomas[58]. The model systems are: spontaneous/induced animal tumors, tissue culture lines (including short term culture in soft agar), and xenografts into immune-suppressed and congenitally athymic ('nude') mice or into immunologically privileged sites.

Animal Tumors

Spontaneous testicular germ cell tumors occur rarely in mouse and man, but are more common in these two species than in others in the animal kingdom. The injection of metal salts into the testes of fowl during their seasonal growth period was the way in which the first experimentally induced tumors were produced, and similar injections into the testes of mice and rats give a low yield of spontaneous tumors[59]. Studies of plant teratomas (galls) suggest that the development of neoplasia is a caricature of normal tissue renewal processes[60].

Spontaneous tumors occur in the inbred mouse strain 129 with a frequency of 1% [61]. This led to the development of strain 129/ter SV with a 30% incidence of tumors[62]. This increase in incidence is thought to be due to the mutation of a single gene. From the outset it was realized that mouse teratocarcinomas (a word used by biologists and not synonymous with the term used in the American histological classification of human germ cell tumors) may provide a good model for human tumors. A wealth of information has been gained about the tumor's biology, the characteristics of stem cells and the mode of turnover differentiation, but convincing parallel studies with human tissue have not been performed[63]. However these tumors have provided good models for the study of primordial germ cells, demonstrating their multi-potentiality for differentiations their surface characteristics, marker protein production and response to chemotherapy[55].

Teratomas have also been induced in the 129 mouse strain by the transplantation of genital ridges to the testes and epididymi[64] and by the syngeneic transplantation of post-gastrulation embryos to extra-uterine sites in other strains of mice and hamsters[65].

Marked similarities exist between the undifferentiated cells of human and animal teratomas and normal uncommitted embryonic cells. The primordial germ cells of the 129 strain teratoma are pleuripotent, being able to differentiate into cells of any of the 3 germ layers (endo-, ecto- or mesoderm). Developmental biologists suggest inter-relationships between the pathways of oncogenesis and ontogenesis[56,57,58].

Tissue Culture Lines

Several cell lines originating from the mouse teratocarcinoma model have been developed and used to investigate marker production, biochemical pathways, and cell surface marker characteristics - one of which (F9) is shared by preimplantation mouse embryos, human sperm and mouse and human embryonal carcinoma cells[56,66]. A number of human tumors have been established in cell culture[55]. These lines have been used to investigate markers, cell surface antigens and other biochemical phenomena, but as yet no radiobiological or chemo-sensitivity tests have been reported. Considerable efforts to isolate the stem cells from these human tumors continues.

The more recently developed technique of in vitro cloning of human testicular cancer cells in soft agar has a potentially brighter future in the field of drug sensitivity testing[67].

Xenograft Techniques

Work with these explant techniques began with the transplant-ation of tumor to immunologically privileged sites such as the hamster cheek pouch. However the establishment of the technique of xenografting human tumors into immune-suppressed animals or congenit-ally athymic mice has added a new dimension to the study of these tumors which is dynamic and quantifiable[56]. Some of the problems with this technique are the relative size of the different animals, and their different host reponses, metabolism and growth rates. More than 20 human germ cell xenograft lines have been reported and ef-forts are being directed towards their characterization. Their use so far appears to be directed more towards biochemical and marker studies[68,69], and attempts at radio-immuno-localization[70], than towards therapy testing.

The majority of studies with animal and xenograft models have concentrated on the more descriptive aspects of germ cell tumor biology and biochemistry as well as similarities between malignant cells from tumors and normal cells from developing tissues.

REFERENCES

1. F. K. Mostofi, Testicular tumors - epidemiologic, etiologic and pathologic features, Cancer, 32:1186 (1973).
2. R. C. B. Pugh and K. M. Cameron, Teratoma, in: "Pathology of the testis", R. C. B. Pugh, ed., Blackwell, Oxford, p.199 (1976).
3. J. A. H. Waterhouse, Epidemiology of germ cell tumors, in: "Germ Cell Tumors", C. K. Anderson, W. G. Jones, and A. Milford Ward, eds., Taylor and Francis, London, p.104 (1981).
4. R. Doll, C. S. Muir, and J. Waterhouse, Cancer incidence in five continents, Volume II, U.I.C.C., Geneva (1970).
5. J. Waterhouse, C. Muir, P. Correa, and J. Powell, Cancer incidence in five continents, Volume III, IARC Scientific Publications No.15, Lyon (1976).
6. P. J. Corbett, R. A. Cartwright, and H. Annett, Testicular tera-toma in Yorkshire - a retrospective review, in: "Germ Cell Tumors, C. K. Anderson, W. G. Jones, and A. Milford Ward, eds., Taylor and Francis, London, p.93 (1981).

7. J. M. Davies, Testicular cancer in England and Wales: Some epi-
    demiological aspects, Lancet, i:928 (1981).

8. J. Clemmesen, A doubling of morbidity from testis carcinoma in
    Copenhagen, 1943-1962, Acta Pathol.et Microbiol.Scandinav.,
    72:348 (1968).

9. J. Clemmesen, Statistical studies in the aetiology of malignant
    neoplasms, Acta Pathol.et Microbiol.Scandinav., Supplement
    247, Munksgaard, Copenhagen (1974).

10. D. Schottenfeld and M. E. Warshauer, Testis, in: "Cancer
    Epidemiology and Prevention", W. B. Saunders, ed., London and
    Philadelphia, p.947 (1982).

11. J. A. H. Lee, M. Hitosugi, and G. R. Petersen, Rise in mortality
    from tumors of the testis in Japan 1947-1970, J.Nat.Cancer
    Inst., 51:1485 (1973).

12. C. S. Muir and J. Nectoux, Epidemiology of cancer of the testis
    and penis, Nat.Cancer Inst.Monog., 53:157 (1979).

13. E. Pedersen and K. Magnus, Cancer registration in Norway 1953-
    1954. The Norwegian Cancer Society, Oslo (1959).

14. L. Lipworth and A. D. Dayan, Rural preponderance of seminoma of
    the testes, Cancer, 23:1119 (1969).

15. S. Graham and R. W. Gibson, Social epidemiology of cancer of the
    testes, Cancer, 29:1242 (1972).

16. A. Talerman, J. G. Kaalen, and W. Fokkens, Rural preponderance
    of testicular neoplasms, Brit.J of Cancer, 29:176 (1974).

17. P. N. Braun, The origin of germ cell tumors of the testis,
    Cancer, 51:1610 (1983).

18. N. E. Skakkebraek and J. G. Berthelsen, Carcinoma-in-situ of
    testis and orchidectomy, (Letter) Lancet, ii:204 (1978).

19. J. Clemmesen, Statistical studies in the aetiology of malignant
    neoplasmas. 3, Acta Pathol.et Microbio.Scand., Supplement 209
    (1969).

20. E. N. MacKay and A. H. Sellars, A statistical review of
    malignant testicular tumors based on the experience of the
    Ontario Cancer Foundation Clinics, 1938-1961, Canad.Med.
    Assoc.J., 94:889 (1966).

21. S. Aristizabal, J. R. Davis, R. C. Miller, M. J. Moore, and
    M. L. M. Boone, Bilateral primary germ cell testicular
    tumors. Report of four cases and review of the literature,
    Cancer, 42:491 (1978).

22. H. J. Hoekstra, D. M. Mehta, and K. H. Schraffordt, Syn-
    chronous bilateral primary germ cell tumors of the testis:
    A case report and review of the literature, J.Surg.Oncol.,
    22:59 (1983).

23. J. G. Berthelsen, N. E. Skakkebraek, P. Morgensen, and B. L.
    Sørensen, Incidence of carcinoma in situ of germ cells in
    contralateral testis of men with testicular tumors, Brit.
    Med.J., ii:363 (1979).

24. M. Sokal and M. J. Peckham, Bilateral incidence of testicular
    tumors, Brit.J.Radiol., 51:477 (1978).

25. C. G. Scorer and G. H. Farrington, Congenital deformities of the
    testis and epididymis, Butterworths, London (1971).

26. D. T. Mininberg, J. C. Rodger, and J. M. Bedford, Ultra-
    structural evidence of the outset of testicular pathological
    conditions in the cryptorchid human testis within the first
    year of life, J.Urol., 128:782 (1982).
27. Th. Wobbes, H. J. Hoekstra, J. Oldhoff, and K. H. Schraffordt,
    Malignant testicular germ cell tumors in father and
    son: a report on two families, J.Urol., 129:152 (1983).
28. R. P. Abratt, Testicular cancer in two brothers one of whom has
    achondroplasia, Brit.J.Urol., 54:427 (1982).
29. H. J. Wilbur, M. W. Woodruff, and M. S. Welch, Concomitant germ
    cell tumors in monozygotic twins, J.Urol., 121:538 (1979).
30. R. J. Izant and H. C. Filston, Sacrococcygeal teratomas:
    analysis of 43 cases, Amer.J.Surg., 130:617 (1975).
31. P. S. Feldman, S. S. Howards, C. Harris, and C. Harris, A geo-
    graphic cluster of testicular seminomas, J.Urol., 129:839
    (1983).
32. M. A. Batata, F. C. H. Chu, B. S. Hilaris, W. F. Whitmore, and
    R. B. Golby, Testicular cancer in cryptorchids, Cancer, 49:
    1023 (1982).
33. D. C. Martin, Malignancy in the cryptorchid testis, Urol.Clin.
    N.Amer., 9:371 (1982).
34. P. Potts, The Chirurgical Works of Percival Pott, F.R.S., London
    (1779).
35. K. Welvaart and J. G. P. Tijssen, Management of the undescended
    testis in relation to the development of cancer, J.Surg.
    Oncol., 17:219 (1981).
36. P. J. Van Cangh, P. Hennebert, and P. Malvaux, Cryptorchidism
    and testicular cancer, (Letter) J.Urol., 125:603 (1981).
37. F. Hinman, Unilateral abdominal cryptorchidism, J.Urol., 122:71
    (1979).
38. K. W. M. Scott, A case of polyorchism with testicular teratoma,
    J.Urol., 124:930 (1980).
39. M. J. Peckham, General introduction: biological diversity and
    predisposing factors, in: "The Management of Testicular
    Tumors", M. J. Peckham, ed., Edward Arnold, London, p.1
    (1981).
40. Directory of on-going research in cancer epidemiology. I.A.R.C.
    Publications No.46, C. S. Muir and G. Wagner, eds., Lyon
    (1982).
41. P. Mustacchi and D. Millmore, Racial and occupational variations
    in cancer of the testis: San Francisco, 1956-65, J.Cancer
    Inst., 56:717 (1976).
42. M. S. Pollack, D. Vugrin, W. Hennessy, H. W. Herr, B. Dupont,
    and W. F. Whitmore, HLA antigens in patients with germ cell
    cancers of the testis, Cancer Research., 42:2470 (1982).
43. P. Aininger, H. P. Schwarz, J. Kuhböck, and W. R. Mayr, HLA and
    testicular cancer, Cancer Immunology and Immunotherapy, 10:
    169 (1981).

44. A. J. Swerdlow, C. A. Stiller, and L. M. Kinnier Wilson, Pre-
     natal factors in the aetiology of testicular cancer: an epi-
     demiological study of childhood testicular cancer deaths in
     Great Britain, 1953-73, J.Epidemiol.and Comm.Health, 36:96
     (1982).
45. J. M. Birch, H. B. Marsden, and R. Swindell, Prenatal factors in
     the origin of germ cell tumors of childhood, Carcinogenesis,
     3:75 (1982).
46. R. J. Fram, M. B. Gamick, and A. Retik, The spectrum of genito-
     urinary abnormalities in patients with cryptorchidism, with
     emphasis on testicular carcinoma, Cancer, 50:2243 (1982).
47. S. Sakashita, T. Koyanagi, I. Tsuji, K. Arikado, and T. Matsuno,
     Congenital anomalies in children with testicular germ cell
     tumor, J.Urol., 124:889 (1980).
48. D. Schottenfield, M. E. Warshauer, S. Sherlock, A. G. Zauber,
     M. Leder, and R. Payne, The epidemiology of testicular cancer
     in young adults, Amer.J.Epidemiol., 112:232 (1980).
49. B. E. Henderson, B. Benton, J. Jing, M. C. Yu, and M. C. Pike,
     Risk factors for cancer of the testis in young men, Int.J.,
     Cancer, 23:598 (1979).
50. W. B. Gill, G. F. B. Schumacher, M. Bibbo, F. H. Straus, and
     H. W. Schoenberg, Association of diethylstilboestrol exposure
     in utero with cryptorchidism, testicular hypoplasia and semen
     abnormalities, J.Urol., 122:36 (1979).
51. J. E. Loughlin, S. J. Robboy, and A. S. Morrison, Risk factors
     for cancer of the testis, (Letter) New Engl.J.Med., 303:112
     (1980).
52. L. J. Kinlen, M. A. Badaracco, J. Moffett, and M. P. Vessey, A
     survey of the use of oestrogens during pregnancy in the
     United Kingdom and of the genitourinary cancer mortality and
     incidence rates in young people in England and Wales,
     J.Obstet.and Gynaecol.of the British Commonwealth, 81:849
     (1974).
53. W. Ehrengut and M. Schwartau, Mumps orchitis and testicular
     tumors, Brit.Med.J., ii:191 (1977).
54. A. J. Coldman, J. M. Elwood, and R. P. Gallagher, Sports activi-
     ties and risk of testicular cancer, Brit.J.Cancer, 46:749
     (1982).
55. B. Nørgaard-Pedersen and D. Raghavan, Germ cell tumors: a col-
     laborative review, Oncodevelopmental Biol.and Med., 1:327
     (1980).
56. D. Raghavan and P. Selly, Testicular tumor xenografts and other
     experimental models, in: "The Management of Testicular
     Tumors", M. J. Peckham, ed., Edward Arnold, London, p.70
     (1981).
57. M. J. Evans, Are teratocarcinomas formed from normal cells? in:
     "Germ Cell Tumors", C. K. Anderson, W. G. Jones, A. Milford
     Ward, eds., Taylor and Francis, London, p.24 (1981).
58. C. G. Graham, Initiation of mouse teratomas, in: "Germ Cell
     Tumors", C. K. Anderson, W. G. Jones, A. Milford Ward, eds.,
     Taylor and Francis, London, p.17, (1981).

59. I. Damjanov, Teratoma and teratocarcinoma in experimental
     animals, Nat.Cancer Inst.Monog., 49:305 (1978).
60. A. C. Braun, The usefulness of the plant tumor system for study-
     ing the basic cellular mechanisms that underlie neoplastic
     growth generally, in: "Cell Differentiation", R. Harris, P.
     Ailin, and D. Viza, eds., Munksgaard, Copenhagen, p.115
     (1972).
61. L. C. Stevens and C. C. Little, Spontaneous testicular teratomas
     in an inbred strain of mice, Proc.Nat.Acad.Sci.(USA)., 40:
     1080 (1954).
62. L. C. Stevens, A new inbred subline of mice (129/ter SV) with a
     high incidence of spontaneous congenital testicular
     teratomas, J.Cancer Inst., 50:235 (1973).
63. M. Evans, Experimental teratomas. Clinics in oncology, 2:77
     (1983).
64. L. C. Stevens, Experimental production of testicular teratomas
     in mice, Proc.Nat.Acad.Sci.(USA)., 52:654 (1964).
65. I. Damjanov, Development of teratomas from embryos transplanted
     into outbred and inbred adult hamsters, J.Nat.Cancer Inst.,
     61:911 (1978).
66. B. L. M. Hogan, Teratomas in culture, in: "Germ Cell Tumors",
     C. K. Anderson, W. G. Jones, and A. Milford Ward, eds.,
     Taylor and Francis, London, p.123 (1981).
67. R. F. Ozols, B. J. Foster, and N. Javadpour, Cloning of human
     testicular cancer in soft agar: potential diagnostic and
     therapeutic applications, in: "Germ Cell Tumors", C. K.
     Anderson, W. G. Jones, and A. Milford Ward, eds., Taylor and
     Francis, London, p.216 (1981).
68. F. Searle, New marker possibilities, in: "Germ Cell Tumors",
     C. K. Anderson, W. G. Jones, and A. Milford Ward, eds.,
     Taylor and Francis, London, p.233 (1981).
69. D. Raghavan, J. Gibbs, N. Costa, J. Kohn, A. H. Orr, A. Barrett
     and M. J. Peckham, The interpretation of marker protein
     assays: a critical appraisal in clinical studies and a xeno-
     graft model, Brit.J.Cancer, 41:191 (1980).
70. V. Moshakis, R. A. J. McIlhinney, and A. M. Neville, Radio-
     labelled monoclonal antibodies for the localization of human
     teratoma xenografts in vivo, in: "Germ Cell Tumors", C. K.
     Anderson, W. G. Jones, and A. Milford Ward, eds., Taylor and
     Francis, London, p.149 (1981).

CELL KINETICS IN HUMAN GERM CELL

TUMORS OF THE TESTIS

R. Silvestrini, A. Costa
S. Pilotti, and G. Pizzocaro

Istituto Nazionale Tumori
Milan, Italy

INTRODUCTION

The high sensitivity to drugs and radiation[1-6] and the availability of biological markers for monitoring the evolution of disease[7-9] have made germ cell tumors of the testis one of the most curable. However, many aspects, such as sensitivity of different histological types to chemical and physical agents and the identification of markers for a large number of seminomas, mainly at early stages, have still to be investigated. Moreover, although the development of experimental models[10-12] has improved the knowledge of the biology of this tumor, and morphological features have been the object of thorough studies, other aspects, such as cell kinetics, have not been well investigated[13,14]. Most of the relevant studies have been performed by _in vivo_ tumor perfusion[15,16] aimed at analysis of intratumor heterogeneity of cell proliferative activity, while intertumor heterogeneity has been completely neglected.

In the present study we wanted to define the cell proliferative activity of the two main seminomatous and nonseminomatous tumors with their different histological subtypes, and to relate this activity to the extent of disease. The purpose was to determine whether cell kinetics, alone or in association with morphological and pathological features, would give additional information of clinical relevance. This type of analysis is justified by earlier evidence that pretreatment proliferative activity has consistently proven to be an indicator of increased risk in the human tumor types thus far considered[17-20]. Such a study requires a large series of tumors, and the feasibility was enhanced by the use of a simple technical approach. We evaluated the proliferative activity by determination of the _in vitro_ [^{3}H]thymidine labeling index (LI) of testicular tumors from 108 patients.

MATERIAL AND METHODS

Eighty-three untreated and 25 previously treated patients with
testicular cancer were entered to the study. For 34 of the untreated
patients, the histology of the primary tumor was seminoma (59% stage
I, 35% stage II, and 6% stage III).

The remaining 39 untreated patients had nonseminomatous tumors,
of which 46 specimens were available: 16 samples were from primary
tumor and 30 were from metastatic sites. Embryonal carcinoma was
present as the only histology in 67% of the cases, or associated with
endodermal sinus tumor and mature or immature teratoma in 33%. Surgical staging, which was performed in these patients by means of retroperitoneal lymphadenectomy, showed that there were 7% at stage I, 60%
stage II, and 33% stage III lesions.

Specimens from previously treated cases were obtained mainly from
those patients who had bulky or inoperable disease, and who had undergone presurgical combined chemotherapy. At the time of surgical excision, part of the pathological material was used for in vitro cell kinetic determinations[21]. Small fragments of tumor tissue were incubated
in culture medium (McCoy's 5a medium + 20% foetal calf serum + antibiotics) for 1 hour at 37°C with agitation in a shaker water bath with
[^{3}H]thymidine. After incubation the fragments were fixed in Bouin's
solution and processed for the usual histological procedures. Auto-
radiographic examinations were carried out on histological sections
using the stripping film (Kodak AR10) technique[22]. The labeling
index (LI) was determined by scoring a total of 3000 to 5000 cells from
different specimens of the same tumor. The counting was limited to the
periphery of the section. No threshold for considering labeled nuclei
was necessary, because the background was less than 1.5 grains per 100
μm^2 and therefore silver grains of the background were only occasionally observed outside the nuclei. A total of 20 grains per nucleus was
required for a cell to be scored as labeled.

RESULTS

The overall analysis of kinetic features on 46 embryonal carcinomas,
34 seminomas, 12 teratomas and 4 endodermal sinus tumors from both pure
and mixed forms, regardless of the tumor site, showed LI values ranging
from 0.10 to 77.3% with a frequency distribution of an exponential type
(Table 1). When seminomas and embryonal carcinomas, which were the
two most frequent types in our series, were considered separately, we
found a similar form of frequency distribution but the LI values were below 30% in seminomas, while embryonal carcinomas had a distribution towards higher values with a maximum frequency of LI between 30 and 60%
(Figure 1).

The overall median LI value was 25.5%, with a significantly lower
value for seminomas (13.9%) than for nonseminomatous tumors (32.2%)
(p < 0.0007). Moreover, within the nonseminomas, markedly different

Table 1. Cell Kinetics in Relation to Histomorphology in Germ Cell
          Tumors of the Testis

	No. of cases	Labelling index (LI) (%)	
		Median	Range
Overall	96	25.5	0.01-77.3
Seminoma } *	34	13.9	0.8 -31.5
Nonseminomatous }	62	32.2	0.01-77.3
Embryonal carcinoma	46	42.0	1.6 -77.3
Endodermal sinus	4	29.6	17.0 -30.5
Teratoma, immature	5	8.5	0.1 -25.0
Teratoma, mature	7	1.4	0.01- 7.0

*Seminomas vs. nonseminomas, $p < 0.0007$.

proliferative rates were found for the various histological types. The
maximum proliferative rate was found for embryonal carcinomas (42.0%),
while a somewhat lower one was found for the few cases of endodermal sinus tumors (29.6%). Very low values were found for proliferating cells
in teratomas. A trend toward higher proliferative activity, which was
not statistically significant, was found in immature (8.5%) vs. mature
(1.4%) forms. However, a high variability of LI values was observed
within each histological subtype. The narrowest range occurred in mature teratomas, intermediate ranges in immature teratoma, endodermal
sinus tumor and seminoma, and wider ranges occurred in embryonal
carcinoma.

The cell kinetics of pure forms of seminoma were determined in
34 cases. Among these, 5 cases were anaplastic (with and without
syncytiotrophoblasts), and 29 were typical seminomas. No relation
was found between proliferative rate and histological subtypes.

For embryonal carcinomas, an adequate number of cases was
examined to study the proliferative rate separately in the pure and
the mixed forms (Table 2). The most frequent associated component in
mixed embryonal carcinomas was mature teratoma (13 of 15 tumors).
Other seminomatous and nonseminomatous histological types occurred in
only 2 cases. The ranges and median values of LI were the same for
both the pure and mixed forms.

The proliferative activity of embryonal carcinomas was examined
by the different tumor sites (Table 3). Owing to the difficulty of obtaining samples from both primary and lymph node sites on the same patients,
the proliferative activity of primary tumors was compared to that observed in lymph node metastases from a different series of patients.
A somewhat higher but not statistically significant median LI value was
found for 29 lymph node metastases (43.0%) compared to 16 primary tumors
(36.0%).

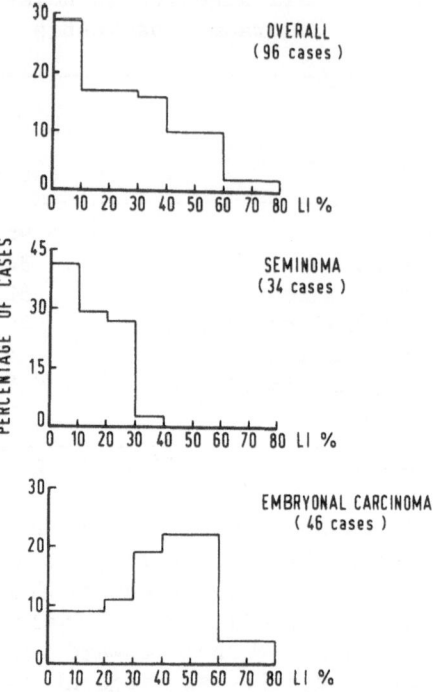

Fig. 1.
Frequency distribution of LI
values in germ cell tumors of
the testis from 96 untreated
patients.

For a small group of seven cases, the analysis of proliferative
activity of the primary tumor vs. that of lymph node metastases was
possible on the same patient (Table 4). Among these, six had the same
histological pattern in primary and lymph node sites, and five of these
had similar proliferative rates for similar histological components.
The only disagreement was observed for a patient with a distant nodal
metastasis. In one case, different cell kinetics observed between the
primary tumor and its lymph node metastasis was associated with different histology (mature teratoma in the primary and embryonal carcinoma in the metastasis.

Another observation deals with the kinetic characteristic of the
same histologies (embryonal carcinoma and mature teratoma) in untreated vs. previously treated patients. A statistically significant decrease in median LI value was found for embryonal carcinomas from
nine previously treated patients at the time of surgical excision
after chemotherapy. Conversely, an increase in LI value was observed for 16 patients with mature teratoma in the same clinical condition. However, this increase was not statistically significant even
though very high LI values were reached, mainly in lymph node sites.

The proliferative activity was also analyzed in relation to the
extent of disease. Analysis of 34 cases of seminoma (Figure 2)
showed an increase in the proliferative rate of the primary tumor
with increased extent of disease. However, the Median LI value found
for stage II (17.5%) was not significantly higher than that observed
for stage I (10.7%). Moreover, a large variability was consistently

Table 2. Cell Kinetics in Pure vs. Mixed Forms of
          Embryonal Carcinoma

Embryonal Carcinoma	No. of cases	Labelling Index (LI) (%)	
		Median	Range
Pure	31	44.0	1.6-77.3
Mixed	15	41.0	7.0-70.0

Table 3. Cell Kinetics in Primary Tumor vs. Lymph
          Node Metastases of Embryonal Carcinoma

Tumor site	No. of cases	Labelling Index (%)	
		Median	Range
Testis	16	36.0	1.6-77.3
Lymph node	29	43.0	9.7-74.0

found within each stage. Among the 46 patients with pure and mixed
embryonal carcinoma (Figure 3), 93% had disease at pathologic stages
II and III. Again, a large variability was observed for these
stages, and the median LI values found for stage II and III were
similar (42.0% and 45.0%, respectively). Differences were not observed even when the proliferative activity of the tumor was analyzed for regional vs. advanced disease (39.0% and 43.5%, respectively).

DISCUSSION

     This study of the kinetics of germ cell testicular tumors is
greatly hampered by their complex histology, by the differences in
histological patterns between primary and metastatic sites, and by
variations in histopathology observed at relapse [23,24]. However,
from this study it can be concluded that nonseminomatous germ cell
testicular tumors have a widespread proliferative activity, and that
this characteristic is consistent within each histological subgroup.
Specifically, mature teratomas are characterized by the narrowest
range of variability and the lowest LI median value. Conversely,
embryonal carcinomas present the widest range of variability and the
highest median LI value. Embryonal carcinoma, which is the fastest
proliferating histological subtype amont the human tumors thus far
studied[13], probably represents a cell population with 100% growth
fraction and has an actual growth pattern of an expoential type.
Its kinetic behavior is the same in both pure and mixed forms and, as
observed for other tumor types [25], in lymph node as well as primary
sites. This latter information, which was obtained from determinations on different series of patients, is confrimed by preliminary data

Table 4.   Cell Kinetics of Primary Tumor vs. Lymph Node Metastasis
           from the same Patient (7 cases)

Case no.	Tumor site	Histology*	LI (%)
1	Testis	ECA (pure)	77.3
	Retroperitoneal node	ECA (pure)	74.0
2	Testis	ECA (mixed)	54.0
	Retroperitoneal node	ECA (pure)	44.0
3	Testis	ECA (pure)	50.0
	Supraclavicular node	ECA (pure)	14.0
4	Testis	ECA (mixed)	34.0
	Abdominal disease	ECA (mixed)	41.0
5	Testis	MT (mixed)	7.0
	Retroperitoneal node	ECA (mixed)	27.0
6	Testis	ECA (pure)	3.27
	Retroperitoneal node	ECA (pure)	9.67
7	Testis	MT (mixed)	0.4
	Retroperitoneal node	MT (pure)	0.7

*ECA, embryonal carcinoma; MT, mature teratoma.

on a small group of patients in which both the primary and nodal samples
were available. However, this point needs to be investigated further in
relation to pathological or clinical stage in a larger series of cases.

In seminomas, an even larger variability of LI values was found,
with an intermediate median value between that observed for embryonal
carcinoma and that for teratoma. No relationship was observed between histological subtypes and cell kinetics, and the number of
mitoses failed to correlate with the number of proliferating cells.
A possible explanation could be that there is intertumor variability
of the ration between duration of S and the mitotic phases. At
present, the small number of cases examined and the still short
follow-up do not permit final evaluation of the progrnostic relevance
of LI in subgroups of patients homogeneously treated.

However, it must be stressed that differences in the degree of
proliferative activity for the different histologies is in agreement
with their different prognostic relevance. Moreover, the lack of a
correlation between cell proliferative activity and stage does not ex-
clued the potential prognostic relevance of this biological variable.
In other tumors throughly investigated, it has been shown that cell
kinetic features are independent of other morphological parameters[19].

ACKNOWLEDGEMENT

Supported in part by Target Project "Control of Neoplastic Growth,"
grant no. 82:01335.96, from the Consiglio Nazionale delle Richerche, Rome.

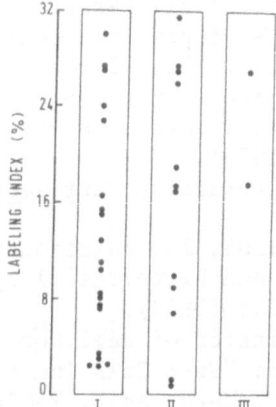
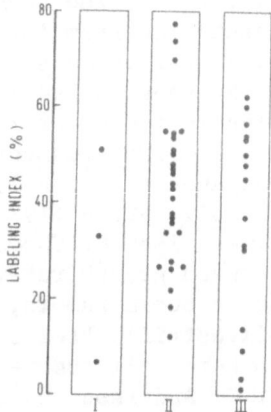

Fig. 2. Scattergrams of LI values   Fig. 3. Scattergrams for II ranges
         in Clinical stages I, II   in pathological stages I, II
         and III for 34 seminomas   and III for 46 embryonal car-
         from untreated patients.   cinomas from untreated pa-
         Note the concentration of   tients. Note the concentra-
         cases in stage I and the   tion of cases in stage II, and
         low labelling index.   the higher labelling index.

REFERENCES

1. D. C. Skinner, Nonseminomatous testis tumors: A plan of manage-
    ment based on 96 patients to improve survival in all stages
    by combined therapeutic modalities, J.Urol., 115:65 (1976).
2. L. H. Einhorn and J. P. Donohue, Improved chemotherapy in dis-
    seminated testicular cancer, J.Urol., 117:65 (1977).
3. C. Williams, Current dilemma in the management of non-seminomatous
    germ cell tumors of the testis, Cancer Treat. Rev., 4:275 (1977).
4. P. T. Scardino and D. G. Skinner, Germ-cell tumors of the testis: Im-
    proved results in a prospective study using combined modality
    therapy and biochemical tumor markers, Surgery, 86:86 (1979).
5. J. E. Pontes, Z. Wajsman, S. Beckley, P. Williams, and G. P.
    Murphy, The treatment of stage III nonseminomatous testicular
    tumors. Roswell Park Memorial Institute results (1970-1979).
    Cancer, 51:1327 (1983).
6. C. D. Callery, E. C. Holmes, S. Verson, J. Huth, W. F. Coulson, and
    D. G. Skinner, Resection of pulmonary metastases from nonsemio-
    matous testicular tumors. Correlation of clinical and histologi-
    cal features with treatment outcome, Cancer, 51:1152 (1983).
7. G. Liekovsky and D. G. Skinner, Significance of serum lactic
    dehydrogenase in stage B and C nonseminomatous testis tumors,
    J. Urol., 123:516 (1980).
8. M. C. Lippert and N. Javadpour, Lactic dehydrogenase in the monitoring
    and prognosis of testicular cancer, Cancer, 48:2274 (1981).
9. E. S. Newlands, R. H. J. Begent, G. J. S. Rustin, D. Parker, and K. D.
    Bagshawe, Further advances in the management of malignant terato-
    mas of the testis and other sites, Lancet, 2:948 (1983).

10. L. C. Stevens, The development of transplantable terato-
    carcinomas from intratesticular grafts of pre- and post-
    implantation mouse embryos, Dev.Biol., 21:364 (1970).

11. L. C. Stevens, A new inbred subline of mice (129/ter Sv) with a
    high incidence of spontaneous congenital testicular tera-
    tomas, J.Natl.Cancer Inst., 50:235 (1973).

12. I. Damjanov and D. Solter, Experimental teratoma, Curr.Topics
    Pathol., 59:69 (1974).

13. E. P. Malaise, N. Chavaudra, and M. Tubiana, The relationships
    between growth rate, labelling index and histological type of
    solid human tumors, Eur.J.Cancer, 9:305 (1973).

14. R. Silvestrini, Proliferative characteristics of cell population
    in germ cell tumors of the testis, in: "Germ Cell Tumors",
    C. K. Anderson, W. G. Jones, and A. M. Ward, eds., Taylor &
    Francis, London, p.145 (1981).

15. U. Rattenhuber, H. M. Rabes, P. Carl, U. Löhrs, R. Lamerz, K.
    Mann, G. Rindfleisch, and G. Staehler, Analysis of prolifer-
    ative compartments in human seminoma (abstract), Urol.Res.,
    8:234 (1980).

16. H. M. Rabes, U. Rattenhuber, P. Carl, U. Löhrs, G. Staehler, R.
    Lamerz, K. Mann, and G. Rindfleisch, Analysis of prolifer-
    ation of tuman testicular teratocarcinomas (abstract), Urol.
    Res., 8:235 (1980).

17. J. S. Meyer and B. Hixon, Advanced stage and early relapse of
    breast carcinomas associated with high thymidine labelling
    indices, Cancer Res., 39:4042 (1979).

18. B. G. M. Durie, S. E. Salmon, and T. Moon, Pretreatment tumor
    mass, cell kinetics, and prognosis in multiple myeloma,
    Blood, 55:364 (1980).

19. A. Costa, G. Bonadonna, E. Villa, P. Valagussa, and R.
    Silvestrini, Labelling index as a prognostic marker in non-
    Hodgkin's lymphomas, J.Natl.Cancer Inst., 66:1 (1981).

20. C. Gentili, O. Sanfilippo, and R. Silvestrini, Cell prolifer-
    ation in relation to clinical features and relapse in breast
    cancers, Cancer, 48:102 (1981).

21. R. Silvestrini, M. G. Daidone, and G. Di Fronzo, Relationship
    between proliferative activity and estrogen receptors in
    breast cancer, Cancer, 44:665 (1979).

22. F. Polvani and R. Silvestrini, Note di tecnica autoradiografica
    con pellicole "stripping", Riv.Istochim.Norm.Patol., 2:239
    (1956).

23. J. J. Bredael, D. Vugrin, and W. F. Whitmore, Jr., Autopsy
    findings in 154 patients with germ cell tumors of the testis,
    Cancer, 50:548 (1982).

24. P. N. Brawn, The origin of germ cell tumors of the testis,
    Cancer, 51:1610 (1983).

25. A. Costa, G. Del Bino, L. Ventura, and R. Silvestrini, Relation-
    ship between cell kinetics of primary tumor and lymph node
    metastases of the same patient (abstract). XIIth Meeting of
    the European Study Group for Cell Proliferation, Budapest,
    May 4-6 (1983).

TUMOR MARKERS IN TESTICULAR CANCER: A REVIEW OF

12 YEARS EXPERIENCE AT THE NCI

N. Javadpour

National Cancer Institute
Bethesda, USA

INTRODUCTION

Among the tumor systems, testicular cancer is unique in having a number of reliable tumor markers. This tumor system has both placental and embryonic components, therefore, any marker protein found in the embryonic or placental portion may also be detected in the blood. A number of such markers have been found in testicular cancer (Table 1) among which, alpha-fetoprotein (AFP) and human chorionic gonadotropin (HCG) have been found to be the most useful. Other tumor markers such as placental alkaline phosphatase (PLAP) gamma-gutamyl transpeptidase (GGT) and lactic dehydrogenase (LDH) are also of value especially in seminoma where other markers are not commonly present.

This review updates our 12 years of experience with these markers with emphasis on those most useful. A number of reviews and updates have previously been published and readers are referred to them for further details[1-8].

BACKGROUND INFORMATION

The diagnosis of cancer relies upon morphological features of tumor cells. Despite the apparent precision of these criteria, the diagnosis of some neoplasms, such as testicular and ovarian tumors, is, at best, unsatisfactory if based only on histologic features. Over recent years it has become apparent that certain cancer cells synthesize special protein moieties which may be identified in these cells and also measured in the patient's serum, using sensitive radioimmunoassay (RIA) and immunocytochemical techniques.

Among the proteins that have been found in the serum of patients with germ cell tumors of the testis are human chorionic gonadotropin (HCG) and alpha-fetoprotein (AFP). Sensitive and specific radio-immunoassays and immunocytochemical techniques for HCG and AFP which have been developed at the National Cancer Institute are capable of detecting minute amounts of these markers in the sera and cancer cells of patients with testicular tumors. When these two glyco-proteins are utilized together, they are the best serologic and cellular markers available in diagnosis, detection of early re-currence, accurate staging, and in reflecting the adequacy of treat-ment of testicular cancer.

In this communication, I shall review the techniques of detec-tion and the role of biologic markers in testicular cancer with special emphasis on the more established tumor markers such as AFP and HCG. These data have been generated at our laboratories and clinical programs in studying testicular tumor prospectively for the past 12 years.

IMMUNOCYTOCHEMICAL TECHNIQUES (ICCT)

Of the immunocytochemical techniques available the most repro-ducible and convenient are the immunoperoxidase (IP) and peroxidase antiperoxidase (PAP) techniques[9,10]. The advantages of these techniques over fluorescent microscopy include: (1) no specialized equipment is necessary; (2) no fresh tissue is required - the paraf-fin block, formaldehyde-fixed tissues that are usually available in departments of pathology are sufficient for prospective studies; (3) storage of material and slides do not pose any particular problems; (4) slides can be filed and kept as permanent records; (5) the unde-sirable background staining is minimized, and the detail of tissue and exact location of a given cell marker may be precisely localized

Table 1. Testicular Tumor Markers.

Specific Markers
    1.   Alpha-fetoprotein (AFP)
    2.   Human chorionic gonadotropin (HCG)
    3.   Placental alkaline phosphatase (PLAP)
    4.   Gamma-glutamine transpeptidase (GGT)
    5.   Placental Proteins number 5, 10, 15
    6.   Placental lactogen

Nonspecific Markers
    1.   Lactic dehydrogenase (LPH)
    2.   Polyamines (putrescine, spermine, spermicline)
    3.   Carcinoembryonic antigen

---

in various parts of a cell by a simple counterstain. Although there
are a number of such procedures, the IP and PAP techniques will be
briefly described here.

Immunoperoxidase Technique

Immunoperoxidase methods have much in common with established
immunofluorescence procedures. Both have the potential for demonstration of specific cell and tissue antigens, with similar limitations demanding rigorous control of specificity. In any study, the
choice of an immunofluorescence method or an immunoperoxidase method
can be made on rational grounds, according to the desired objective,
the degree of morphologic detail required, the material available for
study, and the ease of access to specialized ultraviolet microscopy.
The major advantage of immunoperoxidase is that it can be utilized in
either a prospective or a retrospective study, since the tissue to be
stained can be fixed in formaldehyde as opposed to immunofluorescence
which requires fresh or frozen tumor specimens. This makes immuno-
cytochemistry more convenient and practical.

This technique utilizes a 4 to 6-micron thick section of formaldehyde-fixed tumor that is deparaffinized in xylene and cleared in
the usual fashion. The section is incubated in a humid chamber for
thirty to sixty minutes with appropriate antisera to a given marker.
The second antibody is a gamma globulin that is conjugated with
horseradish peroxidase. This section is washed again and exposed to
diaminobenzidine containing 0.05% hydrogen peroxide for ten minutes.
All the appropriate controls, including sections exposed to normal
serum and absorbed antiserum are included.

Peroxidase Antiperoxidase Technique

We have recently utilized the peroxidase antiperoxidase technique with equally reliable results. Because of scarcity of information in the literature, it deserves a brief comment.

The technique of peroxidase antiperoxidase is similar to immunoperoxidase with perhaps more sensitivity and less background. As in
the IP technique, antiserum is required, for example, a rabbit antiserum to a given marker; the second antibody is usually a goat anti-
rabbit immunoglobulin G (IgG). This technique utilizes a third
antiserum, peroxidase antiperoxidase, which is raised in rabbit. The
remainder of the technique is similar to that of the conventional IP
technique.

Although it was hoped that steroid receptors could be detected
and localized by the use of this immunologic technique with specific
and sensitive antisera to various steroids, it has not worked satis-

factorily at the present time. Hopefully, preparation of specific
antibodies to estrophilin may open the way for further progress in
this area. Utilizing these techniques, we have localized AFP, HCG
and pregnancy β_1 glycoproteins in various cells of placenta and
testicular cancer (Table 2).

FREQUENCY OF HCG AND AFP IN TESTICULAR CANCER

In the study of 389 patients with testicular cancer, the following distributions of HCG and HFP were found[8] (Table 3).

Eleven of 130 patients with seminoma had elevated levels of
serum HCG. One of these patients had an element of choriocarcinoma
in subsequent histologic sections of the primary tumor. He developed
aortocaval metastases that proved to be choriocarcinoma after retroperitoneal lymphadenectomy. One hundred and twenty-nine patients had
normal serum AFP. However, in 1 patient with seminoma, the serum AFP
was elevated, and in serial sectioning, an element of embryonal
carcinoma was found. This patient was proven to have metastatic
embryonal carcinoma after retroperitoneal lymph node dissection. One
hundred and two of 145 (70%) patients with embryonal carcinoma, 36 of
56 patients (69%) with embryonal carcinoma with or without teratoma
(64%), 3 of 4 patients with yolk sac tumors (75%) and none of 5
patients with choriocarcinomas had an elevated level of serum AFP.
Eighty-seven of 145 patients (60%) with embryonal carcinomas, 32 of
56 patients (57%) with embryonal carcinomas with or without teratoma,
one of 4 patients (25%) with yolk sac tumors and 5 of 5 patients
(100%) with choriocarcinomas had elevated serum HCG. When both or
either markers were considered, 7.7% of seminoma, 44% of teratoma,
88% of embryonal carcinoma, 86% of embryonal carcinoma with teratoms,
75% of yolk sac tumors, and all patients with choriocarcinoma had
elevated levels of serum HCG and/or AFP.

Table 2.   Immunohistologic Classification of Germ Cell Tumors.

Tumor	AFP	HCG	SP_1
Placenta	−	+	+
Yolk Sac Tumor	+	−	−
Seminoma	−	−	−
Seminoma with STGC	−	+	+
Embryonal carcinoma	+	+	−
Choriocarcinoma	−	+	+
Teratoma	−	−	−

Table 3. Frequency of Elevated HCG and AFP in Patients with Testicular Cancer.

	AFP		HCG		AFP amd/or HCG	
	No. of Patients	Percent of Patients	No. of Patients	Percent of Patients	No. of Patients	Percent of Patients
Seminoma	0/160	0	14/160	9.0	14/160	9.0
Teratoma	6/16	37.5	4/16	25.0	7/16	43.7
Embryonal carcinoma	102/145	70.3	87/145	60.0	127/145	87.5
Embryonal carcinoma with teratoma	36/56	64.2	32/56	57.0	48/56	85.7
Choriocarcinoma	0/5	0	5/5	100.0	5/5	100.0
Tolk sac tumor	3/4	75.0	1/4	25.0	3/4	75.0

Staging of Testicular Cancer

The effective use of surgery, chemotherapy and radiation therapy
for patients with testicular cancer requires accurate staging for
therapy and/or interpretation of end results. The conventional
staging parameters, including the lymphangiogram, inferior venacavogram and excretory urogram, often yield a considerable staging
error.

With sensitive and specific radioimmunoassays of serum alpha-
fetoprotein and human chorionic gonadotropin in 118 patients with
embryonal carcinoma with or without teratoma undergoing clinical and
surgical staging the staging errors have decreased to 9-14% in stage
I and 5-10% in stage II cases. Various clinical observations have
been made in this group of patients: (1) persistently elevated serum
markers after orchiectomy for testicular cancer invariably indicate
stage II or III disease, (2) persistently elevated serum markers
after positive lymphadenectomy usually suggest stage III disease and
(3) persistently elevated serum markers after lymphadenectomy negative for tumor invariably indicate stage III disease. Therefore,
such determinations are important guides to further therapy and must
be an essential feature of adjuvant trials.

Clinically, when these markers were considered in staging testicular tumors the staging errors decreased to 5-14%, a dramatic
improvement when compared to previously reported staging errors of
35-53%. It is important to consider the biologic half-lives of these
markers (AFP 5 days and HCG 18 to 24 hours) to avoid any confusion
from the progressively decaying markers of the already excised tumor.
Pre-orchiectomy serum markers are not always available but this
should not disturb the proposed system since the original orchiectomy
specimen is usually available and immunohistologic techniques, such
as immunoperoxidase, can determine the presence of cellular markers
when serum is not available.

Important features have been demonstrated when markers are used
in staging testicular cancer. A persistently elevated level of serum
markers after orchiectomy for testicular cancer invariably indicates
stage II or III disease. A persistently elevated level of serum
markers after lymphadenectomy usually indicates stage III disease.
When lymphadenectomy is negative for tumor but serum markers after
lymphadenectomy are persistently elevated, patients invariably have
stage III disease. Persistently elevated serum markers after orchiectomy and/or lymphadenectomy indicate the presence of residual tumor
and, therefore, a need for further therapy. Pathologic staging of
testicular tumors based on markers decreases the staging error to an
acceptable level and directs the physicians to appropriate therapy
(Tables 4 and 5).

Table 4.   Improved Staging of NSTT[1] with Markers.

	Clinical Stage II	Pathologic* Stage II	Staging Error
Without markers	46	40	13%
With markers			
Positive	40	38	5-10%
Negative	6	2	

[1] Nonseminomatous testicular tumors
* 2 patients proven to be Stage III on followup

The important features which tumor markers add to the under-staging of testicular cancer are as follows:-

1. Improved staging based on clinical investigations markers, therapy, and pathological findings.
2. Persistently elevated serum markers after orchiectomy for testicular cancer invariably indicate stage II or III disease.
3. Persistently elevated serum markers after lymphadenectomy indicate stage III disease or an inadequate lymphadenectomy.
4. When lymphadenectomy is negative for tumor but postlymphadenectomy serum markers are persistently elevated, patients invariably have stage III disease. However, surgery still remains the most accurate means of assessing retroperitoneal metastasis.
5. Perhaps the most important applications of these markers are in monitoring of testicular tumor when serially measured.

Radioimmunoassay of Urinary HCG

An improved technique for detecting small amounts of HCG that is 20-fold more sensitive than the conventional RIA for detecting the β subunit of HCG has been reported. This technique utilizes concentrated 24-hour urinary HCG and a highly specific RIA with an antiserum (H93) that is specifically prepared against the carboxyl terminus of urinary HCG.

Table 5.   Improved Staging of NSTT with Markers.

	Clinical Stage I	Pathologic Stage I	Staging Error
Without markers	72	52	28-32%*
With markers	57	52	9-14%*

* Based on three patients considered Stage II on followup

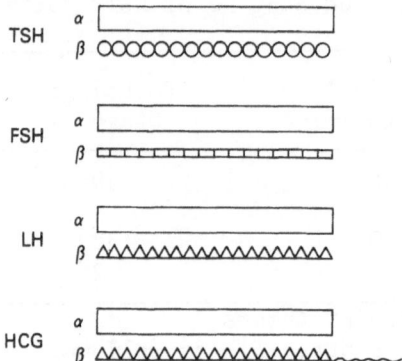

Fig. 1. Structure of 4 human glycoprotein hormones.

By concentrating the HCG in urine specimens, HCG production can
be more sensitively monitored than by measurement of serum levels of
a new carboxyterminal RIA (Figure 1). Hence, the urinary HCG RIA
offers the potential for more sensitively monitoring the tumor burden
and guiding the therapy of patients with testicular cancer.

Simultaneous serum and urinary human chorionic gonadotropin
(HCG) levels were measured in 12 patients with disseminated testicular cancer. Initially these 12 patients who had either seminoma,
embryonal carcinoma, teratocarcinoma, or choriocarcinoma of the testis had elevated levels of serum HCG (Table 6). After treatment with
intensive chemotherapy and/or surgery, the elevated serum HCG values
dropped to undetectable levels. However, the 24 hr. urinary HCG
level measured by radioimmunoassay of urine concentrates was elevated
in 10 of the 12 patients, despite undetectable levels of HCG in the
serum, indicating the persistence of tumor producing this marker. Of

Table 6. Urinary HCG in 15 Patients with Initially Elevated Serum
          HCG.

	No. Patients and Normal Subjects	No. Samples	Serum HCG[a] ng/ml	Urinary HCG[b] ng/ml
Normal	20	30	1	<70
Clinically un-detectable tumor	15	32	1	89–286
After tumors were excised	5[c]	6	1	<70

[a] Measured by Sb6 radioimmunoassay.
[b] Measured by H93 carboxyl-terminal radioimmunoassay.
[c] Five had proved recurrence on no therapy.

these patients 4 were proven to have persistent tumor, which by
histopathology and immunoperoxidase-staining contained HCG. This
highly sensitive urinary HCG radioimmunoassay has improved the detection of persistent tumor burden and has been rewarding in selecting
the patients in whom further therapy is warranted (Figure 2).

Monitoring the Response to Therapy

Serial measurements of serum HCG and AFP by RIA reflect the
efficacy of surgical, radiation, and/or chemotherapeutic regimens in
patients with testicular tumor. When these therapies are effective
they produce an immediate decrease in serum levels of HCG and AFP
that reflects the decrease in tumor size and could be as rapid as the
catabolic rate for these markers (Figure 3). In our series, elevated
markers were found, often months before the patients were symptomatic
or recurrence was detectable by any other clinical tests. Consequently, the markers proved to be sensitive indicators of the presence of otherwise undetectable metastases.

When following a patient with serial marker levels, one may find
elevated marker levels before any tumor is clinically detectable.
The next question to answer, with such a patient, is the location of

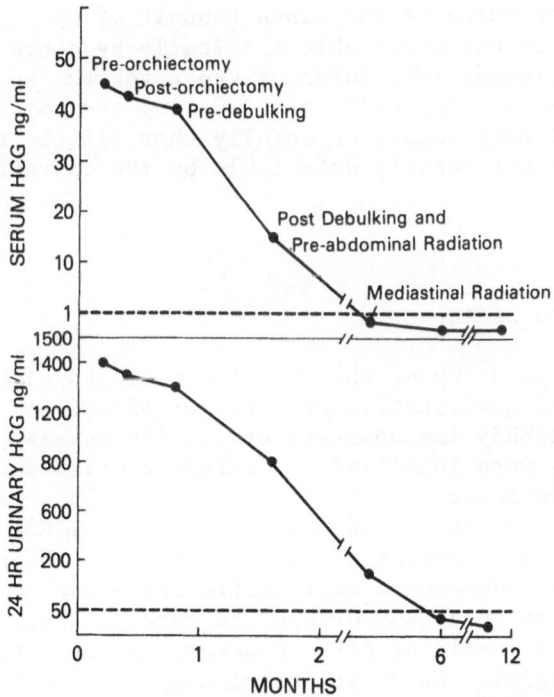

Fig. 2. Simultaneous serum and urinary HCG in a patient with
          testicular cancer.

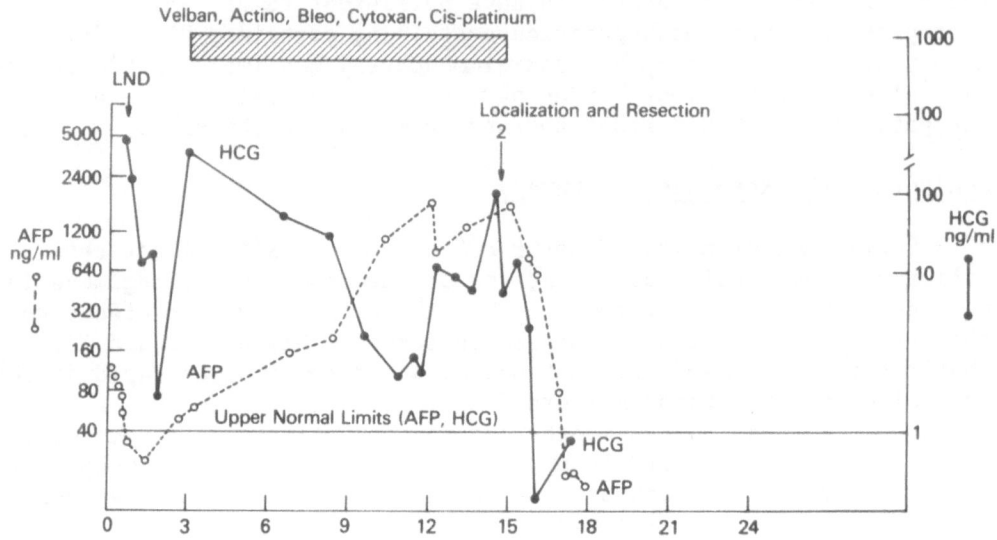

Fig. 3.   HCG and AFP in monitoring the response to various treatments
         in a patient with testicular cancer.

the recurrence. The value of the alpha subunit of HCG is in localizing a tumor that is not detectable clinically by conventional
clinical tests, including IVP, inferior venacavogram, and lymphangiogram. Alpha-HCG has a short half-life (20 min) and may be used in
the localization of metastases, especially those in the retroperitoneal area which are not readily detectable by the conventional tests.

MARKERS IN SEMINOMA

Serum AFP and HCG in Seminoma

     In a prospective study at the NCI, 130 patients with "pure
seminoma" had serial quantitative measurement of HCG and AFP by
specific double-antibody immunoassays originally developed at the
NCI. These markers were localized in different cells using techniques of immunoperoxidase and immunofluorescence on serial sections
of the tumors. Eleven out of 130 had elevated serum HCG. In serial
sectioning of the tumor specimens 1 of 11 patients had an element of
choriocarcinoma and underwent a retroperitoneal lymph node dissection
and chemotherapy; the serum HCG dropped to normal. There were 129
patients with normal levels of AFP. However, in one patient the
serum AFP was 152 ng/ml. On serial sectioning, an element of embryonal carcinoma was found; this patient has also been proven to have
metastatic involvement. A patient with massive bulky retroperitoneal
seminoma and left hydronephrosis underwent debulking and radiation.

In this study, we have observed the following clinical findings.
(1) The frequency of elevated HCG in the serum of patients with
seminoma is about 7.5% (10 of 130).   (2) Although the synctiotropho-
blastic tumor cell occasionally found in pure seminoma is capable of
secreting HCG, one must look for elements of choriocarcinoma, or
embryonal carcinoma, or both. This surely changes the therapeutic
approach.   (3) The elevated serum AFP in patients with seminoma
indicates the presence of an element of embryonal carcinoma that also
changes the therapeutic approach.   (4) The reported cases in the
literature of seminoma with an elevated level of HCG are either
lacking serial sections or localization of cellular HCG or both.
Therefore, we must be cautious in accepting them as pure seminoma.

Multiple Markers in Seminoma

The role of γ-glutamyl transpeptidase (GGT), placental alkaline
phosphatase (PLAP), and human chorionic gonadotropin have been
studied in testicular seminoma. In 89 seminoma patients with negative β-glycoprotein, total serum GGT was measured and values above 30
IU per liter were considered abnormal. Serum PLAP was measured by
enzyme-linked immunoabsorbent assay and values >1.85 mg per ml were
considered abnormal. Serum HCG and AFP were measured by double antibody radioimmunoassays (normals <1 ng per ml and <20 ng per ml, respectively). At the time of this study, 30 patients had detectable
seminoma, 10 were histologically unconfirmed, and the remaining 49
had no evidence of tumor. Only six of 30 patients (20%) with active
tumor had elevated levels of serum HCG. Twelve of 30 patients with
active tumor (40%) had elevated serum PLAP, and 10 of 30 (33%) of
these patients had elevated serum levels of GGT. When these three
serum markers were considered together, more than 80% of the the
patients with clinically active tumors had detectable serum levels of
one or more of these biochemical serum markers. Since the survival
of patients with stage III seminoma treated by radiation is only 28%,
we advocate serial measurements of these serum markers along with
early utilization of new chemotherapeutic regimens in these patients.
However, it should be emphasized that the false positive, false negative rates of these markers, especially false positive rates for GGT,
due to occasional concomitant liver disease and the biologic half-
lives of these markers should be taken in consideration (Table 7).

Other Markers in Seminoma

A common marker that may be useful in the management of seminoma
is lactic dehydrogenase (LDH). Serum lactic dehydrogenase is a
nonspecific enzyme made up of five heterogenous isoenzymes in man
that can be measured electrophoretically. Cancer cells have increased glycolysis leading to an increased synthesis of lactate, and
it may be utilized as a nonspecific tumor marker in several cancers.

Table 7. Incidence of False Positive and False Negative Frequency of Placental Alkaline Phosphatase (PLAP), γ-glutamyl Transpeptidase (GGT), and Human Chorionic Gonadotropin (HCG) in 89 Patients with Seminoma.

Status of 79 patients	Patients No.	PLAP %	GGT %	HCG %	PLAP, GGT and/or HGG
Detectable tumor	30	40	33	20	80
Nondetectable tumor	49	12	4	0	14

Ten patients who had suspected tumor, but in whom it was not confirmed histologically, were excluded for this analysis.

In seminoma LDH may be particularly useful because of several
factors:

1. There is a lower frequency of serum HCG elevation in seminoma as
    compared with nonseminoma.
2. Measurement of LDH is more readily available and simpler than
    radioimmunoassay studies.
3. The majority of patients with bulky stages II and III seminomas
    seen at the NCI had elevated serum levels of LDH which were
    useful in monitoring their therapy. The preliminary results
    suggest that an elevation of LDH may be relatively specific in
    testicular cancer compared with other neoplasms.

IN VITRO SYNTHESIS OF α-FETOPROTEIN AND HUMAN CHORIONIC
GONADOTROPIN IN TESTICULAR CANCER

    In vitro synthesis of α-fetoprotein and human chorionic gonado-
tropin utilizing C-14 lysine and C-14 leucine incorporation in fresh
testicular cancer tissue is demonstrated by radioimmunoassay, immunoprecipitation of ^{14}C-HCG and ^{14}C-AFP, and quantitation of AFP and HCG
(Figures 4 and 5). The in vitro synthesis of AFP and HCG was correlated with immunocytochemical staining in the cancer cells of
patients. The technique of in vitro detection of AFP and HCG requires labeling and immunoprecipitation of these labeled markers.

    A number of observations have been derived from the results of
these studies. First, testicular tumors are heterogenous, and
immunocytochemical studies have demonstrated that not all the cancer

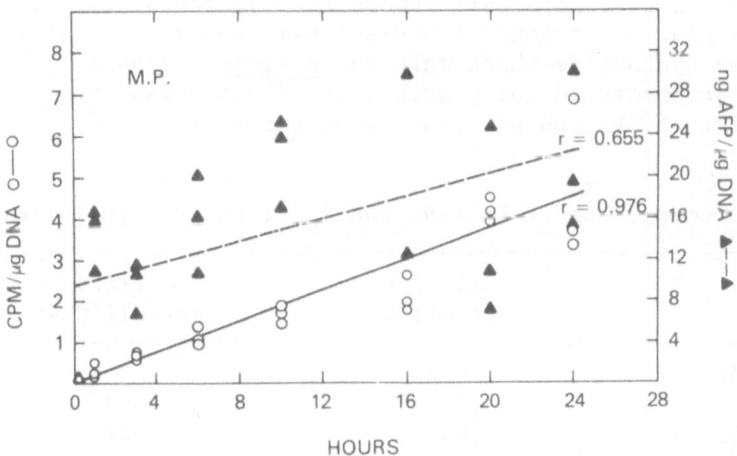

Fig. 4. AFP synthesis by utilizing C-Leucine and lysine in in vitro
         testicular cancer cells.

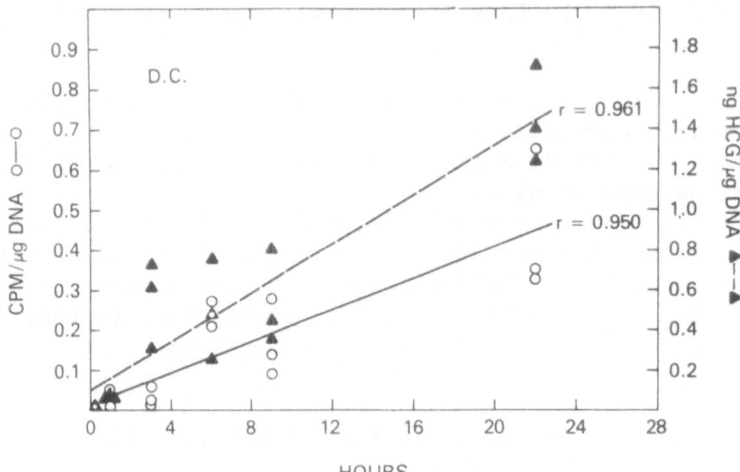

Fig. 5. HCG synthesis by utilizing C-Leucine and lysine in vitro by
         testicular cancer cells.

cells stain for HCG or AFP. It appears that only clones of cells
synthesize these markers in a given tumor. Furthermore, the use of
minced tissue may mean that the number of dead cells unable to produce any markers is underestimated; therefore, the synthetic rates of
these markers may be underestimated. Second, the available data
indicate occasional discordance between serum and tissue levels of
AFP and HCG and in vitro synthetic rates. These discordances may be
due to lack of a sufficiently sensitive technique to measure extremely low levels of these markers or may be due to sampling errors.
Third, it appears that germ cell tumors from different sites, such as
primary testicular, extragonadal mediastinal, and metastatic retroperitoneal may synthesize these markers in vitro. Finally, these
studies have demonstrated the possibility of measuring the in vitro
synthetic rates of AFP and HCG from fresh tumors (Table 8).

Table 8. Comparison of In Vivo and In Vitro AFP Synthesis.

	In Vivo pg/cell/day	In Vitro pg/cell/day
R.M.	.0005	.0004
R.G.	.002	.003
W.Q.	.002	.001
Same Comparison for HCG		
Q.W.	.005	.006

DISCORDANCE BETWEEN AND LIMITATIONS OF TUMOR MARKERS

The discordance between various testicular tumor markers is well known and may be explained on the basis of the findings that different cells produce these various markers. Also, during chemotherapy of a patient with elevated levels of serum HCG and AFP, one may return to normal whilst the other remains elevated. This may occur if some of the cells producing a given marker are resistant to the therapy (Figure 6). Furthermore, we have demonstrated the cellular source of various tumor markers utilizing immunocytochemical techniques.

In conclusion, the utilization of serum AFP, HCG, and serum LDH in conjunction with other studies plays an important role in testicular cancer. In monitoring these patients, monthly measurements of serum AFP and HCG and chest X-ray are essential in following patients in the first 12 months; these measurements should be made every 2 months during the second year and yearly thereafter.

In spite of certain limitations these markers appear to be the best available in any solid tumors. The current practices and recommendations to minimize certain problems and maximize the efficacy

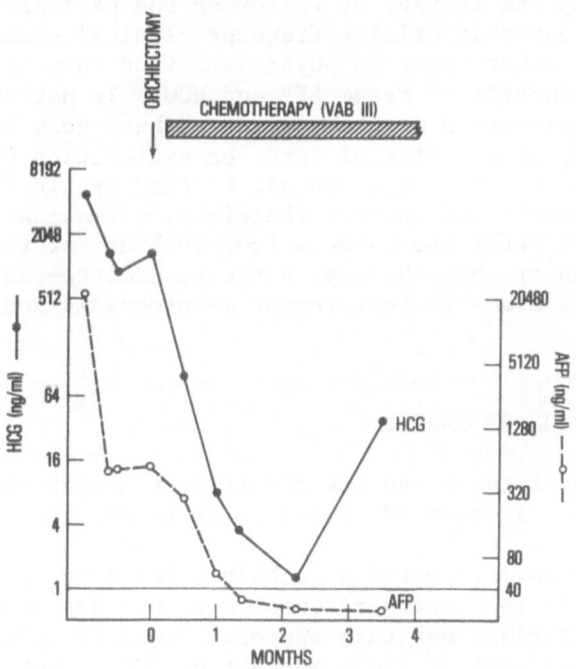

Fig. 6. Discordance between AFP and HCG.

of RIA measurement of serum AFP and HCG from the commercial sources
for testicular cancer are:

1.   The physician should discuss the sensitivity and specificity of
a given commercial assay with the laboratory, and perhaps, occasional
inclusions of normal serum or serum with known levels of AFP and/or
HCG may serve as negative and positive controls when blindly coded.

2.   These markers should not replace scrotal exploration for histopathologic diagnosis of the primary tumor and retroperitoneal lymphadenectomy to detect or exclude the presence of retroperitoneal
metastasis. However, the elevated levels of tumor markers are
indicative of the presence of tumor and the necessity for further
treatment. They are also helpful in monitoring the efficacy of and
the need for changes in therapy.

3.   The problem of impurity of certain antisera against the subunit
of the HCG or the possibility of high levels of luteinizing hormone
(LH) in patients undergoing orchiectomy and/or chemotherapy causing a
false positive result should also be kept in mind. The false positive results may be clarified by the testosterone suppression test,
determination of serum LH, and measurement of HCG on urinary concen-
trate utilizing a carboxy-terminal RIA that is currently available to
all urologists through the NCI laboratories as a courtesy.

4.   In monitoring the therapy or following the patients with testicular tumor, one should utilize frequent clinical examination,
chest X-rays, and other tests as physicians find them necessary,
along with determination of serum AFP and HCG. In patients on chemotherapy, the normalization of these serum markers does not mean
tumor-free status; as a matter of fact, on exploration of the retroperitoneum and chest, it is not unusual to find cystic fibrotic
material with necrosis and tumor. Therefore, normalization of serum
markers should not deter the surgeon from looking for tumor. Appropriate utilization of chemotherapy, surgery, radiotherapy, and tumor
markers can make a dramatic improvement in prognosis and survival of
these patients.

RADIOIMMUNODETECTION OF CANCER

     Localization of tumor and its significant impact in surgical and
radiotherapeutic management of cancer is of increasing significance.

     Recently, we have reported a technique for locating α-HCG-
producing tumors in the retroperitoneal area not detectable with the
conventional modalities, but this approach requires multiple venous
catheterizations and use of alpha subunit of HCG. Another method
recently developed for locating tumors in vivo is based on injecting
radioactive antibodies made against a tumor-associated marker and

then performing total-body scintigraphy to pinpoint foci of abnormal
radioactivity, corresponding in sites of tumor; this procedure has
been termed radioimmunodetection (RID) of cancer. The first extensive application of this approach was antibodies to carcinoembryonic
antigen (CEA). More recent studies have included the use of antibodies to AFP and HCG[13].

Radioantibody Preparation

Hyperimmune goat antiserum is prepared with purified urinary
HCG. The anti-HCG serum is absorbed with urinary protein using an
automated chromatography system with a solid phase (Sepharose 4B,
Pharmacia, Piscataway, New Jersey), immunoabsorbent column. The
immunoglobulin G (IgG) fraction of the absorbed antiserum is chro-
matographically purified as previously discussed. Goat anti-HCG is
labeled with Iodine (Amersham/Searle, Arlington Heights, Illinois)
by the chloramine-T method. Individual preparations of l-labeled
goat anti-HCG IgG are tested for pyrogenicity in rabbits, and for
sterility and acute toxicity. After they are proved to be nontoxic
and pyrogen-free, they are utilized for patients. Goat anti-human
AFP is also labeled with Iodine utilizing the same techniques.

Scintillation Techniques

The radioiodinated anti-HCG or anti-AFP IgGs are administered
intravenously at a total dose of approximately 1 mCi in 20 ml of
sterile normal saline over a period of ten to fifteen minutes. To
subtract the free iodine in bladder and stomach and to suppress the
background caused by antigen-antibody complexes, 500 μCi of Tc-label-
led human serum albumin is injected intravenously before imaging.
Images of the anterior chest and anterior and posterior abdomen were
obtained by gamma scintillation twenty-four and forty-eight hours
after the intravenous injection of the radioantibody. The data
obtained are stored in a computer capable of computing digital images
of the l-labelled antibody alone, Tco4 and Tc-albumin, or l-labelled
antibody minus the technetium components. Cancer radioimmunodetection with antibodies to HCG and to AFP appears to be a useful procedure for the pretreatment and post-treatment evaluation of patients
with testicular cancer and can reveal sites of tumor not detected by
other methods for HCG-producing tumor. This technique may be utilized in RIT by utilizing a therapeutic dosage of antibody labelled
with radioactive material. Experimental design using athymic mice
with xenographic human yolk sac tumor is in progress in our laboratory. The potential for RID (Figures 7 and 8) and RIT in this tumor
system and other tumors is significant.

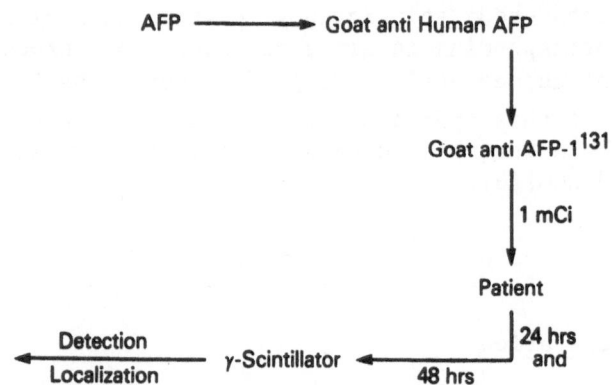

Fig. 7. Radioimmunodetection of AFP producing tumor.

PREPARATION OF MONOCLONAL ANTIBODIES

Immunologic assays have always been plagued by uncertainties and unpredictability in heterogeneity of the immune response[14]. Immunization is still more of an art than a science, and serologists have had to be satisfied with whatever quality and quantity of antibodies an immunized animal will provide. However, hydridoma technology has eliminated some of these technical problems. In order to appreciate the advances that have been made and the difficulties that remain, it is necessary to understand how antibody-producing hybridomas are generated by hybridoma. Mice are immunized with the antigen of interest and are then usually given another injection to obtain a secondary response (Figure 9). Two to 4 days later, the spleen is removed and teased apart to form a suspension of spleen cells. The spleen cells are mixed with mouse myeloma cells that have previously been adapted to grow continuously in culture. Polyethylene glycol is added to the mixture to promote the fusion of cell membranes, and the cells are suspended in tissue culture medium. Because only one of every 2×10^5 spleen cells actually forms a viable hybrid with a myeloma cell, it is necessary to eliminate all the unfused myeloma and spleen cells to allow recovery of the rare hydrid. Such hybrids can be isolated by growth in a selective medium. Variants of mouse

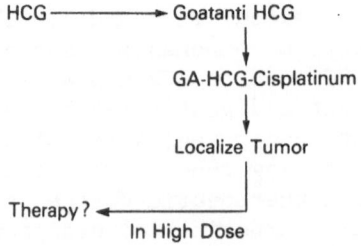

Fig. 8. Affinity therapy of cancer by chemotherapeutic agents labelled antibody.

myeloma cells that lack the enzyme hypoxanthine phosphoribosyltrans-
ferase are used as the myeloma parent, and hypoxanthine, aminopterin,
and thymidine are added to the growth medium. Since the cells with
only the myeloma genetic material lack the enzyme, they cannot use
exogenous hypoxanthine to synthesize purines. The aminopterin blocks
their endogenous synthesis of purines and pyridines, and the myeloma
cells die. However, a hybrid between a myeloma cell and a spleen
cell contains the transferase provided by the normal spleen cell,
uses the exogeneous hypoxanthine and thymidine, and survives. The
normal spleen cells are not killed by this selective medium, but they
do not survive the culture. The hybrids, which double approximately
every 24 to 48 hours, rapidly outgrow the few nondividing but persistent spleen cells. Clones that are producing antibody are grown
to mass culture and are recloned as soon as possible. This procedure
eliminates contaminating non-antibody-producing clones and cells that
have lost the ability to produce antibody as a result of chromosomal
loss. The specificity of the antibody is confirmed by more extensive
testing.

STUDIES OF LACTIC DEHYDROGENASE (LDH)

    LDH is a glycolytic enzyme found in many human tissues and
fluids. The enzyme is released into serum owing to tissue injury due
to inflammatory conditions, degenerative processes, toxicity, or
cancer. Elevations of serum LDH levels have been found to reflect
growth and regression of various malignant neoplasms[15]. Eight
patients with testicular cancer were studied with serial serum LDH.

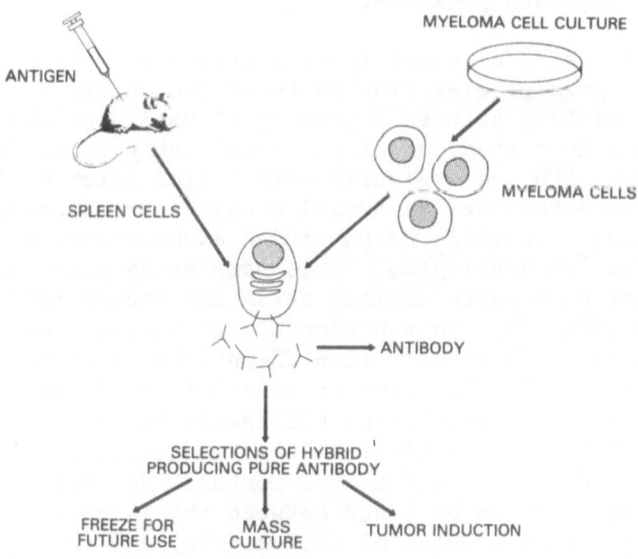

Fig. 9. Techniques of making monoclonal antibodies.

LDH was determined by observing the change in absorption of DPNH as pyruvate was converted to lactate. Normal values were <340 ng/ml before June 1, 1978, <381 for June 1978, and <248 after July 1, 1978, owing to technical improvements[16].

Of the 80 patients, 23.8% (19) were Stage I (tumor confined to the testicle), 23.8% (19) were Stage II (metastatic disease in the retroperitoneal lymph nodes only), and 52.5% (42) were Stage III (visceral or distant metastases). Because the NIH specifically sought patients with bulky metastatic disease during this period for protocol purposes, bulky Stage III is more heavily represented. Eight of the Stage III patients had extragonadal tumors. Eleven patients had seminomas while 69 patients had nonseminomas.

In this study, frequency of elevation of pretherapy levels of LDH in patients with germ-cell testicular tumors was definitely higher in Stage III patients. Only 20% (1/5) Stage I patients had elevated LDH levels (Table 3), in comparison to 26.3% (5/19) Stage II patients, and 62.5% (25/40) Stage III patients. However, the frequency of pretherapy elevated HCG and AFP levels was similar when comparing stages in these patients. Others have also found in testicular tumor patients that although LDH is much more frequently elevated in advanced disease, the same findings are applicable to HCG and AFP.

Currently, no known serum marker is frequently elevated in patients with seminoma. Since seminomas do not have elevated AFP levels and seldom have elevated HCG levels, one may measure LDH that is a simple and easily available test. If elevated, it can be utilized in monitoring such patients.

Although the size of tumor found intraoperatively did not correlate with the degree of elevation in total LDH in other solid tumors, an overall correlation between degree of elevation of LDH and cancer burden was found in a study of 27 patients with germinal testicular tumors. However, LDH was evaluated with serial serum evaluations and tumor burden was determined by serial evaluations of calculated areas of cancer lesions measurable by physician examination, chest roentgenogram, and/or lymphangiogram. This same study also shows an overall correlation between maximum serum LDH concentration and prognosis. Specifically, four patients with maximum serum LDH >5000 IU/liter had a poorer prognosis than 23 patients with maximum serum LDH <5000 IU/liter. In this present study of 80 patients, 43 patients who had normal pretherapy LDH levels had a mean survival time (MST) of 15.5 months while the 26 patients who presented with elevated LDH values had an MST of 9.2 months. However, little significant difference could be found between the 14 patients who had initial LDH elevations <900 ng/ml (9.1 months) and the 12 patients who had initial LDH elevations >900 ng/ml (9.3 months). Therefore, whether initial LDH was elevated or normal played a prognostic role while the amount of elevation did not correlate with prognosis. However, in a study of 204 nonseminomatous germinal testis tumors, initial LDH levels were found to correlate with mean survival times in that for initial LDH values of <225, MST was > eight years; for LDH of 255-600, MST was 14 months; and for LDH >600, MST was 10 months. Hence, LDH can be seen to play a role in pointing to a poorer patient prognosis when it is elevated initially (Table 9).

In conclusion, serum LDH levels are elevated in a number of bulky testicular tumors and correlate well with the course of treatment. Therefore, when elevated, LDH may be utilized as a guide for response to therapy (Figure 10). It is not helpful in diagnosis, or staging of patients with testicular tumor. However, it can be valuable in seminoma patients with no other markers. Of benefit is the fact that it is a simple inexpensive hospital test that is easily available with quick results. Finally, because we have had a fair number of patients with advanced bulky testicular tumors with multiple poor prognostic features, we have correlated prognosis with serum LDH and it appears that elevated serum LDH level is of value as a prognostic indicator. This is, perhaps, reflecting the bulk of tumor since bulky disseminated tumors have a poorer prognosis. Further studies are being performed to evaluate elevation of the various LDH isoenzymes rather than total LDH for possible increased sensitivity and specificity.

OTHER PLACENTAL PROTEINS

Over the past several years, we have studied a number of placental proteins including pregnancy specific β_1 glycoprotein and placental proteins number 5, 10 and 15 utilizing immunoperoxidase. We have localized these markers in syncytiotrophoblastic components of the human placenta, choriocarcinoma and syncytiotrophoblastic giant cells associated with testicular cancer (Table 10).

Table 9. Correlation of Prognosis with Pretherapy LDH Levels in Bulky Stage III Testicular Tumor Patients.

Pretherapy LDH Levels	Surviving Patients	
	No.	%
Normal	7/14	50.0
Elevated	7/24	29.6

LDH = lactic dehydrogenase.

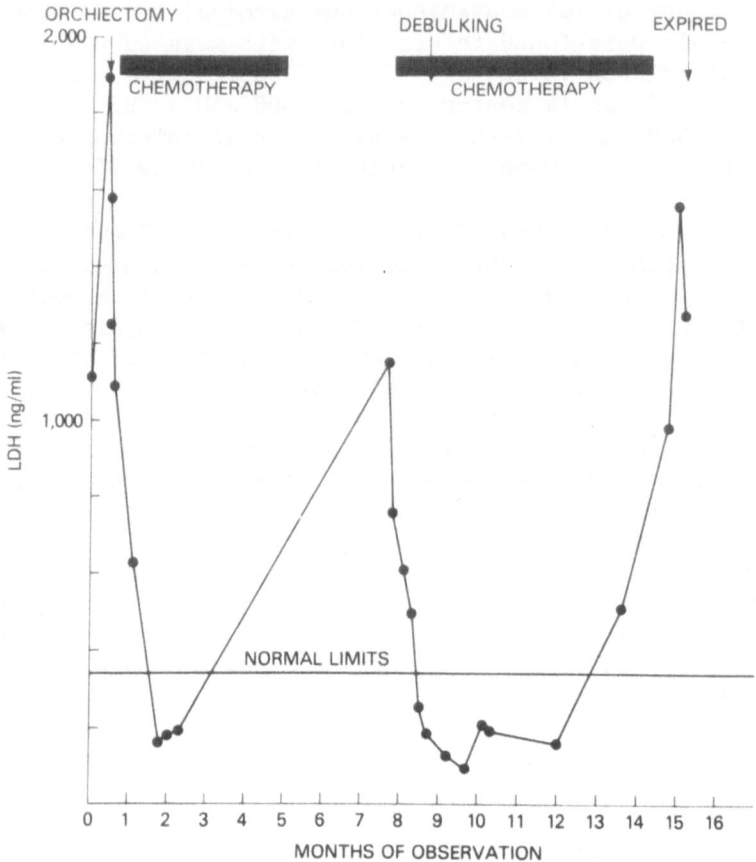

Fig. 10. LDH as monitor of testis tumor.

CONCLUSION

Perhaps the most outstanding progress in cancer immunology has been in the field of immunodiagnosis; mainly the development of specific and sensitive immunocytochemical techniques to measure and localize cell markers in the sera and cancer cells of cancer

Table 10. Other Placental Proteins.

	SP_1	Placental Protein # 5, 10, 15
Placenta	+	+
Syncytiotrophoblast	+	+
Tumor Giant Cell	+	+

patients. In this review, the development and utilization of cell
markers in urologic cancer with newer techniques such as IP, PAP,
RID, (Figures 7 and 8) RIT and monoclonal antibodies are presented.

These findings have exceeded even the wildest dreams of immu-
nologists, and they are revolutionizing serology. Not only is it now
possible to generate a homogeneous antibody, but also the production
of that antibody is immortalized, and the only limitation on the
amount of antibody available is the number of mice to which one is
willing to give injections. In addition, impure antigens can be
used, since the technique for generating antibody-forming hybrids
results in the identification and propagation of cloned cell lines
producing a single antibody that will react with the antigen and will
not react with any contaminating material. Basic researchers in all
areas of biology and clinical investigators studying a wide variety
of systems have recognized the enormous potential of large amounts of
monoclonal homogeneous antibodies and have begun to generate such
reagents. These developments have become an important part of contemporary urologic practice.

GERM CELL AND GONADAL STROMAL TUMORS

Utilizing IP, antibodies to AFP and SP1 have been localized in
various germ cell tumors (Table 2). This immunohistologic classification has helped up to understand the origin of these markers. It
has been shown that the immunoperoxidase technique can be used to
identify steroid hormones in sections of fixed embedded tissues.
This advance paves the way for prospective and retrospective studies
of specific sites of steroid hormone localization in a poorly understood group of gonadal neoplasms, namely those within the sex cord-
stromal category. The ability to localize specifically testosterone,
estrogen, and progesterone had challenged many of the time-honored
concepts of steroid biosynthesis by gonadal stromal tumors. In the
past, specific hormone synthesis was attributed more or less to
specific types of cells. These cells were thought to be responsible
for estrogen synthesis and Leydig cells for testosterone production;
granulosa and Sertoli cells were regarded as inactive generally.
Utilizing highly specific antibodies for testosterone, estradiol, and
progesterone, it has now been shown that all these cells are func-
tionally active and furthermore, that most have the capacity to
synthesize both estrogens and androgens. Testosterone is most frequently localized in Leydig cells, but it may also be present in
Sertoli cells and occasionally in granulosa cells. Estradiol is
found not only in theca cells but also frequently in granulosa,
Sertoli, and Leydig cells, whereas progesterone appears to be localized mainly in luteinized theca cells and less commonly in granulosa
and Leydig cells (Table 11).

Table 11.   Cell Origin of Steroids in Testicular Nongerm
           Cell Tumors.

	Estrodiol	Testosterone	Progesterone
Sertoli Cells	+	+	–
Leydig Cells	+	++	±

REFERENCES

1. N. Javadpour, K. R. McIntire, and T. A. Waldmann, Immunochemical
    determination of human chorionic gonadotropin (HCG) and
    alpha-fetoprotein (AFP) in sera and tumors of patients with
    testicular cancer, Natl.Cancer Inst.Monogr., 49:209–213
    (1978).
2. N. Javadpour and S. M. Bergman, Recent advances in testicular
    cancer, Curr.Probl.Surg., (1978).
3. T. A. Waldmann and R. A. McIntire, The use of a radioimmunoassay
    for alpha-fetoprotein in the diagnosis of malignancy, Cancer,
    34:1510–1515 (1974).
4. N. Javadpour, Serum and cellular biologic tumor markers in
    patients with urologic cancer, Hum.Pathol., 10:557 (1979).
5. R. J. Kurman, P. T. Scardino, K. R. McIntire, T. A. Waldmann,
    and N. Javadpour, Cellular localization of alpha-fetoprotein
    and human chorionic gonadotropin in germ cell tumors of the
    testis using an indirect immunoperoxidase technique. A new
    approach to classification utilizing tumor markers, Cancer,
    40:2136–2151 (1977).
6. J. L. Vaitukaitis, G. D. Braunstein, and G. T. Ross, A radio-
    immunoassay which specifically measures human chorionic
    gonadotropin in the presence of human luteinizing hormone,
    Amer.J.Obstet.Gynecol., 113:751–758 (1972).
7. T. A. Waldmann and K. R. McIntire, The use of a radioimmunoassay
    for alpha-fetoprotein in the diagnosis of malignancy, Cancer,
    34:1510–1515 (1974).
8. N. Javadpour, The role of biologic tumor markers in testicular
    cancer, Cancer, 45:1755–1761 (1980).
9. C. R. Taylor, Immunoperoxidase techniques, Arch.Pathol.Lab.Med.,
    102:113 (1978).
10. N. Javadpour, Immunocytochemical techniques in localization of
    tumor markers in cells and tumor: A potential for radio-
    immunodetection and radioimmunotherapy, Urology, 1:1 (1983).
11. P. H. Lange, K. R. McIntire, T. A. Waldmann, T. R. Hakala, and
    E. E. Fraley, Serum alpha fetoprotein and human chorionic
    gonadotropin in the diagnosis and management of non-
    seminomatous germ-cell testicular cancer, N.Engl.J.Med.,
    295:1237 (1976).

12. N. Javadpour, T. Soares, and G. L. Princler, In vitro synthesis
     of alpha-fetoprotein and human chorionic gonadotropin in
     testicular cancer, Cancer, 45:1755 (1980).
13. N. Javadpour, E. E. Kim, and F. H. Deland, The role of radio-
     immunodetection in the management of testicular cancer,
     J.A.M.A., 246 (1981).
14. C. Millstein, Somatic cell genetics of antibody-secreting cells:
     studies of cloncal diversified and analysis by cell fusion.
     Cold Spring Harbor Symp. Quant.Biol., 41:793 (1977).
15. F. E. Von Eyben, Biochemical markers in advanced testicular
     tumors, Cancer, 41:648 (1978).
16. M. C. Lippert and N. Javadpour, Lactic dehydrogenose in the
     monitoring and prognosis of testicular cancer, Cancer,
     48:2278 (1981).

PLACENTAL-LIKE ALKALINE PHOSPHATASE AS

A TUMOR MARKER IN SEMINOMA

L. Andersson, F. Edsmyr and A. Jeppsson

Department, of Urology and Radiotherapy
Karolinska Hospital, University of Umeå
Umeå, Sweden

Tumor markers have proved to be useful aids in the follow-up
after treatment for malignant tumors of the testis and further
screening tests have become even more necessary following the introduction of potent drugs for the combination of chemotherapy and
surgery with a potential curative effect in patients in whom the
disease is not too far advanced. Alphafetoprotein (AFP) is a reliable marker substance in yolk-sac differentiated tumors, in which
it occurs in about 75% of cases. Human chorionic gonadotrophin
(β-HCG) occurs in approximately 90% of cases in chorionic carcinoma.
Both of these substances are also found in around 60% of embryonic
carcinoma or teratoma[1,2]. HCG also occurs in 10-15% of patients
with seminoma[3]. However, we have not had a marker with such a
specificity in seminomatous tumors.

Fishman et al.,[4] and later Holmgren et al.,[5] identified in
the serum of cancer patients, an alkaline phosphatase immunologically
similar to the phosphatase normally occurring in the placenta. This
marker (PLAP) was found in 7% of patients with embryonic carcinoma
and in higher frequency when very advanced cases were investigated.

Wahren et al.,[6] studied the content of various marker substances in testicular tumors using an immunofluorescence technique
for identification of the markers in smears from tumors, and in
radioimmuno-assays of tumor lysates. They observed that there were
high concentrations of AFP in embryonic carcinoma and that a high
percentage of the tumor cells took the immunofluorescent stain for
anti-AFP. AFP was not found in measurable amounts in seminomas. On
the other hand high levels of placental alkaline phosphatase were
found in many of the seminomas. An even higher concentration of PLAP
was found in a testis with embryonic carcinoma. No component of

seminoma could be found in the sections from this tumor. A testicular lymphoma and five testicles removed for other reasons or at
postmortem were also examined although not with all tests. The
antisera gave no stain with the non-malignant testicular tissues or
the lymphoma and no abnormal marker levels were found.

With the aim of immunolocalizing a PLAP-producing human tumor
and to study the specificity of the antibodies used, Jeppsson et
al.,[7] produced monoclonal and polyclonal antibodies against purified placental alkaline phosphatase. The antibodies were labelled
with 125 I and injected intraperitoneally into mice which had previously been implanted with human tumor cells. One of these tumor
cell lines was subcloned from HeLa cells derived from a cervix carcinoma and was known to produce PLAP. The animals were killed after
2, 4 or 6 days and the concentration ratio of 125 I labelled rabbit
anti-PLAP in the various organs was studied. The tumor had a significantly higher concentration ratio than all the other organs and
tissues studied. (Table 1).

It was also found that there was a correlation between the PLAP
concentration in the tumors and the amount of labelled antibody
localized to the tumor.

The immunofluorescent staining of tumor cells was performed
using rabbit anti-PLAP. With imprints from the implanted PLAP-
producing tumor, 40% of the cells took the stain. On the other hand,
100% of tissue cultured tumor cells took the stain. The implantation of the tumor in the nude mouse apparently changes the mode of
production of PLAP.

PLAP is not a secretory protein like AFP, CEA and HCG but is
bound to the outer surface of the cell membrane.

Table 1.   Mean Concentration Ratios of 125 I-Labelled Rabbit
            Anti-PLAP in Various Organs of Nude Mice 4 Days After
            Implantation with a PLAP-Producing Tumor

Site	Concentration ratio organ/whole mouse	Site	Concentration ratio organ/whole mouse
Tumor	8.5	Spleen	1.5
Blood	4.8	Liver	1.4
Kidney	2.7	Heart	1.2
Skin	2.4	Muscle	0.9
Thyroid	2.0	Intestine	0.9
Lung	2.0		

All these findings indicate that there is at least a hypo-
thetical reason to use the placental alkaline phosphatase for immuno-
detection and perhaps even for the selection of chemotherapy.

METHODS

We have studied the placental-like alkaline phosphatase activity
of serum in 100 patients with seminoma using the rabbit anti-PLAP
radioimmunoassay[8]. Ninety-seven of these patients had seminoma of
the testis and three had seminoma of the mediastinum. Control serum
samples were collected from 51 control subjects, men matched for age.
Of these, 17 had acute mumps, and 34 were healthy blood donors. Sera
were also collected from 10 pregnant women in the third trimester.
The sensitivity of the assay was 12 μg/l. Values above that level
were considered to be abnormal in male sera.

RESULTS

In 21 patients with a primary diagnosis of testicular seminoma
(stages I and II according to the classification of Fraley et
al.,[9]) elevated levels of PLAP were found before the orchidectomy
in 9 cases (43%), at a mean value of 53 μg/l. There was no difference between stage I and stage II cases.

Our routine treatment strategy in seminoma is the prophylactic
irradiation of the retroperitoneal nodes, even in the absence of
clinical evidence of metastases. After the orchidectomy 4 patients
still had slightly elevated levels of PLAP. In two of them who were
checked after irradiation, the PLAP value had fallen to an undetectable level. Of course we can speculate whether they may have had
undetected metastases, but this is not proven.

Of those 68 patients with no evidence of disease at follow-up,
nine had slightly raised PLAP levels on one or more occasions.
Borderline values were also found in two of the controls, 16 and 12
μg/l. The specificity of the assay for the control population was
thus 96%. The pregnant women had high levels of serum PLAP, 260 μg/l
on the average. In 12 patients tumor recurrence or metastases were
found. Nine of them had raised levels of PLAP with an average of
53.2 μg/l. The sensitivity of the PLAP assay for primary disease in
this series was thus 43%, or 55% if both primary and advanced disease
was considered. The test was most sensitive in the cases of advanced
disease, nine of 12.

Figure 1 shows the values for PLAP, HCG, AFP and red blood cell
sedimentation rates in a 41-year-old man with testicular seminoma
with no obvious metastases. He had orchidectomy and radiation
therapy. Later he was found to have asymptomatic mediastinal and

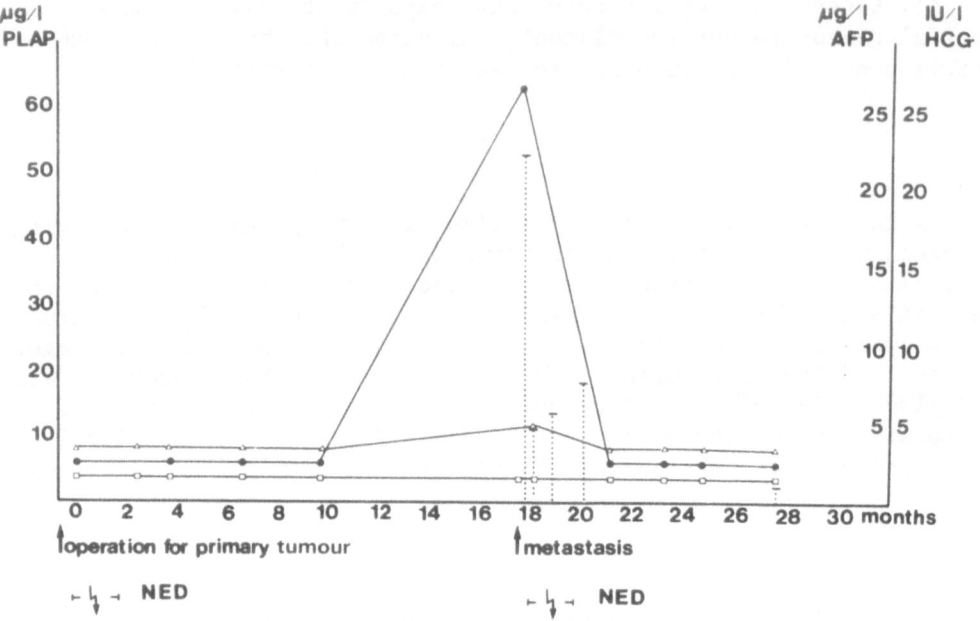

Fig. 1. Serum levels of PLAP (●), HCG (□), AFP (△) and sedimentation
         rates (---) in a 41-year-old man with mediastinal and
         supraclavicular metastases from a testicular seminoma.
         (Courtesy of British Journal of Urology).

supraclavicular metastases. PLAP was elevated and AFP was possibly
slightly raised, though not significantly. The tumor was not at all
reflected in the HCG values. Following irradiation the metastases
disappeared and the PLAP level returned to normal and remained so for
the ensuing 7 months of observation. The test giving the next best
information was the sedimentation rate.

CONCLUSIONS

    Judging from the results so far it appears that the PLAP test is
a useful additional aid in the management of patients with seminoma,
even if it has apparently a lower sensitivity than AFP in non-
seminomatous cancer or HCG in choriocarcinoma. It is now part of our
routine work-up in these cases. For a more conclusive evaluation we
need a somewhat larger patient series and, above all, a somewhat
longer observation period.

REFERENCES

1. G. D. Braunstein, K. R. McIntire, and T. A. Waldmann,
    Discordance of human chorionic gonadotropin and alphafeto-
    protein in testicular teratocarcinomas, Cancer, 31:1065
    (1973).
2. F. Edsmyr, B. Wahren, and C. Silfverswärd, The use of immunology
    and hormonal markers in the treatment of testis tumors,
    Int.J.Radiol.Oncol., 1:279 (1976)
3. N. Javadpour, K. R. McIntire, and T. A. Waldman, Human chorionic
    gonadotropin (HCG) and alpha-fetoprotein (AFP) in sera and
    tumor cells of patients with testicular seminoma. A prospec-
    tive study, Cancer, 42:2768 (1978).
4. W. H. Fishman, N. R. Inglis, C. C. Stolback, and M. J. Kront, A
    serum alkaline phosphatase isoenzyme of human neoplastic cell
    origin, Cancer Res., 28:150 (1968).
5. P. A. Holmgren, T. Stigbrand, M. Damber, B. von Schoultz, and B.
    Wahren, Determinations of placental alkaline phosphatase-
    Regan isoenzyme in cancer sera by sensitive radio-immuno-
    assay. Scand.J.Immunol., 8:515 (1978).
6. B. Wahren, P. A. Holmgren, and T. Stigbrand, Placental alkaline
    phosphatase, alphafetoprotein and carcinoembryonic antigen in
    testicular tumors. Tissue typing by means of cytologic
    smears, Int.J.Cancer, 24:749 (1979).
7. A. Jeppsson, B. Wahren, J. L. Milláan, and T. Stigbrand, Tumor
    and cellular localization by use of monoclonal and polyclonal
    antibodies to placental alkaline phosphatase. Submitted for
    publication 1983.
8. A. Jeppsson, B. Wahren, T. Stigbrand, F. Edsmyr and L.
    Andersson, A clinical evaluation of serum placental alkaline
    phosphatase in seminoma patients, Brit.J.Urol., 55:73 (1983).
9. E. E. Fraley, P. H. Lange, and B. J. Kennedy, Germ-cell
    testicular cancer in adults, New Engl.J.Med., 301:1370
    (1979).

TUMORS OF THE TESTICLE: RECENT ADVANCES IN DIAGNOSIS

J. P. Blandy, R. T. D. Oliver,
R. C. Tiptoft and W. Hately

The London Hospital

The story of testicular tumor therapy is an exciting paradox with wonderful success and discreditable failure. Since the subject was last discussed here in Erice, we have all participated in a revolution in the treatment of advanced disease. No longer are we obliged to watch helplessly while young men die with widespread metastases, and we are all delighted with this success. Sometimes we forget that we have also all been witnesses to a serious failure. Far too many of these young men still present with widespread metastases even today: indeed the proportion of those presenting with advanced disease is, if anything, increasing. And yet - with the exception of the rare Azzopardi tumor that leaves only an undetectable scar in the testis while it marches on to form metastases[1] - all these had a lump in the testis at a stage before it started to spread.

As has already been pointed out earlier in this meeting (Oliver page 131) not only are more patients coming to hospital with metastases, but delay in making the diagnosis - at least in nonseminomas - worsens the prognosis[2].

Why do so many patients come up so late? What can we do to make things better?

We believe we can identify three steps where mistakes occur:

1. Failure of the patient to go to the general practitioner.
2. Failure of the GP to take appropriate action.
3. Failure of the specialist to take appropriate action.

FAILURE OF THE PATIENT

We believe that it is very important to educate young men in
schools, in universities and in the general population to examine
their testicles for a lump, just as we try to educate young women to
examine their breasts for small carcinomas.

FAILURE OF THE GENERAL PRACTITIONER

Although the clinical features of testicular cancers are well
known to all of us, doctors in their surgeries still frequently fail
to refer the patient at once to a surgeon with an interest in the
disease. The only way we can put this right is to make sure that
young medical students are brought into contact with urological
practice, and that urologists get a fair share of the medical
students' time. Unfortunately, in most modern medical schools,
urology forms a very small part of the curriculum.

FAILURE OF THE SURGEON

Not all surgeons realize that there are special groups of men
who have an increased chance of testicular cancer. How often will
they insist on treating the inflammatory form of cancer as if it were
epididymitis? How often do they fail to explore the testicle that
has been the subject of trauma? How often do we see undescended
testicles that have been allowed to lurk, unexplored, in the groin?

Let us be honest: most testicular tumors are easily recognized
by the well-trained fingers of the experienced surgeon. There are a
few difficult areas - small lumps that lie within the body of the
testicle, others that arise near the epididymis and are easily
mistaken for epididymitis, some tumors that lie concealed under a
thick tunica vaginalis, and there are some testicles which are just
difficult to feel properly.

How can the surgeon be helped at this stage? Technetium scanning[3], and Doppler ultrasound examination[4], may be of some value
in making the distinction between epididymitis and torsion, but
neither test is of any value in helping with a solid cancer. Only a
very few hospitals will be able to obtain immediate measurements of
tumor markers, and in any event, they will not always be positive.
Fine-needle aspiration cytology may be useful if you are fortunate
enough to have an expert cytologist, but it is of no value if you do
not know where to put the needle, because you cannot feel a swelling.
Indeed, the only ancillary test which is of practical value, is the
use of ultra-sound.

We have used water-bath ultrasonography at the London Hospital (Tiptaft et al., 1982) in the last 2 years[5]. The principle is simple and elegant. The testicles are immersed in a water-bath containing a battery of transducers. The results are sometimes very helpful. When confronted with a hydrocele through which it is diffi-cult to feel the testicle, or the solitary remaining testis of a neurotic young surgeon, which is difficult to feel, it is very com-forting to know that the underlying testis is homogeneous to ultra-sound. The method can detect small, large or enormous testicular tumours, but it cannot differentiate cancer from granulomatous or-chitis. It is occasionally useful when faced with the normal, but grossly unequally-sized pair of gonads. However, ultrasound is in fact, more often than not, only an extra and additional help in making the decision whether or not to explore the testis.

In fact, what is more important than any other aid to diagnosis, is resolve on the part of the surgeon. There is some curious but very strong reluctance on the part of the - usually male - surgeon, to decide to explore and possibly to remove a testicle. Removal of a breast or an ovary will hardly cost him a moment's thought, but again and again our records betray the hesitation that may cost the young man his life.

Chevassu's manoeure is too little known, and too little prac-tised. Nevertheless it is our belief that it should be used without hesitation whenever there is any doubt that a testicle is not entirely innocent. If the testicle is sliced open, the appearance of a tumor is nearly always obvious: if in doubt, a frozen section may be obtained. If the testis is seen to be entirely normal on gross section, then it can easily be sewn up again, and within 6 months, it will be impossible to distinguish the normal from the operated side[6].

CONCLUSION

Sometimes it seems to take a lot of effort to prove something that has been known for a long time. So it is in cancer of the testicle. It would not be an exaggeration to say that cancer of the testicle could now be eliminated as a cause of death, if only more cases were detected at an earlier stage. Perhaps some of the immense effort that is devoted to research into chemotherapy all over the world, could now be diverted into education of the public, the medical students and the surgical residents of the next generation.

REFERENCES

1.   J. G. Azzopardi and A. V. Hoffbrand, Retrogression in testicular seminoma with viable metastases, J.CLin.Pathol., 18:135-141 (1965).

2.   J. P. Blandy, H. F. Hope-Stone, and A. D. Dayan, Tumors of the
     testicle, William Heinemann Medical Books Ltd., London
3.   K. H. Stage, J. E. Gottesman, D. G. Skinner, and R. M. Ehrlich,
     Testicular scanning: clinical experience with 72 patients,
     J.Urol., 125:334-337 (1981).
4.   R. J. Brereton, Limitations of the Doppler flow meter in the
     diagnosis of the "acute scrotum" in boys, Brit.J.Urol.,
     53:380-383 (1981).
5.   R. C. Tiptaft, Nicholls, W. Hately, and J. P. Blandy, The
     diagnosis of testicular swellings using water-bath ultra-
     sound, Brit.J.Urol., 54:759-764 (1982).
6.   M. Chevassu, Tumeurs du testicule, These de Paris, Paris, No.
     193, G. Steinheil.

TUMORS OF THE TESTIS: LYMPHOGRAPHY AND

OTHER IMAGING TECHNIQUES

R. Musumeci, J. D. Tesoro-Tess, A. Milani,
E. Ceglia, F. Zanoni and G. Pizzocaro

Istituto Nazionale Tumori
Milan, Italy

For the evaluation of the retroperitoneal lymph nodes in cancer
of the testis, a number of radiological investigations can be used.
These diagnostic tools include urography, phlebography, lymphangiography, ultrasound and computed tomography. Each of these procedures
has a different diagnostic accuracy, sensitivity, and specificity.
As a general rule, the more traditional investigations, which are
indirect, are less useful in the identification of retroperitoneal
lymph node metastases. As a rule, the diagnostic value of a radiological procedure, either direct or indirect, is much lower when the
diagnostic tool is used alone than when a combination of two or more
procedures is used.

Urography is the most commonly used of the indirect procedures.
The diagnostic possibilities depend upon the site and size of the
abnormal lymph nodes and upon the possibility that enlarged nodes
will displace the kidney and/or the ureters. For this reason urography is mainly useful in tumors of the left testis in which the
presence of para-aortic metastases can be disclosed by a lateral
displacement of the proximal ureter. The test is less significant
in tumors of the right side, in which metastatic adenopathies are
usually discovered anteriorly to and/or in between the inferior vena
cava and the aorta. Only when the involvement is huge will lateral
displacement of the right ureter be seen.

The "so-called" vascular investigations include opacification of
the inferior vena cava, left renal, and left testicular vein. The
intra-arterial route of administration of the contrast medium is
seldom used and only in selected cases. The diagnostic accuracy of
these procedures correlates with the size and the site of lymph node
metastases (Table 1). When the metastatic node is in a favorable

Table 1.   Inferior Vena Cavographic Interpretation[1].

Histology	Inferior Vena Cavography positive	negative
Positive	24	17
Negative	0	14
Sensitivity	24/41	59%
Specificty	14/14	100%
Accuracy	38/55	69%

Table 2.   Results of Lymphography in Testicular Tumors.
           Data from the literature (1969-80)[2].

No. of cases	2,691
% positive LAG	43.7% (range 28.9-57.5)
% accuracy	85.2% (range 75.7-100)

position, it is possible to see an impression on the walls, a dislocation, or a partial or total obstruction of the opacified veins, with collateral circulation. Only in a few patients will phlebography be the only examination indicating metastases. For this reason, because of its invasiveness and because of the availability of more sophisticated techniques such as computed tomography, its routine use is no longer justified. The procedure may, however, still be useful for demonstrating direct vascular invasion or anomalies of the inferior vena cava and left renal vein prior to retroperitoneal surgery.

At the present time, lymphangiography is the best known diagnostic tool for the direct evaluation of retroperitoneal nodes in cancer of the testis. The published material is abundant but the results are often controversial (Table 2). The percentage of lymphograms abnormal for metastases ranges from 28.9%[3] to 57.5%[4]. According to the literature, the mean value is about 40-42%. Also the diagnostic accuracy of lymphography is controversial with values ranging from 75.7%[5] to 100% presented[4]. The mean value according to the published data is about 85%.

Our experience concerns 807 patients with histologically proven malignant testicular tumors collected between 1965 and 1980 (Table 3). The overall incidence of metastases was 49.7%, higher in germinal (50.6%) than in miscellaneous non germinal cases (31.7%).

Table 3. Results of Lymphography in Tumors of the Testis.

Histology	No. of cases	positive %	Involvement		
			unilateral %	bilateral %	para-aortic %
Seminoma	344	40.7	45.0	34.3	96.4
Carcinoma	422	58.7	43.9	37.1	99.6
	766	50.6	44.3	36.1	98.4
Non germinal	41	31.7	61.5	30.7	100
Totals	807	49.7	44.9	35.9	98.5

This value was also higher in carcinomas (58.7%) than in seminomas
(40.7%). According to our knowledge of the lymphatic drainage of the
normal testis, the ipsilateral para-aortic region is the first site
of involvement. This finding is confirmed in our case material, with
98.5% of para-aortic metastatic diffusion. In about a half of the
pathologic cases, the nodal involvement appeared to be limited to a
single station, usually the ipsilateral para-aortic site. This
occurred most frequently in non-germinal tumors (61.5%). Only one
third of the pathologic lymphograms showed bilateral retroperitoneal
node involvement.

A total of 287 patients (Table 4) underwent retroperitoneal
lymphadenectomy or biopsy with an overall diagnostic accuracy for
lymphography of 86.7%, sensitivity of 85.9% and specificity of 87.9%.
There were 15 false positive and 23 false negative reports. These
data correlate well with those previously referred to in the literature.

The diagnostic contribution of lymphography can be more adequately assessed in carcinomas where an extensive retroperitoneal

Table 4. Lymphographic/Histologic Correlation in Operated Patients.

Histology	No. of operated patients	Lymphography/Histology				Accuracy %
		+/+	+/-	-/+	-/-	
Seminoma	21	18	1	1	1	90.5
Carcinoma	255	118	14	21	102	86.3
Non germinal	11	4	–	1	6	90.9
Totals	287	140	15	23	109	86.7

Sensitivity 85.9% (140/163)   Specificity 87.9 (109/124)

Table 5.   Results of Lymphography in 204 Selected Cases of Clinical
NO Carcinoma.

LAG	No. of cases	Lymphography/Histology				Accuracy %
		+/+	+/-	-/+	-/-	
N_0	120			20	100	83.3
N_1	24	19	5			80.0
N_2	60	53	7			88.3
	204	72	12	20	100	84.3

dissection was the treatment of choice. From the group of patients with carcinoma we selected 204 cases (Table 5) in whom the initial physical evaluation of lymph node areas was negative (N_0). The patients then had lymphography followed by extended or modified retroperitoneal lymphadenectomy. More than half the cases were reclassified after lymphography as negative for retroperitoneal metastases, but in this group histologic correlation was only 83.3% with 20/120 showing false negative readings. Also, in the group with positive lymphography, a number of false positives occurred mainly in patients classified as N_1 with disease limited to a single ipsilateral lymph node. This is probably due to the objective difficulty in assessing the true nature of filling defects in a single node and to the subjective knowledge of the high rate of diffusion to the lymphatics of testicular carcinoma. Furthermore, one must remember that lymphography sometimes understages the disease. This means that, in many cases, even though only one or two radiologically involved nodes are seen, histology discloses more extensive diffusion. At the same time we found that, usually in false negative reports, metastases are so large that these nodes are completely bypassed by the contrast medium and thus impossible to demonstrate. The last important finding to explain the false negative reports is the location of the involved nodes. There are, in fact, node areas impossible to evaluate with lymphography, such as those in the high para-aortic area up to the diaphragmatic crura. The nodes of first drainage of the testis, i.e. the superficial intercavoaortic for right sided tumors and those near the renal vein on the left side, are difficult or even impossible to visualize with standard lymphography. For this reason, when possible, the use of intraoperative funicular lymphography is suggested.

    Among the new imaging procedures echography is recommended as the first diagnostic step since it is inexpensive, widely available and avoids ionizing radiation. However, it must be stressed that lymph nodes with a diameter less than 1.5 cm are almost completely missed or cannot be interpreted with this procedure.

A metastatic lymph node is echographically homogeneous, i.e. transonic or only slightly echogenic, without posterior enhancement if completely solid. Not infrequently a cystic pattern is found if colliquative necrosis has occurred or in the presence of mature teratoma.

The diagnostic value of echography is restricted by the size of the patient. Obese or moderately obese patients are poor candidates for examination due to the physical properties of ultrasound. Further, meteorism influences the picture quality negatively in at least 20% of the patients. In Table 6 published results of echography in the evaluation of the retroperitoneal node chains are shown. The values of overall diagnostic accuracy range from 75 to 90% of the cases in whom the examination was technically possible. Specificity was high (95%) but sensitivity was low. On the basis of the data reported in the literature, only a positive report can be considered of value, while the importance of a negative result appears to be questionable. Further, one must consider that every kind of patient is included in those reports, i.e. patients without disease, those with minimal and those with extensive disease. To assess the true possibilities of echography we performed a retrospective evaluation of our results in a group of 48 patients who underwent retroperitoneal lymphadenectomy. The patients had an early stage disease and lymph node metastases, if present, measuring less than 5 cm and in a half of the cases less than 2 cm in diameter. The results of the review in this group of selected cases are reported in Table 7. Considering the various parameters our results in the early stages of the disease were disappointing.

Computerized tomography performed with machines that permit high resolution and fast scanning times, is considered to be the better non invasive imaging modality for precise identification of a number of abnormal conditions, including diseases of the lymph nodes. The CT criteria for determining nodal involvement are based solely upon identification of enlarged lymph nodes, which some take to be those in excess of 1.5 cm of cross-sectional diameter. Since lymph nodes

Table 6. Results of Echography in the Evaluation of Retroperitoneal Lymphadenopathy.

Author	No. of cases	Sensitivity %	Specificity %	Accuracy %
Winterberger[6]	46	–	–	90
Rochester et al.[7]	17	64	96	–
Brascho et al.[8]	56	–	–	87
Burney and Klatte[9]	39	–	–	75
David et al.[10]	62	79	95	89

Table 7. Results of Echography in Early Pathological Stage
          Testicular Carcinoma.

Sensitivity	8/19	42.1%
Specificity	23/29	79.3%
Overall accuracy	31/48	64.6%
* Accuracy + Report	8/14	57.1%
* Accuracy - Report	23/34	67.6%

are visualized by the presence of surrounding fat, the visualization
possibilities are enhanced in overweight adults. Generally it is
possible to state that metastases in non enlarged or slightly
enlarged lymph nodes go undetected with this procedure. In cancer
of the testis metastases are often large; thus CT can be considered
suitable for the evaluation of the N category in this disease.

According to the literature (Table 8) the diagnostic results of
CT appear comparable with those reported for lymphography. Only in
the paper of Jing et al.[15] are the results of lymphography clearly
better than those of CT in the same group of patients with cancer of
the testis (Table 9).

The main condition that influences the diagnostic accuracy of
the different radiological procedures is the size of the metastases.
It is almost obvious how patients with stage IIC or IID disease are
almost invariably diagnosed, no matter which diagnostic procedure is
chosen. The problem is thus limited to those patients with cancer of
the testis who have stage I, IIA or IIB disease.

To evaluate this approach we selected from our case material a
group of patients who had undergone retroperitoneal lymphadenectomy
and who were pathologically classified as having testicular carcinoma

Table 8. Results of CT in Identification of Lymph Node Metastases
          from Cancer of the Testis.

Author	No. of cases %	Sensitivity %	Specificity %	Accuracy %
Burney[9]	39	–	–	76
Lackner[11]	64	80	79	80
Williams[12]	32	93	82	88
Dunnick[1]	50	66	100	74
Thomas[13]	27	90	83	89
Lein[14]	51	56	96	76

Table 9. Pathologic Correlation in 37 Patients with
          Testicular Carcinoma[15].

	Lymphangiography	Computed Tomography
Positive	28	25
False negative	2	5
Negative	6	6
False positive	1	1
Total	37	37
Sensitivity	28/30 (93%)	25/30 (83%)
Specificity	6/7 (86%)	6/7 (86%)
Accuracy	34/37 (92%)	31/37 (84%)

Table 10. Results of LAG Versus CT in 150 Early
           Pathological Stage Testicular Carcinoma.

Statistical Evaluation	LAG (150 cases)	CT (30 cases)
Sensitivity	70.6%	66.7%
Specificity	79.3%	85.7%
Overall accuracy	75.3%	75.9%
* Accuracy + Report	73.8%	83.3%
* Accuracy - Report	76.5%	70.6%

Table 11. Results of the Combination of LAG (L) + CT (C)
           in a Group of Early Pathological Stage
           Testicular Carcinomas.

Histology	Radiology L-/C- %	L+/C+ %	L-/C+ %	L+/C- %	Correct Interpretation %
N-	72.4	14.3	-	14.3	72.4
N+	13.3	46.7	20.0	20.0	86.7

at stage I, IIA or IIB, i.e. without lymph node metastases or with
metastatic nodes less than 5 cm in diameter; in 50% they were less
than 2 cm. The group includes 150 patients and the diagnostic
results of this evaluation are reported in Table 10. It must further
be stressed that lymphography was performed in all patients while CT
was done in only 20% of the same cases. The results clearly indicate
that it is always difficult to select the true negative patients as
well as those with involvement limited in size and/or extension. The
high published values of accuracy using LAG or CT are probably due to
the combination of the very good results in the patients with bulky
retroperitoneal disease and to the relatively lower diagnostic
possibilities in those patients with or without limited retroperitoneal disease. From this point of view our results can be
considered as valuable. LAG seems a little more adequate in discovering the disease while CT seems better at identifying the
negative cases. The best screening technique is reported to be the
combination of lymphography and CT scan in which CT allows the
exploration of the sites impossible for lymphography and the visualization of the large unopacified metastatic nodes reducing the false
negative rate for lymphography. At the same time lymphography can
eliminate a number of false positives on CT, by diagnosing node
swelling due to non-neoplastic conditions, as well as false negatives
by revealing metastases in non-enlarged nodes. Again the clinical
usefulness of the combination is strictly dependent upon the size of
the metastases. In patients pathologically without metastatic
disease (Stage I) or with disease limited in size and extension
(Stage IIA or B) our experience with the use of the combination LAG +
CT is reported in Table 11. The results of the study are clear. In
the patients without disease at operation (pathologic stage I) the
combination slightly enhances the diagnostic possibilities of the
individual tests. On the other hand, in those patients with histologically proved metastases limited in size and extent (pathologic
stage IIA and IIB), the combination demonstrated abnormal nodes in
86.7%. This represents a real improvement over the diagnostic possibilities of the techniques used separately.

REFERENCES

1. N. R. Dunnick and N. Javadpour, Value of CT and lymphography:
    Distinguishing retroperitoneal metastases from nonseminatous
    testicular tumors, A.J.R., 136:1093-1099 (1981).
2. R. Musumeci and M. Mauri, La linfografia in oncologia, Ed.
    Ilford (1980).
3. S. Wallace and B. S. Jing, Lymphangiography: Diagnosis of nodal
    metastases from testicular malignancies, J.A.M.A., 213:94-97
    (1970).
4. T. de Roo and S. H. van Minden, Lymphographic findings in a
    series of 258 patients with tumors of the testes, Lymphology,
    6:97-100 (1973).

5. W. Zaunbauer, R. Kunz and R. Leuppi, The accuracy of lympho-
    graphy in patients with malignant testicular tumors,
    Fortschr.Roentgenstr., 126:335-340 (1977).

6. A. R. Winterberger, Correlation of lymphography findings in the
    abdomen with B-Scan ultrasonic laminography, in: "Progress in
    Lymphology," pp. 173-178, Plenum, New York (1977).

7. D. Rochester, J. Bouvie, A. Kurzmann, and E. Lexter, Ultrasound
    in the staging of lymphoma, Radiology, 124:483-487 (1977).

8. D. Brascho, J. Durant, and L. Green, The accuracy of retro-
    peritoneal ultrasonography in Hodgkin's disease and non-
    Hodgkin's lymphoma, Radiology, 125:485-487 (1977).

9. B. Burney and E. Klatte, Ultrasound and CT in the staging of
    testicular carcinoma, Radiology, 132:415-419 (1979).

10. E. David, G. van Kaick, U. Ikinger, P. Gerhardt, and P. Prager,
    Detection of neoplastic lymph node involvement in the retro-
    peritoneal space, Europ.J.Radiol., 2:277-280 (1982).

11. K. Lackner, L. Weisback, I. Boldt, K. Scherholy, and G. Brecht,
    Computer-tomographischer Nachweis von Lymphknotenmetastasen
    bei malignen Hodentumoren, Fortschr.Rontgenstr. (RoFo) 130:
    636-643 (1979).

12. R. D. Williams, S. B. Reinberg, L. C. Knight, and E. E. Fraley,
    Abdominal staging of testicular tumors using ultrasonography
    and computed tomography, J.Urol., 123:872-875 (1980).

13. J. L. Thomas, M. E. Bernardino, and R. B. Bracken, Staging of
    testicular carcinoma: Comparison of CT and lymphangiography,
    A.J.R., 137:991-996 (1981).

14. H. H. Lien, A. Kolbenstvedt, K. Talle, S. D. Fossa, O. Klepp,
    and S. Ons, Comparison of computed tomography, lymphography
    and phlebography in 200 consecutive patients with regard to
    retroperitoneal metastases from testicular tumor, Radiology,
    146:129-132 (1983).

15. B. Jing, S. Wallace, and J. Zornoza, Metastases to retro-
    peritoneal and pelvic lymph nodes, in: "The Radiologic
    Clinics of North America," pp. 511-530, vol. 20, n.3 Saunders
    (1982).

THE ROLE OF THE CT-SCAN IN DIAGNOSIS AND FOLLOW-UP

IN PATIENTS WITH METASTATIC TESTICULAR CANCER

K. Scheiber, D. Zur Nedden and G. Jakse

Departments of Urology
Internal Medicine and Diagnostic Radiology
University of Innsbruck, Austria

INTRODUCTION

As malignant testicular carcinoma metastazises with few exceptions primarily into the retroperitoneum, the detection of retroperitoneal metastases is highly important for tumor staging and further therapy. Until recently, retroperitoneal metastases were imaged only by indirect methods such as urography, lymphography or by more invasive methods, such as anigography and cavography[1]

Today sonography and computed tomography allow more accurate examination of the retroperitoneum and can be used for diagnosis as well as for the evaluation of treatment and for follow up after therapy.

Computed tomography seems to be the best method for tumor staging and has been the subject of several reports.[2-7]

PATIENTS AND METHODS

Computed tomography scans were performed with an apparatus of the third generation. Thirty minutes before the examination all patients were given one litre of a diluted suspension of barium sulphate (E-Z-CAT®) to outline the small intestine. Immediately before the investigation each received 2 cc glucagon intravenously to reduce peristalsis.

Between December, 1978 and June 1983, 274 examinations were performed on 76 patients with testicular cancer. In 223 both the abdomen and the lungs were evaluated, 42 examinations were limited to the abdomen or the retroperitoneum and nine were lung tomographies.

The diagnosis was seminoma in 27 of 76 patients, while the other 49 patients had non-seminomatous testicular cancer. According to the nomenclature of Collins and Pugh, 18 patients had malignant teratoma of the intermediate type, 21 patients had an undifferentiated teratoma and ten had trophoblastic teratoma[8].

Computed tomography and ultra-sound were performed immediately after semicastration or after histological diagnoses on 56 patients (19 patients with seminoma and 37 patients with malignant teratoma). Retroperitoneal lymphadenectomy was performed on 31 of 37 patients with nonseminomas.

Table 1 illustrates how clinical stage correlated with tumor histology. The staging system used at our clinic is shown in Table 2.

Unilateral orchiectomy was performed on 20 patients before October 1978, therefore computed tomography was applied only on follow up investigations. Since the introduction of computed tomography at our university hospital, lymphography has been discontinued.

RESULTS

Retroperitoneal lymphadenectomy was performed on 31 patients with nonseminomatous testicular cancer. In 19 of these patients retroperitoneal metastases were resected and 12 patients had no metastases. The comparison between preoperative computed tomography and the results of lymphadenectomy showed correct diagnoses for 16 of 19 patients with positive retroperitoneal lymph nodes. There were three false negative but no false positive diagnoses.

Sonography reports on retroperitoneal metastases were correct for only 10 of 19 patients; eight patients had false negatives and in one the diagnosis was equivocal. In nine of the 12 patients without metastases, sonography was correct, in one it was falsely positive, and in two cases the results were equivocal (Table 3). In our study

Table 1. Histology and Stage in 56 Patients with Testicular Cancer

Stage	Seminoma	MTI	MTU	MTT	Mixed Tumor
I	13	2	7	2	1
IIa		1	3	2	-
IIb	4	2	1	2	1
IIc	2	3	-	1	-
III	-	-	4	3	2

Table 2.   Staging for Testicular Cancer

Stage I	Tumor confined to testis
Stage II	Retroperitoneal metastases or inguinal metastases after transscrotal or previous inguinal operation
IIA	Tumor, smaller than 2 cm
IIB	Retroperitoneal nodes, smaller than 5 cm
IIC	Bulky tumor
Stage III	Metastases outside the retroperitoneum

the sensitivity and specificity for computed tomography are 86% and 100% respectively. Sonography has a sensitivity of 55%, and a specificity of 92%.

In five patients computed tomography revealed distant metastases which were not detected by conventional methods. Two patients had subpleural metastases, two patients had mediastinal involvement and one had hepatic metastases.

In ten patients with initially extensive metastatic disease there was considerable reduction of metastases, after intensive chemotherapy and before the removal of the residual tumor mass. Computed tomography showed four patients with cystic transformation of the residual tumor, three patients with fibrosis, and three patients had metastases which were reduced in size but still homogenous. In none of those ten patients could active cancerous tissue (teratocarcinomatous cells) be found.

Two of the 36 patients who were treated by retroperitoneal lymphadenectomy showed recurrent retroperitoneal tumor on follow-up CT scans and histological proof was obtained by fine needle biopsy. In two patients metastases to other organs were detected postoperatively, a suprarenal metastasis in a patient with bilateral testicular cancer and metastases to the lung in the other.

DISCUSSION

Computed tomography has proved to be an excellent method for the imaging of retroperitoneal metastatic disease. The preoperative diagnosis of malignant testicular cancer was correct in 91% of cases. Formerly lymphography was routinely used at our clinic in patients with testicular cancer, but since the introduction of computed tomography we no longer use this technique for diagnostic purposes. Therefore we cannot compare CT with lymphographic results. However, as can be seen from the literature, the high rate of falsely positive lymphographic diagnoses and the difficulty in discovering metastases at the hilum of the kidney and in the upper lumbar region restricts the value of this method.[7,9,10].

Table 3. Correlation Between CT and Ultrasound in Testicular
          Teratomas

CT Final Status	Correct	False +	False −	Equivocal
Positive–19	16	–	3	–
Negative–12	12	–	–	–
Ultrasound				
Positive–19	10	–	8	1
Negative–12	9	1	–	2

Compared to sonography computed tomography provides more detail and much better reproducibility. Because of abdominal gas, sonographic methods have limitations in the region of the aortic bifurcation and in the true pelvis. Computed tomography images these regions very well if the small intestine is marked.

The demonstration of lymph node metastases with computed tomography depends on the one hand on nodes > 0,5 cm in diameter, and on the other hand on the density of the surrounding fatty tissue.

Isolated enlargements of the lymphatic nodes located in the paraaortic, paracaval, retroaortic and retrocaval region generally are easily visible. They appear as round tissue concentrations or they are indicated by a loss of demarcation of the aorta and the vena cava.

Larger and more involved lymph–nodes cannot be distinguished from the aorta or the vena cava without contrast medium. Sometimes it is very difficult to diagnose moderately enlarged lymph nodes in the region of the mesenteric artery. The differentiation between horizontally or diagonally cut vessels and parts of the intestine is only possible if contrast medium is given as a bolus. The same problems of differentiation exist in the iliac region as enlarged iliac nodes are often obscured by filled loops of the intestine.

Interpretation is especially difficult in patients who have had retroperitoneal lymphadenectomy, because of adherent loops of small intestine in the retroperitoneum. This problem often caused false positive findings on the CT scan[2]. In these cases the use of contrast medium is necessary.

Metal clips which we were using only in exceptional cases are very confusing in computed tomography. If surgical clips are necessary, titanium clips, which reduce artefacts, should be used[11].

   After intensive cytostatic therapy metastases have a different
appearance. Apart from shrinkage, other structural changes can be
observed. In some cases cystic transformation, consisting of necrotic and semiliquid areas, and, in other cases, fibrotic changes are
seen. These visible changes signify increasing benignity of the
metastases. However computed tomography cannot yet replace the
histological examination of these lesions.

REFERENCES

1. K. Lackner, G. Brecht, R. Janson, K. Scherholz, A. Lützeler, P.
    Thurn, Wertigkeit der Computertomographie bei der
    Stadieneinteilung primärer Lymphknotenneoplasien. Fortschr.
    Röntgenstr., 132:1, 21-30 (1980).
2. B. T. Burney and E. C. Klatte, Ultrasound and computed
    tomography of the abdomen in the staging and management of
    testicular carcinoma, Radiology 132:414 (1979).
3. L. Cionini, F. Casamassima, N. Villari, L. Pirtoli, L. Forzini,
    Computed tomography and lymphography of the retroperitoneal
    space in testicular tumors. Acta radiol.oncol. 20:19-24,
    (1981).
4. J. K. T. Lee, B. L. McClennan, R. J. Stanley, S. S. Sagel,
    Computed tomography in the staging of testicular neoplasms.
    Radiology 130:387-390, (1979).
5. G. Marchal, Y. Coenen, G. Wilms, A. L. Baert, The accuracy of
    CT- scan in the diagnosis of retroperitoneal metastases of
    malignant testicular tumors. Fortschr. Röntgenstr. 128,
    1:746-753 (1978).
6. R. G. Rowland, D. Weisman, St. D. Williams, L. H. Einhorn, E. C.
    Klatte, J. P. Donohue, Accuracy of preoperative staging in
    stages A and B nonseminomatous germ cell testis tumors,
    J.Urol. 127:718-720, (1982).
7. R. D. WIlliams, S. B. Feinberg, L. C. Knight, E. E. Fraley,
    Abdominal staging of testicular tumors using ultrasonography
    and computed tomography, J.Urol., 123:872-875, (1980).
8. D. H. Collins and R. C. B. Pugh, The pathology of testicular
    tumors. E & S Livingstone, Ltd., Edinburgh & London, (1964)
9. J. P. Richie, M. B. Garnick, H. Finberg, Computerized
    tomography: How accurate for abdominal staging of testis
    tumors? J.Urol., 127:715-717, (1982).
10. J. K. T. Lee, R. J. Stanley, S. S. Sagel, B. L. McLennan,
    Accuracy of CT in detecting intraabdominal and pelvic lymph
    node metastases from pelvic cancers, Amer.J.Roentgenol.,
    131:633-639 (1978).
11. H. von Holst, M. Bergström, A. Möller, Titanium clips in neuro-
    surgery for elimination of artefacts in computed tomography,
    Acta Neurochir., 38:101-109, (1977).

TUMORS OF THE TESTIS - PROGNOSTIC FACTORS

G. Stoter

Department of Oncology
Free University Hospital
Amsterdam, The Netherlands

The determination of prognostic factors is important in any disease in order to understand the natural history and to optimize the results of therapy, i.e. maximal therapeutic effect in relation to minimal toxicity.

In patients with testicular cancer the putative prognostic factors include histology, extent of the disease, prior treatment, tumor markers, bone marrow status, performance status, and age.

One of the difficulties in the analysis of a patient population with respect to prognostic variables is that the data base often makes it impossible to form subgroups of patients according to arbitrary characteristics. For example, the histology of the primary tumor may include seminoma, malignant teratoma differentiated, undifferentiated, intermediate or trophoblastic (in any combination), but the histological diagnosis will be made according to the presence of the most malignant component, or the predominance of a given component. This may result in inaccurate statements on prognostic significance. Another example is lung metastases, which should be quantitated with regard to number and diameter. However, existing staging systems often have categories such as "more than 5 nodules, each larger than 2 cm" or "more than 3 nodules or any nodule larger than 2 cm". Here again, we lose the continuum of a given variable and have to accept more or less arbitrarily defined categories. Also of importance is whether or not risk factors with regard to survival status or to complete response status are reported. Other reasons which may explain controversies with respect to the importance of prognostic factors may be related to different therapies (even within

one studied patient group), different statistical methods and a bias
due to different factors which have been examined. An example of the
latter is whether one looks at extent of disease by the total number
of sites of metastases or by the total volume of metastases.

Despite these handicaps it appears reasonably possible to define
prognostic variables and their importance.

Univariate analyses by several groups have shown that histology
of the primary tumor, extent of disease, the presence of markers,
previous treatment, performance status and bone marrow status, are
all of significance in predicting the outcome of treatment[1-10].

An important problem in the determination of the relative impact
of a given variable on prognosis arises from the correlation of
prognostic factors. For example, it is reasonable to assume that a
large burden of tumor may contain a high proportion of HCG (Human
Chorionic Gonadotrophin) and AFP (Alpha Feto-Protein) secreting cells
resulting in high serum concentrations of these markers. In other
words, tumor volume and marker levels may be correlated. To solve
this problem, statistical methods for multivariate analysis are
available. By using such models, investigators at Memorial Sloan
Kettering Hospital have defined LDH (Lactate dehydrogenase), HCG and
the total number of metastatic sites as the most important independent prognostic factors[11]. At Charing Cross Hospital, London high
serum levels of HCG and AFP were found to be the most important
prognostic factors, more important than tumor volume[1,2]. A preliminary multivariate analysis of an EORTC study of PVB-treated patients
(cis-Platin, Vinblastine, Bleomycin) with disseminated testicular
non-seminomatous cancer defines trophoblastic teratoma, AFP above
1000 ng/ml and lung metastases more than four in number or larger
than 2 cm as the most important independent prognostic variables. A
central review of the pathology and a separation of number and volume
of the lung metastases will be performed and the analysis will be
repeated thereafter. Scandinavian investigators have performed a
multivariate analysis also including complete response status as a
variable; this has not been done by others. Another difference was
that they looked at immunohistochemical staining of HCG and AFP in
the tumor specimen, whereas they measured only LDH in the serum. The
outcome of their study was that performance status, LDH, HCG in the
tumor and complete response status were the four most important
independent factors which influence survival[12].

In conclusion, multivariate analysis to define independent
prognostic variables with significant impact on remission and survival status is a valuable tool in defining subgroups of patients who
need more intensive therapy or for whom overtreatment can be avoided.
In future studies patients should be stratified for different treatment regimens on the basis of prognostic factors of clear cut relevance.

Currently the serum levels of tumor markers and the extent of the disease are the two most important prognostic variables. The controversy about which is the most important will probably be solved as soon as uniform definitions of prognostic variables are adapted and statistical methods have become standardized.

REFERENCES

1. E. S. Newlands, R. H. J. Begent, G. J. S. Rustin, D. Parker, and K. D. Bagshaw, Further advancement in the management of malignant teratomas of the testis and other sites, The Lancet 1:948 (1983).

2. J. R. Germa-LLuch, R. H. J. Begent, and K. D. Bagshaw, Tumor-marker levels and prognosis in malignant teratoma of the testis, Br.J.Cancer, 42:850 (1980).

3. M. J. Peckham, A. Barrett, T. J. McElwain, W. F. Hendry, and D. Raghavan, Non-seminoma germ cell tumors (malignant teratoma) of the testis, Brit.J.Urol., 53:162 (1981).

4. M. J. Peckham, T. J. McElwain, A. Barrett, and W. F. Hendry, Combined management of malignant teratoma of the testis, The Lancet, 2:267 (1979).

5. G. Pizzocaro, G. De Palo, A. Milani, L. Piva, S. Pilotti, F. Zanoni, and S. Monfardini, Adjuvant chemotherapy related to prognostic subgroups in stage II testicular carcinoma, Proc.Amer.Soc.Clin.Oncol., 1:121, (1982).

6. A. Cockburn, D Vugrin, S. Hajdu, R. Mackey, M. Batata, and W. Whitmore Jr., The prognostic significance of undifferentiated (undif.) histology in seminoma, Proc.Amer.Soc.Clin.Oncol., 1:114 (1982).

7. W. F. Whitmore Jr, D. Vugrin, and R. Golbey: Prognostic significance of choriocarcinoma elements in testis cancer, Proc.Amer.Assoc.Clin.Oncol., 1:115 (1982).

8. M. C. Lippert and N. Javadpour, Lactic dehydrogenase in the monitoring and prognosis of testicular cancer, Cancer, 48:2274 (1981).

9. G. Stoter, C. P. J. Vendrik, A. Struyvenberg, D. Th. Sleyer, R. Vriesendorp, H. Schraffordt Koops, A. T. van Oosterom, W. W. ten Bokkel Huinink, and H. M. Pinedo, The 5-year survival of patients with disseminated non-seminomatous testicular cancer treated with cisplatin, vinblastine, and bleomycin, Cancer, (in press).

10. M. K. Samson, R. Fisher, R. L. Stephens, S. Rivkin, M. Opipari, T. Maloney, and C. W. Groppe, Vinblastine, Bleomycin and Cis-diamminediochloroplatinum in disseminated testicular cancer: Response to treatment and prognostic correlations, Europ.J.Cancer, 16:1359 (1980).

11. G. J. Bosl, N. L. Gelle, C. Cirrincione, N. J. Vogelzang, B. J. Kennedy, W. F. Whitmore, D. Vugrin, H. Scher, J. Nisselbaum, and R. B. Golbey, Multivariate analysis of prognostic vari-

ables in patients with metastatic testicular cancer, <u>Cancer
Research</u>, 43:3403-3407 (1983).

12. F. E. von Eyben, G. K. Jacobsen, H. Pedersen, M. Jacobsen, P. P.
     Clausen, P. C. Zibrandtsen, and B. Gullberg, Multivariate
     analysis of risk factors in patients with metastatic testic-
     ular germ cell tumors treated with vinblastine and bleomycin,
     <u>Invasion Metastases</u>, 2:125, (1982).

AGE - ADJUSTED MORTALITY RATES FOR TESTICULAR

NEOPLASM IN ITALY BETWEEN 1969 AND 1978

F. Di Silverio and G. La Pera*

Department of Urological Pathology
University of Roma and *Division of Urology
Ospedale Provinciale Ascoli Piceno
Italy

INTRODUCTION

Testicular tumors represent 1-2% of all neoplasms and 4-10% of urological tumors[1-3]. The frequency of this lesion has been reported to be approximately 2-3/100,000 inhabitants (4,5) the highest rate being in Denmark with 4.9/100,000 (6-8); in negro and asiatic populations the incidence is low[9-13].

Testicular tumor is the most common neoplasm in males aged between 25 and 34 years of age[14,15]. Since it is an external tumor it can hardly pass unobserved and sooner or later the patient becomes aware of the lesion. Thus the data on mortality must come very close to the real number of deaths occurring from this disease and usually provide more reliable information than morbidity data in the interpretation of time trends[16,17]. In the present investigation the mortality from testicular cancer in Italy during the period 1969 - 1978 was analyzed in order to ascertain the temporal trend.

MATERIALS AND METHODS

Source of data

Data on deaths from this neoplasm, provided by the Istituto Superiore di Sanitá, were taken from the Istituto Centrale di Statistica (ISTAT) published in the Annual Reports of Health Statistics for the years 1969-1978[18] and classified according to the ninth International revision of the causes of death under item 186.0[19]. Data on the Italian population for these years are also

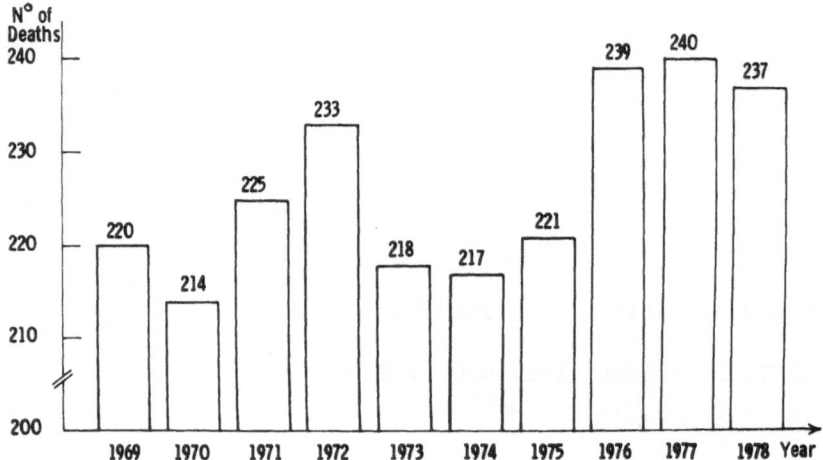

Fig. 1. Crude mortality for testicular neoplasms in Italy between
         1969 and 1978.

taken from Istat[20]. Only deaths related to Italian residents were
recorded in this study.

Analytical Methods

The population and deaths are classified into 17 age groups each
of 5 years: 0-4, 5-9, 10-14, 15-19,- - - - - 70-74, 75-59, 80 and
over.

Rates are reported per 100,000 population and since the Italian
population advanced in age during the 10 years of this study[20],
statistical standardization of age was employed in order to compare
the mortality rate between the various years. This method corrects
the age - specific mortality rates for the percent distribution of
the ages in a given year, which, in our study was 1976.

The result obtained corresponds to the number of deaths expected
for that age group in the reference population if the mortality rate
had been equal to that of the year studied. The mathematical formula
used is:

$$S \quad = \quad \sum_{i=1}^{n} a \; \frac{C}{pop} \quad X \quad \frac{Pop\ 76}{T.Pop\ 76}$$

Where :   a   = the year under study
          C   = number of cases in the age group
        Pop   = Italian population in the age group in the year
                under study

        Pop 76 = Italian population in the same age group of Pop
        T. Pop = Italian population in 1976

The statistical test, reported in the literature to evaluate the time course,[21,22] shows a simple linear correlation between the standardized rates and various years; The statistical significance of the regression coefficient is given by the formula:

$$t = r \sqrt{\frac{n-2}{1 - r^2}}$$

t = Student's t
r = correlation coefficient
n = number of observations
degrees of freedom 8

RESULTS

1. Crude Mortality. Deaths rose from 220 in 1969 to 237 in 1978, an increase of 7.73% (Figure 1).

2. Standardized death rate. The standardized death rate for testicular cancer examined over the 10 year period is given in

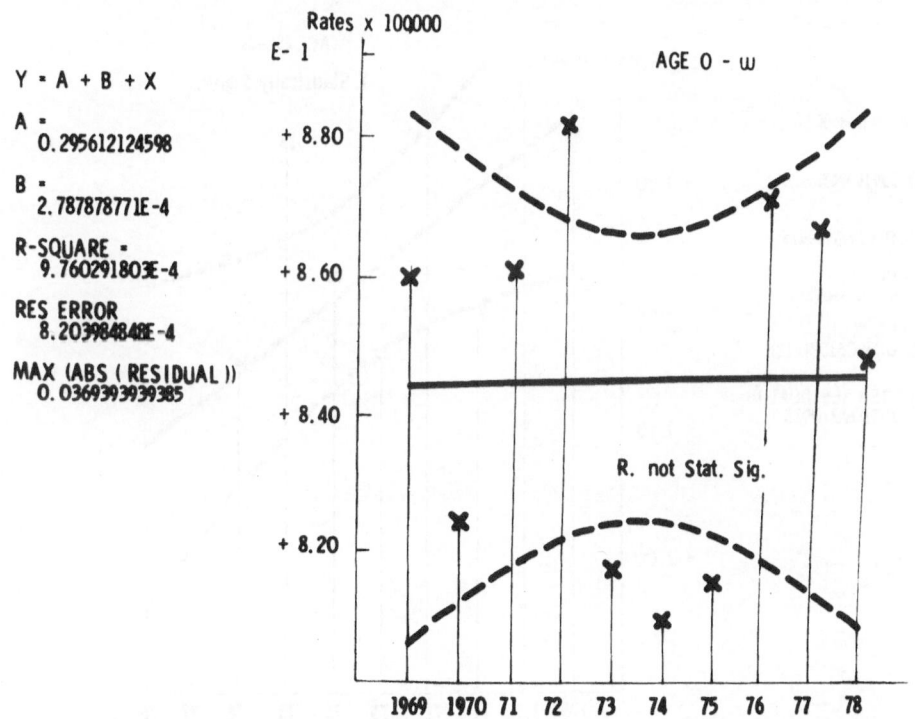

Fig. 2. Standardized mortality rates (Italian population 1976) for testicular neoplasm in Italy between 1969 and 1978. Age 0 - W; E-1 = 1 x 10^{-1}; r = 3.124 x 10^{-2} not statistically significant.

Figure 2. The correlation coefficient r = 3.21 x 10^{-2} is not
statistically significant. However the rates for the 25-39
years group (Figure 3) show a correlation coefficient of r =
-0.70 which is statistically significant with p < 0.05.

DISCUSSION AND CONCLUSIONS

Data from the Italian population for the 10-year period under
study show a low mortality rate and no modification in time trend.

It is not yet possible to offer an explanation for the negative
correlation observed in the group aged 25-39 years. It should not be
forgotten, however, that from a statistical point of view a correlation is not sufficient to suggest a true decrease in death rate.
Further cohort studies on death rate are in fact required and a
better knowledge of the incidence representing the denominator of the
mortality/incidence ratio.

As always in studies of this kind, there are several factors
which could weaken the reliability of the data including 1) incorrect
diagnosis, 2) inaccurate death notification, 3) registration and

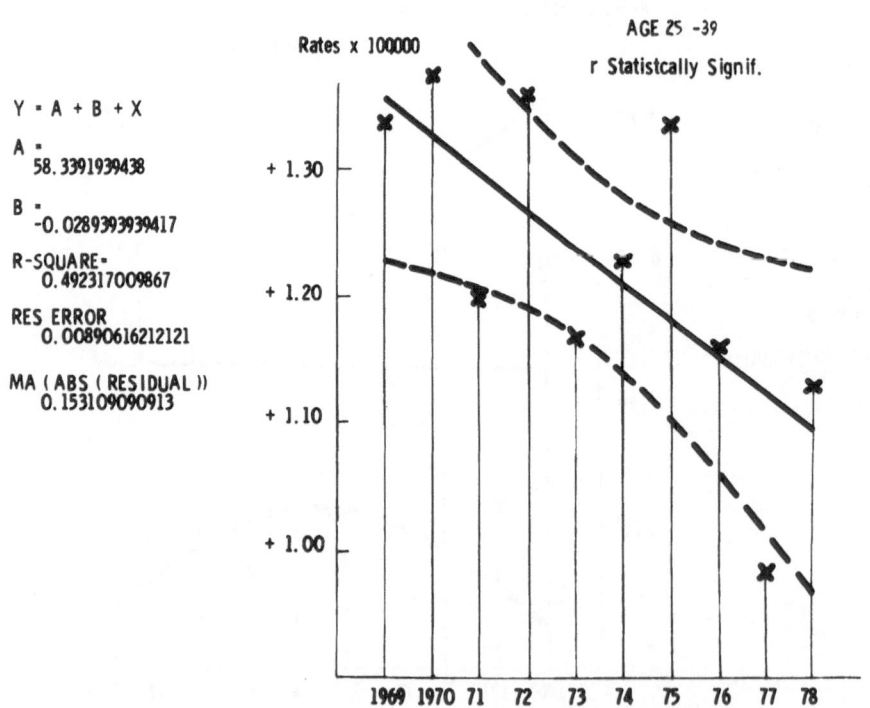

Fig. 3. Standardized mortality rates (Italian population 1976) for
testicular neoplasm in Italy between 1969 and 1978 age
25-39. r = -0.70 statistically significant P < 0.05.

recording errors and 4), inaccuracies in the publication of statistics.[22,24]. The size of this error is not yet known, and whilst
this should not prevent development of meaningful hypotheses for
further study, these factors should not be overlooked in the interpretation of the present data.

Acknowledgement

The authors are grateful to the Istituto Superiore di Sanità for
having provided data used in this study. The valuable collaboration
of dr. Ercole Collazoni and Giovanni Barone is gratefully acknow-
ledged.

REFERENCES

1. G. J. Post and J. A. Bellis, Delayed presentation of testicular
    tumors, Southern Med.J., 73 (I) 33, (1980).
2. F. K. Mostofi, Testicular tumors, Cancer 32:1186, (1973).
3. J. P. Donohue, Surgical management of testis cancer in:
    "Testicular tumors management and treatment," L. H. Einhorn,
    ed., Masson Publishing USA Inc., New York, p.29, (1980).
4. C. S. Muir and J. Nectoux, Epidemiology of cancer of the testis
    and penis, National Cancer Institute, Monograph No.53, p.157
    (1979).
5. G. Piot, J. P. Droz, J. M. Caillaud, D. Bellet, P. Wibault, G.
    Brulé, and J. L. Amiel, Les réchutes des tumeurs malignes
    germinales non séminomateuses du testicule, Journal
    d'Urologie 89, No. 1, p.49, (1983).
6. J. Waterhouse, C. S. Muir, P. Correa, and J. Powell, Cancer
    Incidence in Five Continents, Volume III. IARC Scientific
    Publications No 15; Lyon (1976).
7. J. Clemmesen, A doubling of morbidity from testis carcinoma in
    Copenhagen, 1943-1962, Acta.Pathol.et Microbiol.Scandinav.,
    72:348 (1968).
8. J. Clemmesen, Statistical studies in the aetiology of malignant
    neoplasm, Acta.Pathol.et Microbiol.Scandinav.Supplement 247,
    Munksgaard, Copenhagen, (1974).
9. V. Kaye and C. Isaacson, Genitourinary pathology in urban male
    blacks of South Africa, J.Urol., 123;51 (1981).
10. J. Daniels Jr., R. Stutzman, D. McLeod, A comparison of
    testicular tumor in black and white patients, J.Urol. 125:341
    (1981).
11. T. A. Junaid, Tumors of the testis in Ibadan, Nigeria,
    Brit.J.Urol., 54:411 (1982).
12. J. Higginson and A. G. Oettle, Cancer incidence in the Bantu and
    "Cape colored" races of South Africa: report of a cancer
    survey in the Transvaal (1953-1955), J.Natl.Cancer Inst.,
    24:p 589 (1960).

13. S. Raines, T. G. Hurdle, Tumors of the testis, J.Urol.73:363 (1955).

14. A. Slater, W. K. James, R. Fifield and G. V. Groom, The investigation and treatment of germ cell tumors of the testis, Clin.Radiol., 32:25 (1981).

15. J. M. Davies, Testicular cancer in England and Wales: some epidemiological logical aspects, Lancet i. 928 (1981).

16. C. Osmond, M. J. Gardner, E. D. Analysis of trends in cancer mortality in England and Wales during 1951-80 separating changes associated with period of birth and period of death, Brit.Med.J., 284:1005 (1982).

17. R. Doll, R. Peto, The causes of cancer: quantitative estimates of avoidable risks of cancer in the United States today, J.Nat.Cancer Inst., 66:1192 (1981).

18. ISTAT Annuari di statistiche sanitarie 1969-1978.

19. W.H.O: International classification of diseases and causes of death. IX revision. Geneva 1978.

20. ISTAT Popolazione e movimento anagrafico dei comuni.

21. Armitage, Statistica medica, Feltrinelli, Milano, p49 (1979).

22. M. Trasti, S. Nilsson, L. E. Peterson, Applied diagnostic techniques: a decisive factor in the long-term T-year survival rate in prostatic carcinoma, Brit.J.Urol., 51:135 (1979).

23. F. Di Silverio, G. La Pera, and R. Tenaglia, Age adjusted mortality rate and regional distribution for prostatic carcinoma in Italy between 1969 and 1978, The Prostate 3 631 (1982).

24. J. Staszewski, Cancer of the Urinary bladder: international mortality patterns and trends, World Health Stat. Q. 33(1) 27 (1980).

TUMORS OF THE TESTIS - THE BASIC CONCEPT OF THERAPY

G. Stoter

Department of Oncology
Free University Hospital
Amsterdam, The Netherlands

Germ cell tumors of the testis are divided into seminomatous and non-seminomatous types which have different biological behavior and different response to treatment.

Initially, dissemination to the para-aortic lymph nodes occurs in both, but ensuing haematogenous spread is predominantly to bone and lung in seminomas and to lung and liver in non-seminomas. An important difference between the two groups is that haematogenous dissemination occurs relatively late in seminoma; 90% of the patients are in stage I or stage II at the time of diagnosis[1], which means that the disease can be regarded as a loco-regional problem for a prolonged period of time. In non-seminomatous tumors signs of lymphatic and haematological spread may be found at diagnosis or after a short interval and sometimes lung metastases occur even before lymph node metastases become apparent. This characteristic of the non-seminomas implies that the disease is more often to be regarded as systemic than loco-regional, even if lung metastases are not yet visible.

Seminoma

Seminoma is a very radiosensitive tumor and, since the disease is confined to the testis and the regional lymph nodes, in most patients inguinal orchiectomy is usually followed by retroperitoneal lymph node irradiation. The survival results for stage I are nearly 100% and for stage II the figures vary from 60-90%[1-4]. When stage II is subdivided according to Peckham's classification[3], it appears that patients in stage IIC have a disease-free 5-year survival of approximately 60%[3,4]. In 23 patients with stage IIC at the Royal

Marsden, nine (39%) relapsed. Seven of these had metastases outside the abdomen[3]. This means that there is a rationale to consider patients with stage IIC seminoma as suffering from systemic disease. The same holds for stage III and stage IV patients, in whom the 5-year survival varies from 10-30%[1,2]. The recommendations for therapy therefore should be radiotherapy for patients with stage I-IIB and chemotherapy for patients with stage IIC-V.

In the past, chemotherapy with alkylating agents has been used with poor results in terms of disease-free survival[5]. However, recent results of cisplatin containing chemotherapy show a complete response rate of 74% in 90 patients[3,6-8]. Bearing in mind that there is always a risk that the metastases of a seminoma may also contain non-seminomatous elements, patients with stage IIC-IV semi-noma should receive the same combination chemotherapy as patients with non-seminomatous testicular tumors. Patients with pure seminoma in the primary tumor, but who have elevated levels of AFP as evidence of non-seminomatous metastatic disease, should be treated as such. A moderately elevated HCG level (up to 50 ng/ml) is not proof of non-seminomatous metastases since giant cells in the seminoma may produce small amounts of HCG. The primary tumor of such patients should be carefully examined to detect any area of non-seminomatous tumor tissue. If there is delay in response to treatment, these patients should immediately be switched to a treatment strategy for non-seminomatous germ cell tumors.

Non-Seminoma

Patients with clinical stage I non-seminoma have traditionally been treated by inguinal orchiectomy and retroperitoneal lymph node dissection[9,10,15] or radiotherapy[11,14]. The survival results are in the range of 90-100%, regardless of the treatment modality used. It should be noted, however, that even in pathologic stage I disease (i.e. histologically negative lymph nodes), a 10% rate of metastases mostly outside the retroperitoneal region, is reported[9.10,16]. Following radiotherapy 15% of patients may show disease progression, of which 25% will be in the retroperitoneum[11]. We know that approximately 20% of the patients who are in clinical stage I may have occult micrometastases[12,13]. This means that for about 80% of stage I patients radiotherapy and lymph node dissection are superfluous.

With the development of curative chemotherapy it is now necess-ary to study patients in clinical stage I by careful observation, including CT-scanning and tumor markers, with the institution of chemotherapy as soon as there is any evidence of progression. At the Royal Marsden 53 patients in clinical stage I have been enrolled in such a surveillance study. Nine (17%) have shown progression during a follow-up period of 6-40 months (median 15 months)[14]. All

patients showed progression within 12 months. Five out of eight
patients developed metastases in abdominal lymph nodes. All patients
had small metastases and were salvaged with chemotherapy. If these
data are confirmed, radiotherapy and lymph node dissection should no
longer be used in patients who have clinical stage I disease, provided that patient and doctor have the discipline to follow the
requirements of the close surveillance procedure.

Patients with stage II disease with lymph node metastases <2 cm
(IIA) have a disease-free survival of approximately 85% following
radiotherapy. However, when lymph node metastases are larger than 2
cm the results are much worse with 2/3 of the patients relapsing,
predominantly in the irradiated area[11]. Clearly, the use of radiotherapy should be abandoned in patients with stage IIB and stage IIC
and chemotherapy should be used.

In some European Institutions and in most American centers,
radical lymph node dissection is performed in stage II disease.
There is a shift in policy from attempted dissection in all stage II
cases towards resections in cases of stage IIA and stage IIB only
[15]. If we exclude stage IIC, the progression rate - even in the
best surgical series - varies from 20-45% [9,10,16]. The majority of
relapses will occur in patients with stage IIb[16]. Non-aggressive
adjuvant chemotherapy including drugs such as actinomycin-D, vinblastine and bleomycin does not improve these results[10,16]. Since
relapsing patients nearly always have their metastases outside the
retroperitoneal region, one may assume that micrometastases were
already present at the time of surgery and that the non-aggressive
adjuvant chemotherapy did not produce complete malignant cell kill.
Therefore, in stage IIB and stage IIC, retroperitoneal lymph node
dissection should not be used and patients should be primarily
treated with cisplatin-based combination chemotherapy.

Before the era of modern chemotherapy the 3-year survival of
patients with stage III and IV was only 10%[17]. With cisplatin-
containing combination chemotherapy, the survival has increased to
approximately 70%[18,19]. Therefore, all patients with stage III and
IV need chemotherapy first followed by debulking surgery of residual
masses[15]. The efficacy of chemotherapy is reflected in the decline
of the mortality rate of testicular cancer from 8.2 to 5.4 deaths per
million males in the USA between 1973 and 1978, a decrease of
34%[20].

REFERENCES

1. W. L. Caldwell, M. T. Kademian, Z. Frias, and T. W. Davis, The
       management of testicular seminomas, Cancer, 45:1768 (1980).
2. J. G. Maier and M. H. Sulak, Radiation therapy in malignant
       testis tumors, Cancer, 32:1212 (1973).

3. D. Ball, A Barrett, and M. J. Peckham, The management of metastatic seminoma testis, Cancer, 50:2289 (1982).

4. J. F. Doornbos, D. H. Hussey, and D. E. Johnson, Radiotherapy of pure seminoma of the testis, Radiology, 116:401 (1975).

5. L. H. Einhorn and S. D. Williams, Chemotherapy of disseminated seminoma, Cancer Clin.Trials, 3:307 (1980).

6. M. Samuels, C. Logothetis, M. D. Trindade, and M. D. Johnson, Sequential weekly pulse-dose cis-platinum for far advanced seminoma Proc.Amer.Soc.Clin.Oncol., 21:423 (1980).

7. D. Vugrin, W. F. Whitmore, and M. Batata, Chemotherapy of disseminated seminoma with combination of cis-diamminedi-chloroplatinum (II) and cyclophosphamide, Cancer Clin.Trials 4:423 (1981).

8. A. T. Van Oosterom, S. D. Williams, H. Cortes Funes, C. P. J. Vendrik, and W. W. ten Bokkel-Huinink, The treatment of advanced seminomas with cisplatin, velban and bleomycin. (Abstract) UICC Conf.on Clin.Oncol.Lausanne, 127 (1981)

9. D. G. Skinner and P. T. Scardino, Relevance of biochemical tumor markers and lymphadenectomy in management of non-seminomatous testis tumors, current perspective, J.Urol., 123:378 (1981).

10. S. D. Williams and L. H. Einhorn, Clinical stage I testis tumors: The medical oncologist's view, Cancer Treat Rep., 66:15 (1982).

11. C. J. Tyrrell and M. J. Peckham, The response of lymph node metastases of testicular teratoma to radiation therapy, Brit. J.Urol., 48:363 (1976).

12. F. Cavalli, S. Monfardini and G. Pizzocaro, Report of the International Workshop on staging and treatment of testicular cancer, Eur.J.Cancer, 16:1367 (1980).

13. W. Dewys, F. M. Muggia and E. M. Jacobs, Staging of testicular cancer, a proposed clinical-surgical schema, Cancer Treat Rep. 64:669 (1980).

14. M. J. Peckham, A. Barrett, J. E. Husband, and W. F. Hendry, Orchidectomy alone in testicular stage I non-seminomatous germ-cell tumors, The Lancet, 2:678 (1982).

15. J. P. Donohue, L. H. Einhorn, and S. D. Williams, Cytoreductive surgery for metastatic testis cancer, considerations of timing and extent, J.Urol., 123:876 (1980).

16. D. Vugrin, W. F. Whitmore, E. Cvitkovic, and R. B. Golbey, Adjuvant chemotherapy in non-seminomatous testis cancer; mini-VAB regimen, long-term followup, J.Urol., 126:49 (1981).

17. J. G. Maier and M. H. Sulak, Radiation therapy in malignant testis tumors, Cancer, 32:1217 (1973).

18. L. H. Einhorn and S. D. Williams, Chemotherapy of disseminated testicular cancer, A random prospective study, Cancer, 46:1339 (1980).

19. G. Stoter, C. P. J. Vendrik, A. Struyvenberg, D.Th. Sleyfer, R. Vriesendorp, H. Schraffordt Koops, A.T. van Oosterom, W.W.ten Bokkel Huinink, and H. M. Pinedo, The 5 year survival of

patients with disseminated non-seminomatous testicular cancer
treated with cisplatin, vinblastine and bleomycin, Cancer,
(in press).

20. S. M. Hubbard and J. S. Macdonald, An introduction to current
controversies in cancer management, Stage I testicular
cancer, a case in point, Cancer Treat Rep., 66:1 (1982).

CURRENT CONTROVERSIES IN THE MANAGEMENT OF

TESTICULAR GERM CELL TUMORS

R. T. D. Oliver, J. P. Blandy and H. F. Hope-Stone

Departments of Genito Urinary Oncology
Urology and Radiotherapy
The London Hospital, London

The last decade has witnessed major changes in the management of
Testicular Germ Cell Tumors with the development of toxic but curative chemotherapy for patients with advanced metastatic disease, such
that many previously accepted dogmas have had to be revised. This
paper will review some aspects of the changing attitudes to this
disease in the light of the experience of the London Hospital during
the past 30 years.

AETIOLOGY

Previous studies from this hospital have demonstrated that in
patients with testicular germ cell tumors in addition to the recognized increased incidence of cryptorchidism these patients have also
had a high incidence of previous serious trauma to the testicle,
herniorraphy, genital infection and of mumps after puberty in their
past history[1]. As there was no information as to the incidence of
these events in individuals without testicular tumors, an attempt has
been made to investigate this in an age and sex matched control
population of Outpatient and Casualty attenders without testicular
tumors. This study demonstrated that the incidence of hernia and
severe testicular trauma in patients with testicular tumor was not
different from the control population but two possible new aetiological factors have emerged[2], an increased incidence of malignant
disease in the parents of patients with testicular tumors whose
disease proved drug-resistant (Table 1) and an increased incidence of
relapsing herpes genitalis in the total patient population (Table 2).
Because of the small number of patients studied it will be important
that these factors are investigated in other series, and that more
objective methods of documenting exposure to the Herpes virus such as

131

Table 1. Family Hisotry in Control
and Testis Tumor Population
(Modified from Ref.2)

	N	Parents with History of Cancer
Casualty and Outpatient Controls	82	11
Patients Currently Alive	77	12[a]
Patients who have died with Drug-Resistant Disease	18	9[b]

a vs. b p<.005

serology and investigation of tumor cell DNA for evidence of Herpes
virus DNA sequences be investigated before concluding that this virus
is involved.

INVESTIGATION OF CAUSES OF TREATMENT DELAY

Despite the impressive improvement in results of chemotherapy in
the last decade, there has been no change in the time from first
symptom to initial treatment. Average delay in the patients treated

Table 2. Incidence of Sexually Transmitted
Disease in Patient and Control
Population. (Modified from Reference 2)

	N	Herpes Genitalis	Other
Casualty and Outpatient Controls	82	0[a]	5
Patients	77	8(5)*[b]	8

a vs. b p < .005

* Figures in brackets relate to patients with relapsing form

during the last 5 years has been 3 months and as can be seen from
Table 3, delay is an important determinant of the chance of long term
cure by chemotherapy. This delay occurs predominantly in patients
who have small primary tumors and present to their physicians with
symptoms from metastases (Table 4). However, as all except three of
this sub-group of patients had a lesion palpable in their testicle it
is clear that there is a need to pay more attention to educating all
doctors to palpate the testicles carefully in routine examinations of
young men with unexplained symptoms and in educating all males of the
importance of self-examination of the testicle.

Table 3. Impact of Atypical Presentation on Response to Chemotherapy
          of Testicular Teratoma

	N	Median duration of delay in diagnosis	Current NED > 12 months
Patients presenting with swollen testicle and then found to have metastases	41	4 months	71% (29)
Patients presenting with symptoms of metastases and subsequently found to have testicular teratoma	15	9 months	47% (7)

NED = No evidence of disease

Table 4. Impact of Delay in Diagnosis on Occurrence of Metastases
          and Response to Chemotherapy of Testicular Teratoma

	N	Median duration of delay (months)	>12 months' delay
Patients without metastases at presentation and continuously NED > 12 months	10	2	0%
Patients found to have metastases; currently NED > 12 months	36	4	12%
Patients found to have metastases and now dead or dying from drug-resistant tumor	20	7	35%

NED = No evidence of disease

NEW APPROACHES IN STAGING AND DIAGNOSIS

Since 1950 there has been a continuous changing methodology for
staging patients, beginning with the use of the Intravenous Pyelogram
and biological testing for chorionic gonadotrophin in urine, progressing through the use of the lymphangiogram and whole lung
tomography and ending in 1978 with the introduction of CT scanning
and alpha fetoprotein (AFP) and B human chorionic gonadotrophin
(BHCG) tumor marker estimation. In the London Hospital series[3]
this has been associated with a decrease in the incidence of Stage
1 teratoma from 60% to 38% between 1950 and 1979 coincident with an
increase in survival of such patients from 40% to 90% (Table 5a).
In addition, there has been a change in pathological practice with
an increasing number of sections being cut in seminomas to try to
identify patients with occult teratoma. In addition, the use of AFP
as a tumor marker has enabled better definition of patients with pure
seminoma. This changing pattern of staging has led to the incidence
of seminoma as a proportion of all germ cell tumors of the testicle
falling from 69% to 34%, coincident with the increase in survival of
patients with pure seminoma from 55% to 94% (Table 5b).

Table 5. Changes in Proportions and Survival of Various
          Types of Germ Cell Tumor After Radiotherapy
          (Modified from Reference 3)

	Pre 1950 n = 11	1950–59 n = 48	1960–77 n = 123	1978–79 n = 27
a)				
Stage 1 Teratoma as a proportion of all treatments	81%	60%	50%	37%
Stage 1 Teratoma Disease-free survival > 2 years	45%	70%	75%	90%
	n = 36	n = 124	n = 212	n = 41
b)				
Seminoma as % of all Germ cell Tumors	69%	61%	42%	35%
Seminoma Disease-free survival >2 years	55%	68%	88%	94%

MANAGEMENT OF EARLY STAGE TERATOMA

The results discussed in the previous section nullify the validity of all the past information on the survival of patients with clinical Stage 1 tumor (either seminoma or teratoma) after orchidectomy alone reported from the era prior to the introduction of adjuvant lymph node dissection or prophylactic radiotherapy. The severe problems arising from using chemotherapy in patients who relapse after radiotherapy and the psychological aspects of ejaculatory impotence which occurs in patients after lymph node dissection has led to the need to re-examine the survival of patients determined to be without any metastases detectable by modern staging methods. In common with other authors we have begun to examine this prospectively by intensive radiological and biochemical surveillance. In our series[3] 4 of 16 patients followed for 12 months have relapsed to date and all have so far been salvaged with combination chemotherapy, though as yet, follow up after chemotherapy is still short (3,5, 11 & 27 months). This series complements the much larger series recently reported by Peckham et al.,[4] whose current relapse rate is 19% in a series of 53 cases and Reed et al.,[5] whose relapse rate is 22% in a series of 46 cases.

These observations confirm that there has been a substantial improvement in our ability to define patients without metastases compared to the 30% cure reported when Whitmore reviewed the results in 1970[6], and that it is safe to consider this approach as an alternative to prophylactic radiotherapy or retroperitoneal lymph node dissection in the management of patients.

However, there are three major disadvantages of this approach. It puts severe restrictions on the mobility of the patient for the period of intense follow-up. Though at present all relapses have occurred within 12 months of diagnosis, the recent report of late relapses, up to and beyond 5 years after chemotherapy for metastatic disease means that close follow-up will need to be maintained for longer than was originally suggested. Undoubtedly, some patients will default from follow-up because of a failure to understand the seriousness of the situation. The second disadvantage of this approach is that the need for close contact with these patients is very labor intensive. The final disadvantage is that the seeming simplicity of the method may lead the surgeon to follow his one to two patients a year without adequate experience of all the problems which may occur these thus reducing the referral of these patients to specialist centers. Because of these problems the ultimate aim of surveillance studies must be to define patients with the highest risk of relapse so that they may benefit from prophylactic chemotherapy.

Table 6. Incidence of Detectable
          Placental Alkaline
          Phosphatase in Patients with
          Various Sub-Types of Germ
          Cell Tumor of Testis
          (Modified from Reference 18)

	N	Positive PLAP
Seminoma	15	54%
Mixed Seminoma and Teratoma	13	38%
Malignant Teratoma	19	16%

Table 7. Metastatic Seminoma

	Single Agent Platinum		BVP/BEP	
	Untreated n=10	Relapse Post Radiotherapy n=4	Untreated n=8	Relapse Post Radiotherapy n=5
Complete Response	5	3	6	3
NED Post Surgery	4	–	1	2
Relapse	0	1	–	–
Currently NED	10	3	7	4
	Median Follow-up 15 months		Median Follow-up 30 months	

MANAGEMENT OF METASTATIC TERATOMA

Since the report of Einhorn and Donohue[7] which was the first to demonstrate that close to two thirds of patients with metastatic teratoma could be cured by treatment with Bleomycin, CisPlatinum and Vinblastine (BVP), two other regimes have achieved as good or possibly better results. However, direct comparison is difficult as each study uses slightly different criteria for defining the poor prognosis groups and each group has demonstrated 10-15% improvement in

results in the four years since they first reported their results
(Tables 8 and 9). The London Hospital and Institute of Urology
experience has mirrored this change with the two-year survival increasing from 38-80% when the results of patients receiving BVP
between 1978 - 80 are compared to those treated in between 1980 and
1982[16]. Following these studies, the main endeavour of the department has been to attempt to incorporate VP16-213 into standard BVP as
a fourth drug. There was a considerable increase in the toxicity of
the regime without any clear cut increase in cure rate. This included treatment related deaths which continued despite decreasing
the VP16-213 dosage to $120mg/m^2/3$ days instead of 5 days. With the

Table 8. Metastatic Testicular Teratoma Chemotherapy Results

Duration of Treatment	Combination	N	Overall Survival	N	Small Volume	N	Large Volume
3 months	BVP[7] (1974-77)	47	64%	–	–	–	–
3 months	BVP[8] (1976-78)	78	73%	–	–	–	–
6 months	Marsden[9] BV+DXT (1976-78)	39	51%	21	81%	18	17%
3 months	BEP[10] (1980-82)	52	83%	26	96%	26	69%

Table 9. Metastatic Testicular Teratoma Chemotherapy Results

Duration of Treatment	Combination	N	Overall Survival	N	Small Volume	N	Large Volume
2 years	VAB II[11] (1974-76)	50	30%	–	–	–	–
1 year	VAB IV[12] (1976-78)	41	60%	–	–	–	–
1 year	VAB VI[13] (1978-79)	25	80%	–	–	–	–
6 months	CXI[14] (1977-79)	43	66%	13	92%	30	70%
4-6 months	CX2[15] (1979-82)	69	83%	47	96%	22	56%

demonstration by Peckham et al.,[12] that removal of Vinblastine did
not decrease the initial response rate, the BEP regime has now become
the standard for future studies.

MANAGEMENT OF PATIENTS WITH SEMINOMA OF THE TESTICLE

     Until recently, radiotherapy has been the principal modality of
treatment for all stages of Seminoma. Today Stage 1 seminoma is as
selected a group of patients as Stage 1 Teratoma (Table 5). With the
cure rate being higher than 95% and the treatment associated with
minimal side effects, the use of surveillance will require justification. The observation of Fossa[17] (these proceedings) that prophylactic radiotherapy affects the patients' sperm counts provides
the first justification for considering re-investigation of the
survival of Stage 1 Seminoma using surveillance.

     In addition, two recent discoveries have made it possible to
consider patients with seminoma for sruveillance. First, the demonstration that placental alkaline phosphatase is a marker of metastatic seminoma (Table 6) and second, the demonstration of the ex-
quisite sensitivity of seminoma to chemotherapy, with the initial
cure rate from single agent CisPlatinum (Table 7) being as good as
that achieved with the intensive combination regimes[16] which became
standard because of their success in the management of patients with
metastatic malignant teratoma.

CONCLUSION

     The last decade has witnessed a revolution in the treatment of
germ cell tumors of the testis. Part of this has been due to the
changes in radiological, biochemical and histopathological staging
resulting in metastases being detected in about 50% of patients
previously considered to be Stage 1 and occult teratomas being discovered in approximately 50% of patients previously considered to be
Seminoma. These improvements mean that patients defined as having no
metastases have a less than 20% chance of relapse and given the high
salvage rates of chemotherapy if current preliminary results are
confirmed, it will be unnecessary in the future, to advise prophylactic radiotherapy or retroperitoneal lymph node dissection for
these patients.

     The principal lesson concerning the management of patients with
metastatic disease is that the more advanced the metastases, the
lower the chance of long-term cure. The demonstration of a direct
relationship between delay and extent of disease suggests that
further improvement in survival of patients with metastatic disease
will come more easily from encouraging earlier recognition of testicular tumors both by patients and clinicians.

For the future, three priorities have been defined. First, to balance the discomforts and long-term sequelae of surgery and cyto- toxic chemotherapy in patients with early metastasis to the retro- peritoneum. Second, to discover less toxic regimes which can be used prophylactically in Stage 1 patients with a high risk of relapse. Third, to find ways better to define when it is safe to stop chemo- therapy in patients with bulky necrotic tissue which is currently being removed at post treatment surgical staging operations.

REFERENCES

1. R. T. D. Oliver, Progress in the management of testicular germ cell tumors, The Practitioner, 226:1903-1915 (1982).

2. R.T. D. Oliver and M. H. Pancharatnam, Risk factors associated with occurrence of germ cell tumors of the testis, (in preparation).

3. R. T. D. Oliver, H. F. Hope-Stone, and J. P. Blandy, A justification for the use of surveillance in the management of stage 1 germ cell tumors of the testis, Brit.J.Urol., (in press).

4. M. J. Peckham, A. Barrett, J. E. Husband, and W. F. Hendry, Orchidectomy alone in testicular stage 1 non-seminomatous germ-cell tumors, Lancet II: 678-680 (1980).

5. G. Read, R. J. Johnson, P. M. Wilkinson, and B. Eddlestone, A prospective study of follow-up only in stage 1 Teratoma of the testis, Brit.Med.J., (in press).

6. W. F. Whitmore, Germinal tumors of the testis. Proceedings of the 6th National Cancer Conference, Philadelphia, J. B. Lippincott, p.219 (1970).

7. L. H. Einhorn and J. P. Donohue, cis-diamminedichloroplatinum, vinblastine and bleomycin combination chemotherapy in dis- seminated testicular cancer, Ann.Intern.Med., 87:293-298 (1977).

8. L. H. Einhorn and S. D. Williams, Chemotherapy of disseminated testicular cancer: a random prospective study, Cancer, 46: 1339-1344 (1980).

9. M. J. Peckham, A. Barrett, T. J. McElwain, and W. F. Hendry, Combined management of Malignant teratoma of the testis, Lancet, I: 267-270 (1979).

10. M. J. Peckham, A. Barrett and K. H. Lieu, The treatment of Metastatic germ cell testicular tumors with bleomycin, etopo- side and cis-platin (BEP), Brit.J.Cancer., 47:613-619 (1983).

11. E. Cheng, E. Cvitkovic, R. E. Wittes, and R. B. Golbey, Germ cell tumors II: VAB II in metastatic testicular cancer, Cancer., 42:2162-2168 (1978).

12. D. Vugrin, E. Cvitkovic, W. F. Whitmore Jr, E. Cheng, and R. B. Golbey, VAB-4 Combination chemotherapy in the treatment of metastatic testis tumors, Cancer, 47:833-839 (1981).

13. D. Vugrin, W. F. Whitmore Jr, and R. B. Golbey, VAB-6
     Combination chemotherapy without maintenance in treatment of
     disseminated cancer of the testis, Cancer, 51:211-215 (1983).
14. E. S. Newlands, R. H. J. Begent, S. B. Kaye, G. J. S. Rustin,
     and K. D. Bagshawe, Chemotherapy of advanced malignant tera-
     tomas, Brit.J.Cancer., 42:378-384 (1980).
15. E. S. Newlands, R. H. J. Begent, G. J. S. Rustin, D. Parker, and
     K. D. Bagshawe, Further advances in the management of malig-
     nant teratomas of the testis and other sites, The Lancet, I:
     948-951 (1983).
16. R. T. D. Oliver, J. P. Blandy, W. F. Hendry, J. P. Pryor, and J.
     P. Williams, and H. F. Hope-Stone, Evaluation of radiotherapy
     and/or surgico-pathological staging after chemotherapy in the
     management of metastatic germ cell tumors, Brit.J.Urol., (in
     press).
17. S. Fossa, Radiotherapy in treatment of seminoma (These
     proceedings, 1983).
18. R. T. D. Oliver, D. Tucker, P. Travers and W. Bodmer, Use of a
     monoclonal antibody to assess placental alkaline Phosphatase
     as a marker of seminoma (in preparation).

SURVEILLANCE AFTER ORCHIDECTOMY FOR CLINICAL

STAGE 1 TESTICULAR NON-SEMINOMA

M. J. Peckham

Institute of Cancer Research and
The Royal Marsden Hospital
London and Surrey

In 1979 post-orchidectomy lymph node irradiation for Stage I testicular non-seminoma was discontinued in favor of a policy of close observation. This policy was introduced in recognition of the important haematogenous component to tumor spread in malignant teratoma. With the development of effective chemotherapy it had become clearly essential to reappraise the roles of radiation and node dissection.

Surveillance has only become feasible and indeed ethical in recent years since before curative combination chemotherapy had been evolved the ability of radiotherapy and surgery to deal with metastatic disease was limited. Two major recent developments have made it possible to study the natural history of Stage I disease, to evaluate the usefulness or otherwise of radiotherapy and surgery and to determine whether a deliberate policy of deferred chemotherapy is effective. These are:

INCREASING ACCURACY OF CLINICAL STAGING PROCEDURES

Computerised axial tomographic (CAT) x-ray scanning is a powerful tool in the management of testicular cancer. Refinement of clinical staging is achieved primarily by the ability of CAT scanning to detect small lung metastases[1] (Table 1). A second major contribution to accurate staging is the prognostic significance of the evolution of serum beta human chorionic gonadotrophin (beta HCG) and alphafetoprotein (AFP) levels after orchidectomy[2].

Table 1. Modification of Clinical Stage by CT Scanning in Patients
with Disease Apparently Localized to the Lymphoid System.

Stage	Number of patients	Abdominal nodes	Previously unsuspected metastases demonstrated by CT scanning	
			Supra-diaphragmatic nodes	Pulmonary metastases
I	21	1	—	4
II	22		1	5
III	12			7
Total	55	1 (1.8%)	1 (1.8%)	16 (29%)

ACTIVITY AND VOLUME DEPENDENCY OF CHEMOTHERAPY

Tumor volume has an important influence on chemotherapy outcome
with virtually uniform curability of patients with small metastases
(Table 2). This means that chemotherapy can be deferred in patients
without clinical evidence of metastases in the expectation that
current non-invasive staging methods will permit early detection of
tumor spread. Furthermore, it has proved possible to reduce the
toxicity of chemotherapy in this good prognosis group without compromising therapeutic effect[3].

THE EVENTUAL OBJECTIVES OF SURVEILLANCE

Based on historical data it was postulated that approximately
20% of patients would relapse after orchidectomy in optimally staged
testicular non-seminoma patients who had clinical evidence of metastases. Based upon experience of the tempo of relapse after radiotherapy it appeared probable that relapses would be rapid and that
late relapse would be unlikely[4].

The eventual aim of the surveillance study is to define two
groups of patients:

a) Those with a high risk of tumor dissemination (>50% chance of
    relapse) in whom immediate post-orchidectomy chemotherapy would
    be justifiable, and
b) those with a low risk of metastases (<10%) where radical node
    dissection would hardly be justified and where a surveillance
    policy would be continued.

Table 2. Advanced testicular non-seminoma: outcome of treatment in
relation to extent of disease.

Staging subgroup	Number of patients	Alive & disease-free
Small volume metastases	63	60 (95%)
Bulky abdominal disease ± small metastases at other sites	62	52 (84%)
Bulky generalized disease	39	14 (36%)
Total	164	126 (76.8%)

CRITERIA FOR ENTRY TO STUDY

These include:

1. Histologically proven testicular non-seminoma or seminoma with
    raised serum alphafetoprotein levels.
2. Absence of metastases after clinical staging.
3. Prompt normalization of serum HCG and AFP levels after
    orchidectomy ($T\frac{1}{2}$: approximately 2 and 6 days respectively).
4. Absence of tumor at the cut end of the spermatic cord or direct
    infiltration through the tunica into the scrotal sac.

Intratumor vascular invasion, involvement of distal cord, rete
testis or epididymis, high pre-orchidectomy HCG and AFP serum levels
and various types of surgical violation of the scrotum, including
biopsy with replacement of the tumor bearing testis in the scrotal
sac, transcrotal orchidectomy and percutaneous needle aspiration were
not contraindications to inclusion in the study.

PROCEDURE

Pre-orchidectomy serum HCG and AFP levels are measured wherever
possible. At least twice weekly levels are desirable in the post-
orchidectomy period. This is not always practicable and the patient
may be seen for the first time two or three weeks after removal of
the primary tumor.

The primary tumor should be processed for histology, evidence of
lymphatic and vascular invasion, local extent, particularly cord
involvement and immunocytochemical tissue stains for beta HCG and
AFP.

   Clinical staging includes lymphography, intravenous urography,
CAT scans of abdomen and thorax, ultrasonic scan of liver and retroperitoneum, as well as routine biochemistry including liver function
tests.

   Surveillance patients are seen at monthly intervals for the
first year, two monthly for the second year and three monthly for the
third year. CAT scans of the chest and abdomen are carried out on
alternate visits for the first year. Initially scans were performed
during the second year but their routine use has been discontinued.
Serum AFP and HCG levels are measured at each visit.

RESULTS

   In a recent update of the study 84 men had been observed from
six to 52 months (mean 22.6 months) after orchidectomy and 16 (19%)
had relapsed[5].

HISTOLOGY AND RELAPSE

   As shown in Table 3 there is a significant difference between
the relapse rate in embryonal carcinoma and teratocarcinoma; 40.7%
and 6.8% respectively.

SITES OF RELAPSE

   As shown in Table 4 13/16 patients had raised AFP and/or HCG at
relapse and three showed an elevated marker level as the only evidence of relapse. Of the 16 relapsing patients ten had abdominal
metastases either alone (six patients) or associated with deposits at
other sites (four patients).

INTERVAL BETWEEN ORCHIDECTOMY AND RELAPSE

   As shown in Table 5 relapses occurred 2-8 months after orchidectomy (mean 5.3 months).

PRE-ORCHIDECTOMY MARKER STATUS AND RELAPSE

   As reported elsewhere 6/9 relapsing patients who had preorchidectomy levels measured had raised AFP titres (13-430 ng/ml, mean
120)[5]. This compares with 25/34 for the non-relapsing group (range
10-10,300 ng/ml, mean 889). For beta HCG 3/8 of the relapse group
had raised preorchidectomy levels (range 14-8090, mean 2721). In the
no relapse group 15/27 had raised levels (range 5-24,760, mean

Table 3. Surveillance study of Stage I testicular non-seminoma
(The Royal Marsden Hospital, 1979-1982).

Histology of primary	Number of ° patients	Number relapsing	%
MTU*	27	11	40.7)
MTI	44	3	6.8)
MTT	3	1	(33)
TD	4	0	
YSC	3	0	
Sem AFP+	3	1	(33)
Total	84	16	19

) P < 0.01

°Observation time since orchidectomy 6-52 months, mean 22.6,
median, 20.

*MTU   - malignant teratoma undifferentiated (embryonal
             carcinoma)
MTI   - malignant teratoma intermediate (teratocarcinoma)
MTT   - malignant teratoma trophoblastic
TD   - differentiated teratoma
YSC   - yolk sac carcinoma
Germ AFP+ - seminoma with initially raised serum alphafetroprotein
             levels

(data from Peckham et al., 1983)[5]

Table 4. Surveillance study of stage I testicular non-seminoma:
patterns of relapse.
(The Royal Marsden Hospital, 1979-1982)

Total patients relapsing	Raised marker(s) at relapse	Raised marker(s) as only evidence of relapse	Abdo. nodes	Abdo. + S/D° nodes	Abdo. + lungs	Lungs
16	13 (81%)	3 (19%)	6 (38%)	1 (6%)	3 (19%)	3 (19%)

° S/D = supradiaphragmatic

(data from Peckham et al., 1983)[5]

Table 5.   Surveillance Stage I Testicular Non–Seminoma Study:
           Interval Between Orchidectomy & Relapse.
           (The Royal Marsden Hospital, 1979–1982)

Histology	Number relapsing	Time to relapse
MTU	11	2–8 months mean 5 months
MTI	3	5, 6, 7
MTT	1	3
SEM AFP+	1	6
Total	16	2–8 months (mean 5.3)

(data from Peckham et al., 1983)[5]

1779). These observations suggest that preorchidectomy marker levels
are not of prognostic significance so far as the risk of relapse is
concerned.

     It is of interest that of seven relapsing patients with raised
serum marker levels prior to orchidectomy six showed a marker rise at
the time of relapse. However, of three relapsing patients with
negative markers before orchidectomy two did not show a marker rise
with relapse. This suggests that markers are particularly useful as
monitors when preorchidectomy levels are raised but that the marker
negative patient needs to be watched predominantly by means of physical tumor detection methods.

INFLUENCE OF THE EXTENT OF THE PRIMARY TUMOR AND RELAPSE

     This aspect has not been completely elucidated although pre-
liminary data suggest that in patients with embryonal carcinoma the
presence of vascular invasion and/or lymphatic permeation may confer
a high risk of subsequent relapse with 6/8 of patients in this group
developing metastatic disease.

TREATMENT OF RELAPSING PATIENTS

     All 16 patients are alive, 13 having completed chemotherapy.
Three patients are receiving treatment. These data are summarized in
Table 6.

Table 6.   Surveillance Stage I Testicular Non-Seminoma Study: Details
           of Treatment and Outcome in 16 Patients who Developed
           Metastases.   (The Royal Marsden Hospital, 1979-1982).

Patient (histology)	Stage[°°] in relapse	Chemo-therapy[+]	Involved field ir-adiation	Surgery (histo-logy)	Outcome (months)
1 (MTU)	IIb	VB, PVB	+		[°]NED 43
2 "	IIb	PVB	+		NED 38
3 "	IIA	PVB	+		NED 39
4 "	$IVAL_2$	PVB			NED 32
5 "	IM	BEP			NED 24
6 "	IIA	EP			NED 16
7 "	$IVOL_1$	BEP			NED 9
8 "	IIA	BEP			NED 4
9 "	$IVBL_2$	BEP			On treatment
10 "	$IVOL_2$	BEP			On treatment
11 "	Im	BEP			NED 4
12 (MTI)	$IVAL_1$	BEP			NED 16
13 "	$IVOL_1$			Lung (TD)	NED 12
14 "	Im	BEP			On treatment
15 (MTT)	IIB	BEP			NED 17
16 (Sem AFP+)	IIB	EP		Abdo(-ve)	NED 18

[°]   NED = No evidence of disease
[°°] The Royal Marsden Hospital staging
[+]   V   = vinblastine
      P   = cis-platin
      B   = bleomycin
      E   = etoposide

CONCLUSIONS

     The observed relapse rate in carefully staged non-seminoma
patients without marker or other evidence of metastases is close to
the anticipated figure and experience has demonstrated the feasibility of detecting relapse at a stage when the tumor volume is low
and the patient curable with chemotherapy. Late relapses after
orchidectomy are unlikely to occur since the tempo of relapse observed in the surveillance study is closely similar to that observed
in Stage I patients treated with lymph node irradiation. Serum
markers are not a foolproof method of monitoring patients since 3/16
did not show marker elevation at the time metastases were detected.

     The longer term objective of identifying patients at high risk
of harboring occult metastases in whom immediate chemotherapy would

be justifiable has not yet been achieved, although it is probably
that a detailed study of vascular invasion and lymphatic permeation
in conjuction with histology will eventually allow such a group to be
identified.

REFERENCES

1. J. E. Husband, Computed tomography in testicular tumors, in:
    "The Management of Testicular Tumors," M. Peckham, ed.,
    Edward Arnold Limited, London pp 119-133 (1981).
2. D. Raghavan, M. J. Peckham, E. Heyderman, J. S. Tobias, D. E.
    Austin, Prognostic factors in clinical stage I non-
    seminomatous germ-cell tumors of the testis, Br.J.Cancer
    45:167-173 (1982).
3. M. J. Peckham, A. Barrett, K. H. Liew, A. Horwich, B. Robinson,
    H. J. Dobbs, T. J. McElwain, W. F. Hendry, The treatment of
    metastatic germ-cell testicular tumors with bleomycin, etopo-
    side and cis-platin (BEP), Br.J.Cancer 47:613-619 (1983).
4. M. J. Peckham, A. Barrett, Radiotherapy in testicular teratoma,
    in: "The Management of Testicular Tumors," M. Peckham eds,
    Publ. Edward Arnold Ltd., London, pp 174-201 (1981).
5. M. J. Peckham, A. Barrett, A. Horwich, W. F. Hendry, Orchi-
    ectomy alone for stage I testicular non-seminoma, A progress
    report on the Royal Marsden Hospital study, British Journal
    of Urology, 55:754-759 (1983).

UNILATERAL VERSUS BILATERAL RETROPERITONEAL LYMPH NODE

DISSECTION IN NON-SEMINOMATOUS TESTICULAR GERM CELL TUMORS

S. D. Fossa, S. Ous, H. H. Lien and A. E. Stenwig

The Norwegian Radium Hospital
Oslo
Norway

INTRODUCTION

For many years bilateral retroperitoneal lymph node dissection (RLND) has been the standard surgical treatment for patients with non-seminomatous germ cell tumors (NSTGCT)[1-5]. In clinical stage I disease[6] the operation is mainly a diagnostic procedure, whereas, in stage II patients with retroperitoneal metastases, it may have therapeutic significance. The most frequent side effect of bilateral RLND is loss of semen ejaculation and emission, leading to infertility in the majority of patients[7-9].

In an attempt to achieve the diagnostic advantages of retroperitoneal surgery (identification of patients with retroperitoneal lymph node metastases) without producing infertility, unilateral RLND is performed at The Norwegian Radium Hospital in patients with NSTGCT, clinical stage I, who peroperatively appear to be tumor-free. Bilateral RLND is undertaken only if metastatic growth in lymph nodes is demonstrated during the operation. The present report deals with the experience in the first 53 patients treated according to the above protocol.

PATIENTS AND METHODS

From May 1979 to March 1982 53 orchiectomized patients with NSTGCT, clinical stage I, underwent RLND. The preoperative staging procedures consisted of computerized tomography (CT) of the abdomen (53 patients) and pedal lymphography (51 patients). Postorchiectomy serum levels of alpha-feto protein (AFP) and beta-human choriogonadotropin (β-HCG) were analyzed before retroperitoneal surgery in all but one patient.

Figure 1a and Figure 1b show the extent of the unilateral RLND. If there was evidence of retroperitoneal lymph node metastases per-operatively by using multiple frozen sections, a bilateral RLND was undertaken but without removal of the contralateral iliac lymph nodes. The number of metastatic lymph nodes was recorded separately for each of the following regions (Figure 1): Cavo/Inter-cavo-aortic region (C.ICA)[1], Pre/para-aortic region (PA)[2], right iliac region (R.IL)[3], left iliac region (L.IL)[4]. Patients with retro-peritoneal lymph node metastases received 3 or 4 courses of CVB chemotherapy: cis-platinum (20 mg/m^2, Day 1-5), vinblastine (0.15 mg/kg, Day 1 and 2) and bleomycin (30 mg, Day 2, 5 and 16), repeated every 3rd week. Relapsing patients received the same treatment as salvage chemotherapy.

All patients were followed until April 1st 1983 (minimum follow up: 1 year; median follow up: 31 months).

Twelve to eighteen months after the RLND 52 patients were inter-viewed considering ejaculation disturbances.

RESULTS

Thirty-four patients who were found to be tumor-free, (patho-logical stages IPS) were operated unilaterally. Nineteen patients were classified as PS II, of whom 15 were stage IIa and four were

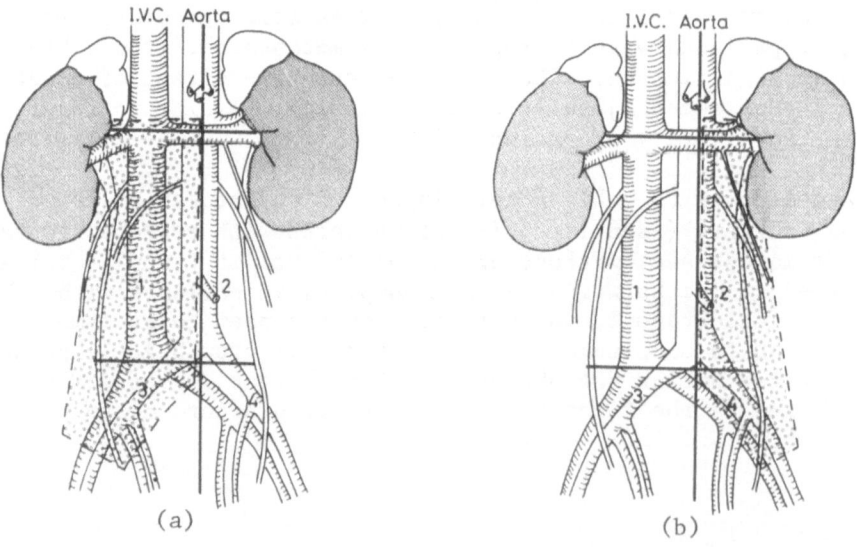

Fig. 1.   Extent of right-sided (a) and left-sided (b) unilateral
           retroperitoneal lymph node dissection (RLND) in patients
           with non-seminomatous testicular cancer, clinical stage I.
           (I.V.C.: inferior vena cava). (1) Cavo/Inter-cavo-aortic
           region; (2) Pre/para-aortic region; (3) Right iliac region
           (4) Left iliac region.

stage IIb. Sixteen of these were operated bilaterally, whereas unilateral RLND was done in the three patients whose metastases were found only in the definitive histological sections.

A total of 867 lymph nodes were resected, 61 of which were metastatic (Table 1). In patients with PS II about every 6th lymph node was metastatic. On the average, the diameter of lymph nodes with metastases measured 14.1 mm (range 3–25 mm). A solitary lymph node metastasis was found in four patients with right-sided and one patient with left-sided tumor.

The 16 bilaterally operated patients were evaluated separately: Contralateral metastases (metastases beyond the limits of ipsilateral RLND) were observed in one of six patients with a left-sided primary tumor (Table 2), and in three of 10 patients with right-sided testicular cancer (Table 3).

Postoperative disease activity was observed in three patients. Two of these patients had a PSI: one patient with a left-sided tumor had persistant β-HCG elevation postoperatively and, in one patient, increased β-HCG levels were found two years after a left-sided RLND. The 3rd patient with a PS II developed a scrotal recurrence 6 weeks after bilateral RLND before he had started adjuvant CVB. These three

Table 1. Number, Localization and Extent of Lymph Node Metastases in Patients with Non-Seminomatous Testicular Cancer, Clinical Stage I.

	C/ICA	PA	IL right	IL left	Total
No of resected lymph nodes					
PSI	212	202	37	27	478
PSII	173	182	20	14	389
Total	385	384	57	41	867
No of metastatic resected nodes (PS II pts)	30	30	0	1	61
(Percentage)	(17.3)	(16.8)	(0)	(7.1)	(15.7)
Diameter (mm) of largest met. node per pt.	16,9	10,0	0	15,0	14,1

(Abbreviations) C/ICA: Cavo/inter-cavo-aortic region, PA: Pre/para-aortic region, IL: Iliac region, and PS: Pathological stage

Table 2. Localization and number
of Lymph Node Metastases
in Bilaterally Operated
Patients with Left-sided
Tumor.

	Total	PA	C/ICA
No of patients	6	5	1
No of metastases per patient (range)	1-4	1-4	4

(Abbreviations) PA: Pre/para-aortic
region, C/ICA: Cavo/inter-cavo-
aortic region

Table 3. Localization and Number of Lymph Node Metastases in Ten
Bilaterally Operated Patients with Right-sided Tumor.

	Total	C/ICA alone	C/ICA + PA	C/ICA + LIL
No of patients	10	7	2	1
No of metastases per patient (range)		1-5	6-11	4

(Abbreviations) PA: Pre/para aortic region, LIL: Left iliac region,
and C/ICA: Cavo/Inter-cavo-aortic region.

patients were all cured by salvage chemotherapy. Thus all 53
patients with clinical stage I are alive with NED (median observation
time 31 months).

"Dry ejaculation" was reported by 13 or the 16 bilaterally
operated patients (Table 4) and in eight of 18 evaluable patients
after unilateral left-sided RLND. After unilateral right-sided RLND
no persistent ejaculatory disturbances were reported.

DISCUSSION

Unilateral RLND or modified bilateral RLND has been recommended
for several years[6,10] in an attempt to preserve fertility in
patients without retroperitoneal lymph node metastases. However,
most centers still prefer to do bilateral RLND, even in patients
without retroperitoneal metastatic spread. Two recent reports
[11,12] however indicate that a more limited RLND may be sufficient

Table 4.   Ejaculatory Status (12-18 months
           after RLND)

RLND	Tumor side	Normal	Dry ejaculate	Total
Unilat.	right	18 (1)		18
	left	10 (3)	8 [2]	18
Bilat.	right	2 (1)	8 [1]	10
	left	1	5	6
Total		31	21	52*

( )   Number of patients with dry ejaculate
      immediately after RLND but with gradual
      improvement during the first post-
      operative year
[ ]   "partial" dry ejaculation.
*   1 patient not evaluable

in clinical low stage disease. Our own observations are consistent
with these authors' conclusions: Unilateral RLND seems to be a
sufficient diagnostic procedure in patients in whom no metastases are
found peroperatively. None of the 3 patients with postoperative
recurrence relapsed with metastases in the retroperitoneal space.
Furthermore 12 of the 16 bilaterally operated patients could have
been rendered tumor-free by unilateral RLND. This means that, even
in many patients with early stage II disease, unilateral RLND seems
to be an appropriate therapeutic procedure.

     Recently, a "wait and see" policy was introduced into the treat-
ment of clinical stage I disease in patients with NSGCT[13]. Only
those patients who relapse during follow-up (about 20-30%) will
receive chemotherapy, which is curative in the majority of patients.
Thus unnecessary treatment and ejaculatory disturbance are avoided in
most of the patients with clinical stage I (70-80%). However, this
treatment philosophy demands a very high degree of diagnostic accuracy and a strict follow up routine. In agreement with other recent
reports[14,15] our false negative diagnostic rate in clinical stage I
disease is about 30%. Furthermore, for geographic, social and health
economy reasons, a very close and frequent follow-up of all patients
with clinical stage I cannot always be guaranteed in our country. We
therefore feel it is safer to improve the diagnostic accuracy by
performing a unilateral RLND in pre- and peroperatively tumor-free
patients and thus identify those patients with a minimal risk of
reactivation of the disease. Furthermore, patients with a PS II will
have the benefit of adequate surgery and adjuvant chemotherapy at an
early phase of the disease.

The advantage of unilateral retroperitoneal lymph node dissection is the maintenance of semen emission/ejaculation and preservation of fertility in the majority of patients. Our study clearly
shows that unilateral RLND preserves ejaculation ability in most of
the patients. The fact that about 40% of the patients are apparently
hypofertile after orchiectomy and before further treatment[16] does
not exclude subsequent normal fertility and fatherhood only 1-1½ year
after the primary treatment, if ejaculation is intact[17].

Our observations suggest that new methods of retroperitoneal
surgery, i.e. more limited RLND, should be developed, even for
patients with early stage II disease. This type of limited surgery,
combined with modern chemotherapy should allow maintenance of fertility in many patients without lowering the cure rate.

CONCLUSIONS

1)   Thirty to 35% of patients with stage I non-seminomatous testi-
     cular cancer have retroperitoneal lymph node metastases.

2)   Unilateral RLND provides excellent results in peroperatively
     tumor-free patients.

3)   No permanent ejaculatory disturbances are observed after uni-
     lateral RLND in the majority of patients.

REFERENCES

1. R. C. Walsh, J. J. Kaufman, W. F. Coulson and W. E. Goodwin,
    Retroperitoneal lymphadenectomy for testicular tumors. JAMA,
    217:309 (1971).
2. D. E. Johnson, Retroperitoneal lymphadenectomy: Indications,
    complications and expectations, Rec.Res.Cancer Res., 60:221
    (1977).
3. J. P. Donohue, L. H. Einhorn, J. M. Perez, Improved management
    of non-seminomatous testis tumors. Cancer, 42:2903-2908,
    (1978).
4. D. G. Skinner, Management of non-seminomatous tumors of the
    testis, in: "Genitourinary Cancer", D. G. Skinner and J. B. de
    Kernion, eds., W. B. Saunders Company, Philadelphia, London,
    Toronto, 470 (1978).
5. W. F. Whitmore, Jr., Surgical treatment of adult germinal testis
    tumors, Sem.Oncol., 6:55-68, (1979).
6. M. J. Peckham, A. Barrett, T. J. McElwain, W. F. Hendry, D.
    Raghaven, Non-seminoma germ cell tumors (malignant teratoma)
    of the testis Brit.J.Urol., 53:162-172, (1981).
7. E. Leiter, H. Brendler, Loss of ejaculation following bilateral
    retroperitoneal lymphadenectomy, J.Urol., 98:375-378, (1967).

8. R. B. Bracken, D. E. Johnson, Sexual function and fecundity
    after treatment for testicular tumors. Urology, 7:35-38,
    (1976).
9. W. B. Waters, M. B. Garnick, J. P. Richie, Complications of
    retroperitoneal lymphadenectomy in the management of non-
    seminomatous tumors of the testis. Surg.Gynec.Obstet., 154:
    501-504, (1982).
10. E. Fraley, C. Markland, P. H. Lange, Surgical treatment of stage
    I and stage II non-seminomatous testicular cancer in adults.
    Urol.Clin.North Am., 4:453-463, (1977).
11. J. P. Donohue, J. M. Zachary, B. R. Maynard, Distribution of
    nodal metastases in non-seminomatous testis cancer. J.Urol.,
    128:315-320, (1982).
12. P. Hermanek, A. Sigel, Necessary extent of lymph node dissection
    in testicular tumors. A histopathological investigation,
    Eur.Urol., 8:135-144, (1982).
13. M. J. Peckham, A. Barrett, J. E. Husband, W. F. Hendry,
    Orchiectomy alone in testicular stage I non-seminomatous germ
    cell tumors, Lancet, 25:678-680, (1982).
14. J. P. Richie, M. B. Garnich, H. Finberg, Computerized
    tomography, How accurate for abdominal staging of testis
    tumors? J.Urol., 127:715-717, (1982).
15. R. G. Rowland, D. Weisman, S. D. Williams, L. H. Einhorn, E. C.
    Klatte, J. P. Donohue, Accuracy of preoperative staging in
    stage A and B non-seminomatous germ cell testis tumors,
    J.Urol., 127:718-720, (1982).
16. S. D. Fossa, O. Klepp, K. Molene, A. Aakvaag, Testicular
    function after unilateral orchiectomy for cancer and before
    further treatment, Intern.J.Androl., 5:179-184, (1982).
17. Fossa, S. D., O. Klepp, S. Ous, H. H. Lien, A. E. Stenwig, T.
    Abyholm, O. Kaalhus, Unilateral retroperitoneal lymph node
    dissection in patients with non-seminomatous testicular tumors
    in clinical stage I, Eur.Urol., submitted (1983).

TUMORS OF THE TESTIS - THE ROLE

OF SURGERY AFTER CHEMOTHERAPY

G. Stoter

Department of Oncology
Free University Hospital
Amsterdam, The Netherlands

Experience acquired in the nineteen-seventies, which includes
one randomized trial of surgery before chemotherapy, has led most
centers to advocate the treatment of advanced disease with chemotherapy first, followed by surgical resection of residual masses
after the completion of induction chemotherapy[1-6].

At Indiana University 12 (23%) of 52 patients treated with
debulking surgery after chemotherapy and who were marker negative at
the time of surgery, showed residual viable cancer. If immature
teratomas are included in the residual malignant disease category,
there was residual disease in 16 (31%) of the 52 patients[8].

At Memorial Sloan Kettering Cancer Center (MSKCC) 11 (30%) of 37
patients, who were marker negative at the time of surgery, showed
residual malignant tissue in the resected specimen[2].

In a Dutch multicenter study of 91 patients, 48 patients underwent debulking surgery after the completion of induction chemotherapy. Seven patients who were still marker positive at the time
of surgery, had viable cancer and later died. Of 41 patients who
were marker negative at the time of surgery, 5 (12%) had viable
cancer in the tissue removed, 12 (29%) had mature teratoma, and 24
(59%) had either fibronecrotic changes (17 patients) or normal archi-
tecture (7 patients). It is difficult to explain the relatively low
percentage of persistent malignant disease and conversely the high
rate of patients with benign histology.

There are two important questions concerning the surgical
results:

1) what proportion of patients can be rendered free of disease by adding surgery to chemotherapy; and
2) how does surgery affect the prognosis of patients with residual malignant tumor, mature teratoma, and fibronecrotic changes, respectively.

The combined experience of the investigators at Indiana University[8], MSKCC[2], and the Dutch Testicular Cancer Study Group (DTCSG)[7] demonstrate a number of important findings:

1) Fifteen per cent of the whole patient population can be rendered free of disease by surgery after induction chemotherapy.
2) Patients with positive tumor-markers at the time of surgery invariably have residual cancer and nearly all of them die eventually.
3) Patients with positive markers are unlikely to have a complete resection.
4) Patients with an incomplete resection have a higher chance of relapse, even when they have negative markers at the time of surgery.
5) Patients with resected mature teratoma and fibronecrotic tissue have a very good prognosis (5% relapse rate) and do not need further therapy.
6) Patients with completely resected residual cancer and negative tumor markers have a high (80%) chance of persistent disease-free status with additional induction chemotherapy.

These findings can be translated into the following treatment strategy: Patients with advanced disease should receive induction chemotherapy. After completion of the chemotherapy they should undergo debulking surgery of residual disease only if the markers have become negative. If markers remain elevated, the patient should be switched to different chemotherapy. Surgery should be aimed at complete resection. If the resected specimen appears to contain viable cancer cells, the patient should be given an additional two courses of induction chemotherapy.

In conclusion, the role of surgery after chemotherapy is: 1) to resect completely the residual disease, and 2) to identify those patients who need further induction chemotherapy.

REFERENCES

1. J. P. Donohue, L. H. Einhorn, and S. D. Williams, Cytoreductive surgery for metastatic testis cancer: considerations of timing and extent, J.Urol., 123:876 (1980).
2. D. Vugrin, W. F. Whitmore, P. C. Sogani, M. Bains, H. W. Herr, and R. B. Golbey, Combined chemotherapy and surgery in treatment of advanced germ-cell tumors, Cancer, 47:2228 (1981).

3. C. Merrin, H. Takita, and R. Weber, Combination radical surgery
    and multiple sequential chemotherapy for the treatment of
    advanced carcinoma of the testis (stage III), Cancer, 37:20
    (1976).
4. C. Merrin, H. Takita, and S. Beckley, Treatment of recurrent and
    widespread testicular tumors by radical reductive surgery and
    multiple sequential chemotherapy, J.Urol., 117:291 (1977).
5. D. J. Mathisen and N. Javadpour, En bloc resection of inferior
    vena cava in cytoreductive surgery for bulky retroperitoneal
    metastatic testicular cancer, Urology, 16:51-56 (1980).
6. N. Javadpour, R. F. Ozols, A. Barlock, D. Anderson, and R. C.
    Young, A randomized trial of cytoreductive surgery followed
    by chemotherapy versus chemotherapy alone in bulky stage III
    (poor prognosis) testicular cancer, Proc.Amer.Soc.Clin.
    Oncol., 22:473 (1981).
7. G. Stoter, C. P. J. Vendrik, A. Struyvenberg, D. Th. Sleyfer,
    R. Vriesendorp, H. Schraffordt Koops, A. T. van Oosterom,
    W. W. ten Bokkel Huinink, and H. M. Pinedo, The 5-year
    survival of patients with disseminated non-seminomatous
    testicular cancer treated with cisplatin, vinblastine and
    bleomycin, Cancer, (in press).
8. L. H. Einhorn, S. D. Williams, I. Mandelbaum, and J. P. Donohue,
    Surgical resection in disseminated testicular cancer
    following chemotherapeutic cytoreduction, Cancer, 48:904
    (1981).

COMBINED CHEMOTHERAPY AND SURGERY IN TREATMENT OF BULKY GERM-CELL TUMORS: COMPARISON OF TWO INDUCTION REGIMENS AND EVALUATION OF POOR PROGNOSTIC FACTORS

J. E. Altwein, E. Kreuser, N. Jaeger,
L. Weißbach, and W. Schreml

Departments of Urology BWK Ulm, and University of Bonn
Department of Haematology/Oncology
University of Ulm, West Germany

Patients with bulky testicular cancer have a complete response (CR) rate from 20 to 50%.[1] The optimal timing of cytoreductive surgery has been an area of frequent debate. Data from Javadpour et al.,[1] based on a prospective randomized trial suggest that cytoreductive surgery prior to chemotherapy is not indicated in patients with advanced testicular cancer. The question arises as to whether the optimal induction therapy should be nonintensive or intensive. The present study was undertaken to evaluate the efficacy of two different nonintensive chemotherapeutic regimens and to examine how pretreatment factors may affect complete remission rates, duration of remission, and survival.

PATIENTS AND METHODS

Patients with bulky germ cell carcinoma (Royal Marsden stages, IIc, III, IV, H+) were examined retrospectively. Of these 65 patients suffered from primary bulky disease (PBD), and 19 patients had secondary bulky disease (SBD). All patients had clearly measurable metastatic disease as determined by modern imaging techniques. During treatment monthly hematologic and X-ray follow-ups were performed. Once complete response (CR) was achieved patients were followed every 2 months for the first year, and every 3 months for the second year. Particular consideration was given to the location of the hemoclip as seen on the scout abdominal film.

Cytoreductive surgery, or debulking[2], was done through a midline incision and was facilitated by the use of the Smith self-retaining or the Sigel-thorax aperture rectractor. The retroperitoneum was approached as suggested by Skinner[3].

Drug Regimens

Two different regimens were used. One group of 38 patients were treated with combination chemotherapy consisting of vinblastine 6 mg/m^2 i.v. days 1 and 2, bleomycin 12 mg/m^2 i.v. days 1-5 and cis-platinum 20 mg/m^2 i.v. days 1-5. Cis-platinum was administered as an 8-hour infusion following an initial six hours of hydration with 5% glucose and normal saline (500 ml/h) and the administration of 40 g of mannitol as the prehydration was being completed. Bleomycin was given continuously over 5 days. Cycles were repeated at 4 week intervals.

A second group of 37 patients received an induction program consisting of two sequential combination regimens (II) consisting of vinblastine 0,2 mg/kg i.v. days 1 + 2, bleomycin 30 U i.v. days 1-5 (A) and adriblastin 60 mg/m^2 i.v. and cis-platinum 100 mg/m^2 i.v. day 1 (B). Cycles were repeated at 4 week intervals in the sequence AABBA. When necessary, drug doses were modified according to the degree of myelosuppression observed with previous courses. No maintenance therapy was given in either regimen I or II.

Response Criteria

Partial remission (PR) and progression (P) were defined according to the WHO criteria[4]. Patients with fibrosis, necrosis, or cystic changes at surgery were classified as CR. When active carcinoma was found but could be completely resected the patient was classified as PR. NED (no evidence of disease) was assigned, according to Einhorn et al.,[5]. The designations for dead of disease (DOD), alive with disease (AWD), dead without disease (DWD) were assigned according to the WHO criteria. In addition to bulky disease the following unfavorable prognostic characteristics (cf. Ozols et al.,[6] 1983) were also examined: extragonadal primary, visceral metastases, pure or partial chrorionic carcinoma, extensively elevated markers, and age over 40 years.

RESULTS

Of the 84 unselected patients, with bulky testis tumor, (including seven protocol violaters and two refusers, 38% achieved a CR, 26% had PR, and 36% progressed (P) (Table 1).

In 14 patients (24%) the resected material at delayed or second look retroperitoneal lymph node dissection (RLND) revealed mature teratoma, and 11 of these patients later had NED. In 21 patients with residual carcinoma 6 patients later had NED (Table 2).

Table 1. Primary and Secondary Bulky Testis Tumors: Results of 84
          Unselected Patients

Response	Classification of cases	Median (range) survival (months)
CR = 32 (38%)	NED = 37 (44%)	18 (3 – 47)
PR = 22 (26%)	AWD = 11 (13%)	10 (3 – 51)
P = 30 (36%)	DOD = 34 (41%)	12 (3 – 42)
	DWD = 2 (2%)	3,5

PVB-Regimen:	38 Patients	Protocol-Violation:	7 Patients
VB-AC-Regimen:	37 Patients	No Chemotherapy:	2 Patients

In PBD, the rate of CR achieved after surgery was 37% (24/65)
while patients with SBD achieved CR in 32% (6/19) and post surgical
NED in 37% (7/19). Median duration of remission in patients with PBD
is 19 months ((range 3–47) and in SBD 36 (range 8–41). In both
groups the median duration of the complete response (CR) had not been
attained (Table 3).

Regimen I produced CR in 42% (16/38) with a NED-rate of 55%
(21/38). Patients treated with regimen II achieved CR in 43% (16/37)
and a NED rate of 43% (Table 4). There is no significant difference
between these two groups. In regimen I the mean number of cycles was
3.94 and in regimen II it was 5.37. The subgroup with the best
results was treated by primary chemotherapy and delayed RLND. In
this group, those receiving regimen I achieved a CR rate of 70%
(16/23). At RLND, 6/16 (38%) of these patients were found to have
residual malignant disease. Regimen II resulted in a CR rate of 75%
(9/12), with residual malignant disease found in 44% (4/9).

Tumor Markers

Serum AFP and HCG levels were measured. Elevation of at least
one of the markers was observed in 74% of cases. BHCG was elevated
in 50% and was over 1000 mU/mL in 13%. AFP was elevated in 60% with
an excessive level (> 1000 ng/ml) in 23% (Table 5).

Surgical Results (Table 6)

Primary RLND was performed in 29/84 (35%) cases while delayed
RLND was done in 28. Thoracoabdominal LND was performed in 5
patients, and thoracotomy for debulking in 6: of these 11/3 (27%)
achieved CR and now have NED. Second look LND and delayed LND was

Table 2. Postchemotherapy - Histology of 58 Patients with PBD + SBD

Histology	N	%	Classification of cases	Median (range) survival (months)
Fibrosis	23	40	NED = 18	24 (3 - 47)
Necrosis			AWD = 0	0
Cystic changes			DOD = 4	28 (18 - 42)
			DWD = 1	5
Mature Teratoma	14	24	NED = 11	30 (9 - 47)
			AWD = 3	30 (8 - 56)
			DOD = 0	0
			DWD = 0	0
Carcinoma	21	36	NED = 6	16 (8 - 44)
			AWD = 3	36 (10 - 51)
			DOD = 12	12 (4 - 26)
			DWD = 0	0

Table 3. Primary Bulk Disease (PBD): Results for 65 Patients (1976-1983)

Response	Classification of cases (N)	Median months of (range) survival
CR = 24 (37%)	NED = 31 (48%)	19 (3 - 47)
PR = 21 (32%)	AWD = 7 (11%)	23 (3 - 56)
P = 20 (31%)	DOD = 25 (38%)	14 (3 - 36)
	DWD = 2 (3%)	2,5

performed in 58 patients (69%). Of these 17/39 (43%) with PR after chemotherapy achieved CR by delayed or second look RLND. There was one therapy related death, a surgical mortality of 1%.

Prognostic Factors

The CR rate of patients with an extragonadal primary was 27% (3/22), and for the remainder it was 53% (34/64). For patients with visceral metastases the CR rate was 17% (2/12) vs 56% (35/63) for those without visceral metastases. Patients with pure or partly choriocarcinoma achieved CR in 20% (6/30) vs 77% (27/35) for those without. Patients with retroperitoneal tumor bulk > 10 cm achieved a CR rate of 30% (16/54) vs 42% (10/23) for those whose metastases were < 10 cm. Patients over 40 years had a CR rate of 37% (3/8) vs

Table 4. Primary and Secondary Bulky Disease:
          PVB vs VB/AC - Regimen

	PVB n = 38	VB/AC n = 37
CR	16 (42%)	16 (43%)
PR	16 (42%)	6 (16%)
P	6 (16%)	15 (41%)
NED	21 (55%)	16 (43%)
AWD	5 (13%)	5 (14%)
DOD	11 (26%)	14 (38%)
DWD	1 (3%)	2 (5%)

50% (34/67) for those under 40 years. Patients with moderate or high
HCG-elevation achieved a CR rate of only 37% (14/38) vs 74% (20/27)
in patients with normal values.

DISCUSSION

     The combination of cytoreductive surgery and chemotherapy re-
presents a rational approach for the treatment of advanced or bulky
testicular tumors. This approach is based upon the biology of solid
tumors. Large tumor masses have a great proportion of cells which
are in the rest period of their cell cycle. Most of the chemotherapeutic agents are cycle, or phase, specific and are unable to destroy
these cells during this rest period. This may explain the failure of
chemotherapy to control large tumor masses. Cytoreductive surgery
has been utilized in the management of advanced or bulky testis
tumors in two distinct clinical settings:

1) Initial cytoreductive surgery followed by chemotherapy
2) Initial chemotherapy followed by cytoreductive or diagnostic
    surgery.

The optimal timing for cytoreductive surgery in the treatment of
bulky testis tumor is still unclear. Data from Ozols et al.,[6]
suggest that cytoreductive surgery prior to chemotherapy is not
indicated in patients with advanced testicular cancer: 39 patients
with stage III-bulky testis tumor were treated in a prospective
randomized trial comparing cytoreductive surgery followed by chemotherapy vs chemotherapy alone. There was no statistically significant improvement in CR and survival between the groups. Furthermore,
negative experience, with primary debulking has been reported[7].
Prolonged postoperative recoveries delayed the onset of needed chemotherapy, thus complicating the chemotherapy in these severely com-

Table 5. Marker Values of 77 Patients with Bulky Testis
          Tumors

AFP ng/ml \ β–HCG mU/ml	Normal 0–6	Medium 6–1000	High >1000	Total %
Normal 0–15	26%	5%	9%	40%
Medium 15–1000	16%	18%	3%	37%
High >1000	9%	13%	1%	23%
Total %	51%	36%	13%	100%

promised patients. Excessive morbidity, accelerated tumor growth
after cytoreductive surgery, and tumor manipulation induced metastases are other arguments against surgery.

Our results in 84 patients with bulky testis tumors indicate
that primary cytoreductive surgery produced a CR rate of 50% (7/14),
while primary chemotherapy followed by diagnostic or therapeutic RLND
produced a CR rate of 70% (23/33). The difference is statistically
significant. In 58/57 (77%) postchemotherapeutic surgery was carried
out; in 40% (35/58) of patients "unnecessary" or diagnostic RLND was
done i.e. residual masses showed necrosis, fibrosis, or cystic
changes. Other reports have shown 29–80% of patients having post-
chemotherapy lymphadenectomy with a negative histology[8–11]. On the
other hand, 60% (35/58) of our patients revealed residual disease;
14/58 (24%) had mature teratoma and 21/58 (36%) had carcinoma.
However, in our study 17/39 (43%) of patients achieving only PR by
chemotherapy were cured by delayed cytoreductive surgery. These
results underscore the importance of the combined approach in bulky
testis tumor.

Several authors have tried to evaluate potential prognostic
factors which affect the CR rate, the duration of remission and
survival rates in disseminated testicular cancer. All the prognostic
factors mentioned, with the exception of bulky disease, were utilized
in these analyses. The unfavorable effect of these 5 prognostic
factors in the CR rate was proven in 75 patients in order to provide
a basis for management of chemotherapy in advanced stage testicular
cancer i.e. nonintensive, intensive or salvage chemotherapy.

The results obtained in this study emphasize the efficacy of
both regimens (I and II) in bulky testis tumor as they produced a CR
rate of 42%, and 43% respectively. Furthermore, remission duration

Table 6.   Surgery 84 Patients with Bulky Testis Tumors

Surgery	First Op. N	Second Op. N	Third Op. N
None	2	20	64
Orchiechtomy	71	0	0
Primary RLND	1	28	0
Delayed RLND	0	26	2
Cervical LND	3	3	1
Laparotomy	5	3	1
Second look	0	2	9
Thoracotomy	0	1	5
Thoracoabd. LND	2	1	2

and survival do not differ significantly. It is of interest, that in patients treated by regimen I the mean number of cycles given was only 3.94 compared to 5.37 in regimen II. In regimen I the rate of patients with residual malignant disease at RLND was 38% vs 44% for regimen II. These results are comparable to those of other authors in the treatment of bulky disease[12-14]. Levi et al.,[15] achieved a CR rate of 73% in advanced germ cell tumors with a 4 drug regimen, PVB plus actinomycin D. Further studies must define the efficacy of more intensive induction regimens.

SUMMARY

In order to study timing of surgical debulking and efficacy of two different chemotherapy protocols we analyzed 84 patients with bulky testis tumor retrospectively, using a modified staging class- ification of the Royal Marsden. During a 6-year period 75 of 84 patients with bulky disease received either a PVB-protocol (n=38) or vinblastine/bleomycin-adriblastin-cisplatinum (37) sequentially. Primary bulk disease (PBD) occurred in 65 (77%) and secondary (SBD) in 19 (23%). Thirty five of the former received initial chemotherapy followed by secondary retroperitoneal lymph node dissection (SRLND), 14 had primary retroperitoneal lymph node dissection (PRLND) and 16 were treated with chemotherapy only. In the group of SBD with primary chemotherapy 16 had a second look retroperitoneal lymph node dissection and in 3 primary surgical debulking was carried out. The overall response rates were: CR 34/38 (40%) with a median disease free survival of 30 months (range 4-47), PR 26/84 (31%) and P 24/84 (29%). The relapse rate in the total group was 3/34 (9%) of those patients who initially had CR. Analysis by subgroups showed that the best results were achieved in patients with primary chemotherapy followed by delayed retroperitoneal lymph node dissection with a CR rate of 70%. Ten patients who had residual disease after chemo- therapy (adult teratoma) had a CR rate of 86%. Initial prethera-

peutic factors, which significantly influenced the response to
therapy were: extragonadal site, pure or partial choriocarcinoma,
massively elevated AFP and HCG, and retroperitoneal bulk greater than
10 cm in diameter. It is pointed out that in primary or secondary
bulk disease evaluation of prognostic factors is necessary to choose
between intensive vs nonintensive induction therapy.

REFERENCES

1. N. Javadpour, R. F. Ozols, A. Barlock, T. Anderson, and R. C.
    Young, A randomized trial of cytoreductive surgery followed
    by chemotherapy chemotherapy alone in bulky stage III, Cancer
    50: 2004-2010 (1981).
2. A. W. Silverman, Surgical debulking of tumors, Surg.Gynecol.
    Obst., 155:577-585 (1982).
3. D. G. Skinner, Advanced metastatic testicular cancer: The need
    for reporting results according to initial extent of disease,
    J.Urol., 128:312-314 (1982).
4. WHO Handbook for reporting results of cancer treatment WHO
    offset Publication No 48 Geneva, (1979).
5. L. H. Einhorn, ans S. D. Williams, Chemotherapy of disseminated
    testicular cancer. A random prospective study, Cancer, 46:
    1339-1344 (1980).
6. R. F. Ozols, A. B. Deisseroth, N. Javadpour, A. Barlock, G. L.
    Meserschmidt, ans R. C. Young, Treatment of Poor Prognosis
    Nonseminomatous Testicular Cancer with a "high dose" platinum
    combination chemotherapy regimen, Cancer, 51:1803-1807
    (1983).
7. N. Jaeger, L. Weißbach, J. E. Altwein, and E. D. Kreuser,
    Primary lymphadenectomy or primary chemotherapy in bulky
    tumor and disseminated cancer, Europ.Urol., 1983 (in press).
8. L. H. Einhorn, S. Williams, J. Mandelbaum, and J. P. Einhorn,
    Surgical resection in disseminated testicular cancer fol-
    lowing chemotherapeutic cytoreduction, Cancer, 48:904-908
    (1981).
9. J. P. Donohue, L. H. Einhorn, ans S. Williams, Cytoreductive
    surgery for metastatic testicular cancer: considerations of
    timing and extent, J.Urol., 123:879-879 (1979).
10. D. Vugrin, W. F. Whitmore, P. Sogani, M. Bains, H. W. Herr, and
    R. Golbey, Combined chemotherapy and surgery in treatment of
    advanced germ-cell tumors, Cancer, 47:2228-2231 (1981).
11. C. Merrin, H. Tatzita, R. Weber, Z. Wajsman, G. Baumgärtner, and
    G. P. Murphy, Combination radical surgery and multiple se-
    quential chemotherapy for the treatment of advanced carcinoma
    of the testis (Stage III), Cancer, 37:20-29 (1976).
12. T. Anderson, T. A. Waldmann, N. Javadpour, and E. Glatstein,
    Testicular germ-cell neoplasms: Recent advances in prognosis
    and therapy Ann.Intern.Med., 90:373-385 (1979).

13. M. K. Samson, R. Fisher, and R. L. Stephens, Vinblastine, bleo-
    mycin and desdiamminedichloroplatinum in dissemiated testic-
    ular cancer: Response to treatment and prognosis cor-
    relations, Eur.J.Cancer, 46:1359-1366 (1981).
14. R. W. Sonntag, H. J. Senn, and F. Cavalli, Treatment of meta-
    static testicular cancer A preliminary report of induction
    chemotherapy followed by maintenance chemotherapy or radio-
    therapy, Cancer Treat.Rep., 63:1669-1674 (1979).
15. J. A. Levi, R. S. Armey, and D. N. Dalley, Significant factors
    in the optimal management of advanced stage germ cell carci-
    noma Aust.N.Z.J.Med., 12:147-152 (1982).

THE ROLE OF CYTOREDUCTIVE SURGERY AND CHEMOTHERAPY IN ADVANCED DISSEMINATED NONSEMINOMATOUS TESTICULAR CANCER - A RANDOMIZED STUDY

Nasser Javadpour

National Cancer Institute
National Institutes of Health
Bethesda, Maryland 20205, USA

INTRODUCTION

Chemotherapy has been shown to be more effective in experimental animals when the size of the tumor is small. In spite of the availability of effective chemotherapeutic agents, the prognosis of patients with massive bulky disseminated testicular cancer is not favorable. In man cytoreductive surgery has been advocated in testicular cancer, Wilms' tumor, rhabdomyosarcoma and Burkitt's lymphoma. The features of bulky testicular cancer rendering it a suitable model for cytoreductive surgery include clarification of histologic type by immunocytochemical techniques, early and sensitive detection of residual postoperative disease by use of alpha-fetoprotein and human chorionic gonadotropin assays, improvements in staging techniques using ultrasound and computerized axial tomography, clarification of staging definitions with improved prediction of potential disease behavior, and effective therapy for advanced disease coupled with early evidence that patients with small tumor burden may be even more responsive.

To date, no prospective randomized trial has been done to test whether the results of effective chemotherapy can be enhanced by cytoreductive (debulking) surgery in patients with tumors not curable by surgery. Although such an approach has a scientific rationale, and results from treatment programs involving some patients support its possible role, the ethical and technical aspects of such a study have been difficult to overcome.

Patients with testicular carcinoma who have multiple poor prognostic features are appropriate for such a study because, compared with patients with minimal tumor burden, their prognosis is rela-

tively poor; ability to achieve a complete remission can be expected
to alter subsequent survival significantly. The patient is usually
otherwise healthy, without other significant medical problems, a
major proportion of the metastatic disease can be surgically removed,
the known chemotherapeutic regimens have excellent anti-tumor activity in patients with small amounts of tumor and the levels of
circulating tumor-associated proteins, that is, alpha-fetoprotein and
human chorionic gonadotropin, can be used to assist in quantitatively
evaluating efficacy of cyto-reductive surgery and intensive chemotherapy[1,2]

MATERIALS AND METHODS

Eligibility Criteria

     The patients eligible for this study all had pathologic confir-
mation of the diagnosis of nonseminomatous testicular cancer by the
Laboratory of Pathology at the NCI. The histologic cell types that
were included in this study were embryonal carcinoma, teratocarcinoma
(embryonal carcinoma with teratoma), or mixed tumors. Patients with
either pure seminoma or pure choriocarcinoma were not eligible.

     Only patients with advanced Stage III testicular cancer were
eligible. The definitions of advanced pulmonary and abdominal
disease are presented in Table 1. Patients were required to have
metastases in areas accessible to surgical resection, i.e., the
retroperitoneum, lung, or supraclavicular masses, although the presence of additional inaccessible metastases, e.g., the liver, mediastinum did not preclude participation. It was not required that the
entirety of the patient's tumor be accessible for surgical resection.

     Patients were not eligible if they had any of the following:
central nervous system metastases or massive hepatic involvement, any
previous chemotherapy or radiotherapy, any medical contraindications
to an extensive surgical debulking procedure, the presence of Stage
III disease with a minimal tumor volume as defined in Table 1, or
impaired renal, hepatic or cardiovascular function.

     Prior to entry into the protocol all patients had a complete
medical evaluation including history, physical examination, complete
blood count, blood chemistries, liver and kidney function tests,
urinalysis, creatinine clearance, radionuclide liver scan, and chest
x-ray. In order to assess the volume of disease, a number of other
radiologic or laboratory studies were performed including chest
tomograms (conventional and computed tomograms), lymphangiogram,
inferior venocavogram (IVC), intravenous pyelogram (IVP), gallium
scan, computed tomogram of the liver, and serial determinations of
serum alpha-fetoprotein (AFP) and human chorionic gonadotropin (hCG)
utilizing specific and sensitive double antibody radioimmunoassays.

Table 1. Extent of Disease in Stage III Testicular Cancer (after Samuels et al.,[5]).

Minimal Disease

- Pulmonary disease with 5 or less metastases per lung field with the maximum diameter of each nodule less than 2.0 cm.

- Pulmonary disease as described above plus a positive lymphangiogram with no ureteral displacement.

Advanced Disease

- Pulmonary disease with any mediastinal mass, hilar mass or intrapulmonary nodule greater than 2.0 cm or the presence of a pleural effusion.

- Abdominal disease with any of the following: a palpable mass, liver metastases, obstructive uropathy, or inferior vena cava distortion from metastatic nodes.

Serum AFP and the hCG levels greater than 20 ng/ml and 1ng/ml, respectively, were considered abnormal. The patients also underwent pulmonary function tests and audiograms before randomization.

Randomization

The eligible patients with advanced Stage III disease were offered the opportunity to participate after extensive interview and discussion of various treatment options. After informed consent was obtained, the patients were randomized to the two treatment groups using randomization decks provided by the Biometric Research Branch of the NCI (Figure 1)[3].

Cytoreductive Surgery

The cytoreductive surgery consisted of the removal of as much tumor bulk as safely possible. The surgical technique involved a long midline vertical or a thoracoabdominal incision through the 9th or 10th ribs exposing the accessible tumor bulk (all the abdominal surgical procedures were performed by a single urologic oncologist)[N.J]. Upon entering the retroperitoneal and abdominal compartments, a thorough exploration was performed. Distal and proximal control of the inferior vena cava was accomplished. The kidneys and ureters were dissected free of tumor. When the tumor was obstructing

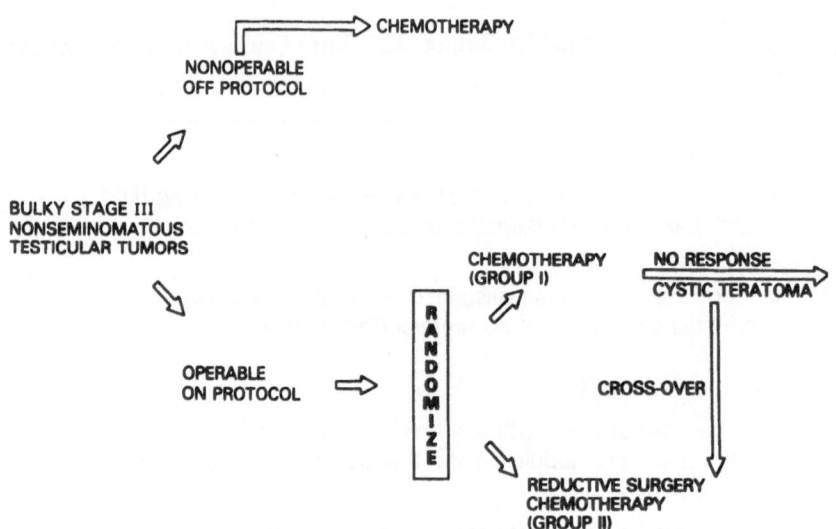

Fig. 1. Scheme of the randomized clinical trial.

the inferior vena cava, an en bloc resection of the inferior vena
cava and common iliac artery was performed. If the kidney was massively involved with tumor it was included in the en bloc resection.
If the tumor encased the aorta, it was carefully dissected free prior
to the en bloc resection of the inferior vena cava. The lumbar
vessels were ligated to facilitate the resection. Isolated hepatic
metastases were resected. If there was encasement of the retroperitoneal duodenum with tumor, a resection and anastamoses was performed. A thoractomy was performed to remove bulky lung metastases.

Recent advances in the chemotherapy of advanced testicular
cancer and the availability of specific serum tumor markers (AFP and
hCG), have made a dramatic improvement in the management of patients
with nonseminomatous testicular cancer. Cis-platinum-containing
combination chemotherapy regimens have produced overall response
rates of 85-95% in metastatic disease[4,5] Fully two-thirds of
patients with Stage III disease (metastases outside the abdomen) can
be expected to achieve a complete response to chemotherapy[6,7]. The
relapse rate in these patients is only 12-13% with 90% of the relapses occurring in the first two years. Since the median follow-up
in these series is greater than two years, it is likely that the
majority of patients who achieve a complete response to chemotherapy
are cured of their disease.

However, within the same group of patients with disseminated
disease, two distinct subsets of patients can be identified with
markedly different prognoses. Patients with minimal disseminated
disease, as defined by Samuels et al., (Table 1), have a complete
response rate of 90% to chemotherapy. In contrast, in patients with

advanced Stage III disease, the complete response rate to chemotherapy has been reported to be similar to that of other malignancies
such as oat cell carcinoma of the lung and ovarian cancer. These
clinical observations as well as similar results in experimental
systems have frequently been used to support a role for cytoreductive
surgery in patients with bulky tumors. However, there has been no
previously reported prospective randomized trial in which their role
of cytoreductive surgery in any human malignancy has been critically
analyzed.

RESULTS

      The results of this prospective randomized trial indicate that
cytoreductive surgery followed by chemotherapy is not superior to
treatment with chemotherapy alone in patients with advanced bulky
testicular cancer. There was no statistically significant difference
in the percentage of complete responders in the two groups or in
their overall survival. To our knowledge, this is the first reported
prospective randomized trial in any malignancy which has evaluated
the role of cytoreductive surgery. Cytoreductive surgery, prior to
chemotherapy, was effective in reducing the tumor volume by 70-90%
(Table 2). The marker data also indicates that surgery was effective
in removing substantial tumor bulk. Even though cytoreduction was
technically feasible in patients with advanced disease, it did not
result in improvement in the subsequent response to chemotherapy or
in overall survival (Table 3; Figure 2 and 3). In fact, the life
table analysis of all 39 patients (Figure 2) suggests an advantage to
those patients who did not receive the debulking surgery. However,
these 39 patients include one patient randomized to surgery who died
prior to any debulking and six patients with extragonadal primaries,
five of whom received debulking surgery. It has recently become
evident that patients with extragonadal primary tumors have a poor
prognosis, and when a life table analysis is performed without including these seven patients, there is no difference in survival
between the two groups (Figure 3).

      The reason why cytoreductive surgery did not improve a subse-
quent response to chemotherapy is not apparent. It is possible that,
even though the surgery was successful in removing a substantial bulk
of tumor, the necessary delay before chemotherapy could be instituted
(mean of 16 days) while the patient was recovering from surgery
allowed a regrowth of residual tumor to such an extent that any
potential benefit from the initial cytoreduction was negated.
However, the marker data do not suggest any rapid regrowth of tumor
during the interval between surgery and chemotherapy.

      The data in this trial suggests that cytoreductive surgery prior
to chemotherapy is not indicated in patients with advanced testicular
cancer. However, surgery may still have an important role in the

Table 2. Cytoreductive Surgery

Extent of Surgery	Number of Patients	Surgical/radiographic estimate of tumor resected
Retroperitoneal exploration	17	70–90%
Partial liver resection	1	80%
En bloc resection of IVC	6	70–90%
Resection of pulmonary metastases	2	80%

Table 3. Results of Therapy

Results	Cytoreductive Surgery followed by Chemotherapy	Chemotherapy as initial treatment
Patients fully restaged	19/20	19/19
No. with response		
Complete response	10/20 (50%)	7/19 (37%)
Partial response	5/20 (25%)	9/19 (47%)
No response	5/20 (25%)	3/19 (16%)

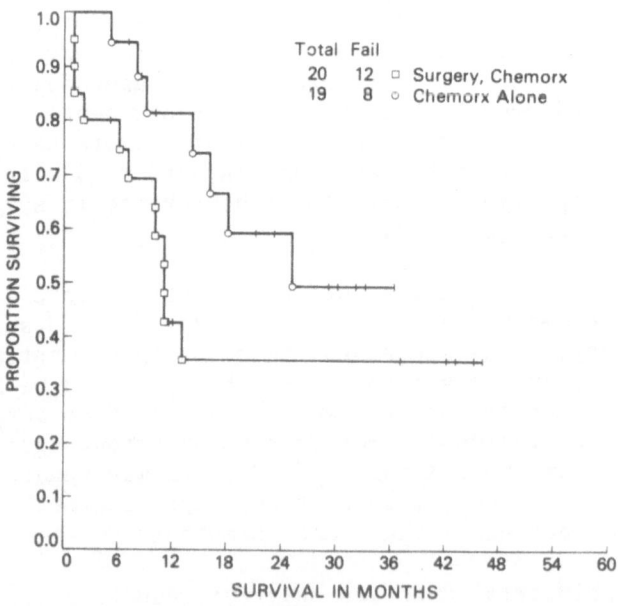

Fig. 2. Life table analysis for 39 patients with disseminated bulky testicular cancer with prior prognostic features.

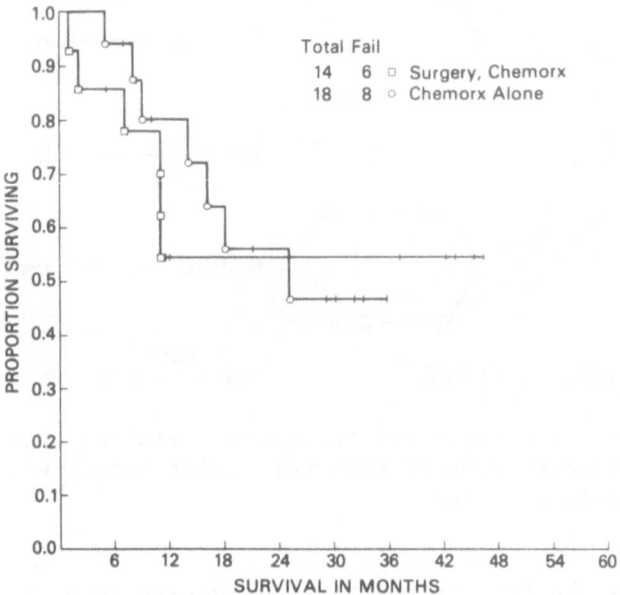

Fig. 3.   The above life table analysis when the extragonadal tumors
          are excluded.

management of patients with advanced disease. It has recently been
reported that intensive chemotherapy followed by excision of any
residual tumor masses can be associated with prolonged survival. In
our study, three patients who had residual masses following chemotherapy had excision of mature teratomas and have been in a prolonged
complete remission.

     The response rates to chemotherapy in these two groups of
patients is similar to what has been previously reported for poor
prognosis patients treated with cis-platinum-containing combinations,
i.e., while overall response rates are between 80-90%, the percentage
of patients with bulky disease achieving a complete response is only
29-50%. The results of the current study suggest that other modalities of therapy such as high dose chemotherapy with autologous
bone marrow infusion or new intensive combination chemotherapy regimens, (instead of prechemotherapy surgical debulking) will be required to improve the response rate in advanced Stage III testicular
cancer patients.

     It is also concluded that patients with bulky tumor do better if
treated with a 3-4 cycles of intensive chemotherapy followed by
resection of any residual disease. In a pilot study, we have utilized an intensive regimen of high dose platinum (P) combined with
vinblastine (V), bleomycin (B) and VP-16 because of encouraging

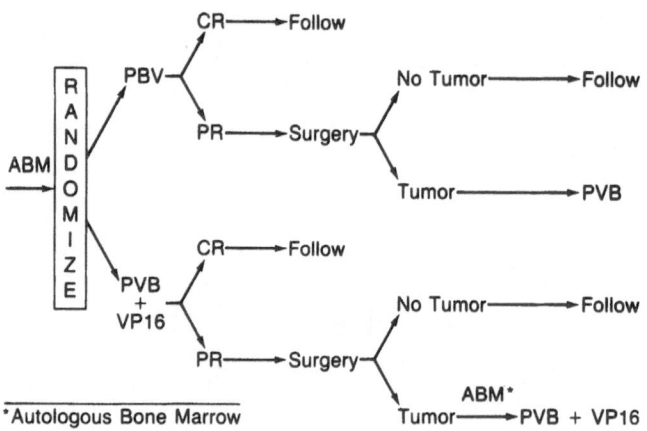

Fig. 4. Current NcI prospective randomized clinical trial for
         massive bulky disseminated testicular cancer with poor
         prognostic features.

initial results. We have an ongoing prospective randomized study
(Figure 4) utilizing 3-4 courses of this chemotherapy versus the
conventional PVB. Although the results are yet too early to judge,
the new regimen gives a complete remission of 80% as opposed to 57%
with the conventional PVB regimen.

REFERENCES

1. N. Javadpour, K. R. McIntire, and T. A. Waldmann, The role of
    radioimmunoassay of serum alphafetoprotein and human chorionic
    gonadotropin in the intensive chemotherapy and surgery of
    metastatic testicular tumors, J.Urol., 119:759-762 (1978).
2. N. Javadpour, The role of biologic tumor markers in testicular
    cancer, Cancer, 46:1755-1761 (1980).
3. E. Gehan, A generalized Wilcoxon test for comparing arbitrarily
    single-censored samples, Biometrika, 52:203-223 (1965).
4. L. H. Einhorn and S. D. Williams, Chemotherapy of disseminated
    testicular cancer: a random prospective study, Cancer, 46:
    1339-1344 (1980).
5. M. L. Samuels, P. Y. Holoye, and D. E. Johnson, Bleomycin combin-
    ation chemotherapy in the management of testicular neoplasia,
    Cancer, 30:318-326 (1975).
6. D. Vugrin, M. Dukeman, and W. Whitmore, VAB-16: Progress in
    chemotherapy of germ cell tumors (GCT), Proc.Am.Assoc.Clin.,
    Oncol., 21:426 (1980).
7. D. Vugrin, E. Cvitkovic, W. F. Whitmore, Jr., VAB-4 combination
    chemotherapy in the treatment of metastatic testis tumors,
    Cancer, 47:833-839 (1981).

ROUND TABLE:

SURGERY OF TESTIS TUMORS: CONTROVERSIAL ASPECTS

J. P. Blandy, N. Vahlensieck, A. Jardin,
P. Scardino, G. Pizzocaro and S. Fossa

The question was put to the panel that, when it came to evaluating the results of contemporary management of tumors of the testis that there was no difference between the almost 100% long-term cure-rate obtained by the advocates of retroperitioneal node dissection, whether used as a staging procedure or in the hope of effecting a cure and the results obtained by Peckhams. Oliver and others who deferred chemotherapy in Stage 1 disease, and only offered surgical removal of retroperitoneal tissue to those patients in whom a mass remained after chemotherapy: both methods of management offered almost complete success. The only difference was that, in Peckham and Oliver's hands, only 20% of the patients were submitted to the major operation of node dissection, while in the hands of Scardino and his colleagues, about 70% were operated on at the cost of the loss of the ability to ejaculate.

There was a short interchange of opinions between Scardino and Peckham which threatened to turn into a dialogue as each cited the statistics which supported his view. Scardino and Pizzocaro pointed out that when there was only small volume disease, lymphadenectomy might be curative, or, at least, might require only two, rather than four, courses of chemotherapy. Scardino pointed out that many patients would willingly undergo surgery rather than suffer the discomforts of unnecessary chemotherapy.

At this stage Pizzocaro introduced results from his own series which showed very clearly that scrotal violation gave rise to a much higher incidence of retroperitoneal node metastases as well as of distant metastases. This prompted a valuable discussion on the correct management of the patient whose testicular tumor had been operated on through the scrotum. Peckham's experience suggested that

179

scrotal violation made little difference to the outcome, but Pizzocaro pointed out that there was an important difference between the testicle explored in error through the scrotum, and perhaps incised, so that tumor was spilt – and the testis merely subjected to a needle biopsy, or removed cleanly albeit through an incision in the scrotum. When challenged, the members of the panel mostly felt that it was perfectly reasonable to wait and see whether or not inguinal metastases appeared. For such metastases radiation for seminoma or chemotherapy for teratoma could be given. Only Vahlensieck took a different attitude: he would perform a node dissection in the groin. Most of the panel and all the audience thought that this was going too far and could not justify the needless edema of the lower limb that must ensue, though Vahlensieck had never seen this complication in his own experience.

None of the participants was likely to change his opinion or his practices and at the end of a very busy cut-and-thrust discussion, it was clear that the advocates of node dissection would continue, for the time being, to employ this operation, if only as a staging procedure, while their opponents kept to the more conservative position of limiting the role of surgery to the removal of the original lump, and to the salvage of the retroperitoneal mass that failed to respond to chemotherapy.

20 July 1983
THE TREATMENT OF CHORIOCARCINOMA: DISCUSSION

Definition

Dr Mostofi began the discussion at the request of the panel, by defining for the participants what was to be understood by the term "choriocarcinoma". Clearly there were two distinct entities – pure choriocarcinoma and choriocarcinoma associated with other cell types, especially embryonal cell carcinoma, yolk sac tumor, and seminoma. To these he added a third – foci of choriocarcinoma in-situ, i.e. within seminiferous tubules:-

1. Pure: cyto- and syncytio-trophoblast only.
2. Choriocarcinoma associated with other cellular elements.
3. Foci in-situ in tubules.

In each of these categories it was necessary to find both cyto- and syncytio-trophoblast for which, Dr Mostofi admitted, special immuno-peroxidase staining might be necessary.

Natural History

Dr Fossa described the experience of the Norwegian Radium Hospital and of the Scandinavian combined study. Of a total of 800 patients there were 5 'pure' chorioncarcinomas (0.6%), and 29 (3.6%) in the mixed group: both were rare, the 'pure' choriocarcinomas exceptionally so. Many seemed to arise in extragonadal sites (5/17 = 29%) and their pattern of spread was largely systemic - findings reinforced by Dr Javadpour whose experience had been much the same.

Treatment

Contrary to the prevailing opinion, treatment in both groups had been by no means hopeless: 24/24 (70.6%) of all types were cured by VACAM or CVB and 2/4 of the pure group, in Fossa's experience. One particular sub-group - those with brain metastases - posed a particular problem in this, essentially systemic, disease. Even so there was room for encouragement: Oliver reported 4/8 such patients with brain metastases that had been cured. Opinions differed as to the best way of effecting such a cure: Torti & Bagshaw preferred to use radiotherapy as an adjunct in treatment: Fossa referred her patients to the neurosurgeon, but Stoter's experience led him to mistrust a surgical approach - in his case the tumor had been found to be inoperable, attempts to remove it left a large cavity, and a drain had been left in for the purpose of irrigating the cavity with a chemotherapeutic agent.

Summary

Taken altogether this exchange of views gave the participants (and the moderator) grounds for renewed hope in a field hitherto governed by gloom and despair. One question was not put to the panel by the audience, but would have received no answer - Why does choriocarcinoma in the male behave so much less favorably than in the female, when treated with modern chemotherapy? Perhaps the answer will never be known, since this latest review of international experience suggests that in fact, it does not do too badly after all.

RADIATION THERAPY OF TESTICULAR GERMINOMAS,

STANFORD UNIVERSITY SERIES (1956-1980)

Malcolm A. Bagshaw and L. Douglas Graham

Departments of Radiology and Radiation Therapy
Stanford University School of Medicine
Stanford, California, USA

ABSTRACT

Over the twenty-five years of experience with pure testicular germinomas reported in this paper, dramatic changes have occurred, both in survival of patients and in the understanding of the natural course of the disease. Current results of our experience with 128 patients who were treated with radiation therapy during this period show an 84% overall actuarial survival ten years after treatment. This high ten year actuarial survival rate is distributed by stage as follows: Stage I (89.5% survival); Stage II (81% survival); Stage III (31% survival). These results compare well with other series of over 100 patients treated by radiotherapy. Stage IV patients may benefit from palliative radiotherapy to metastases of bone or brain. Clinical experience with these tumors has shown that attention to careful staging can optimize both survival and life extension to a significant degree.

INTRODUCTION

Testicular germinomas formerly were considered one of the most lethal cancers in men. With advances in diagnostic and therapeutic radiology, as well as in chemotherapy and diagnostic laboratory tests, this neoplasm is now considered usually curable. This paper deals with the role of radiotherapy in this success story; while surgical removal (inguinal orchiectomy) is the treatment of choice for the primary tumor, radiotherapy has provided an adjuvant treatment capable of sterilizing regional lynphadenopathy and other potential sites of metastatic disease, resulting in improved survival with reduced morbidity.

THE STANFORD LINEAR ACCELERATOR CAPABILITY (1956-1980)

In January 1956, a 4.8 million electron volt (Mev) linear accelerator was installed at the Stanford-Lane Hospital in San
Francisco. The beam was characterized by a maximum dose at 1.2 cm
beneath the skin surface. It had two advantages over conventional
200-400 Kv orthovoltage equipment: skin sparing due to more efficient deposition of ionizing radiation at varying depths beneath
the skin, and a more sharply defined radiation field and target
volume with attendant sparing of surrounding normal tissues. This
unique home-made accelerator was later replaced by a 6 Mev, com-
mercially manufactured, linear accelerator (Varian Clinac 6), with
which the maximum dose occurred 1.5 cm beneath the skin. Source-to-
skin distances could be increased from 100 cm to 140-165 cm, as
larger and longer fields were used. Patients could be treated daily,
receiving a total dose of from 150 to 220 rads through opposing
anterior and posterior fields.

Patient Selection

During the period 1956-1980, 176 patients were referred to the
Division of Radiation Therapy with a diagnosis of testicular seminoma
(more recently we have adopted "germinoma" to replace "seminoma").
Certain of these data we have published previously[1,2,3]. In reviewing and updating our treatment results we excluded 40 patients
for the reason that 13 were seen for consultation only, 16 had had
previous treatment elsewhere, and 11 had pathology other than pure
testicular germinoma, thus leaving 136 patients for study. The
median age was 36 years (Figure 1). Initial evaluation included
physical examination, chest x-ray and an intravenous pyelogram.
Lymphangiography was added in the late 1960's. In the late 1970's,
tumor markers, beta-human chorionic gonadotropin (β-HCG) and alpha-
fetoprotein (AFP) were determined for each patient. Chest tomography
or CT scans were used, as indicated. Of the original 136 eligible
patients, 8 were lost to follow-up, leaving 128 to be described in
this report. Ninety patients were classed at Stage I, 29 as Stage
II, and 9 as Stage III (Table 1). Thus, about 2/3 of these patients
had Stage I, while nearly another third had Stage II disease, leaving
a small group of 9 patients who had serious Stage III disease with
distant metastases.

Treatment Methods

Stage I. These patients received 3000 to 3500 rads* over 3 to
3-1/2 weeks via a single field encompassing the para-aortic and

---

*1 rad = 100 ergs/gm unit of absorbed dose.

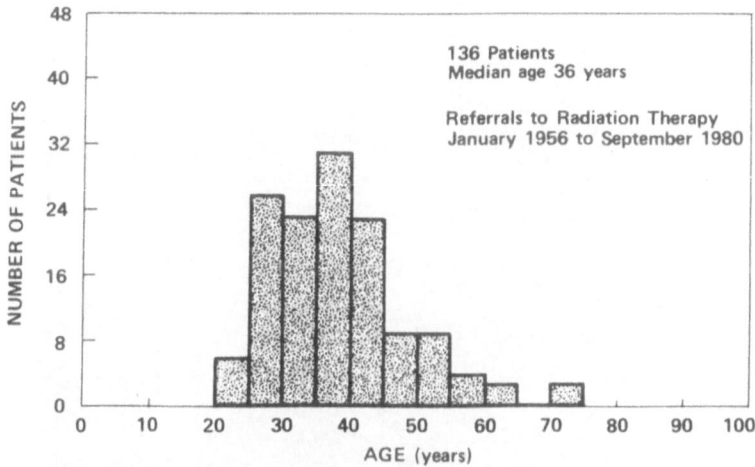

Fig. 1. Age distribution of 136 patients with pure germinoma of the
testis.

ipsilateral pelvic lympth nodes (Figure 2). Previous experience
with orthovoltage equipment had taught us that lower doses were
tumoricidal, but the ease and accuracy of delivery with the liner
accelerator (using an anteroposterior-posteroanterior opposed pair
technique) led to a gradual escalation of absorbed dose. Presently,
however, we treat only to 3000 rads in order to spare bone marrow for
future chemotherapy, should it be needed. The hemiscrotum was
treated to 2000 rads either by extending the inferior aspect of the
field or by separate treatment with orthovoltage equipment if a
transcrotal orchiectomy had been performed, or if the primary tumor
had penetrated beyond the tunica albuginea.

Table 1.   Clinical Staging System used in this Report.
(Modified from Boden and Gibb[19])

Stage		No. Patients
Stage I	Tumor limited to the testis and spermatic cord, but not present at resected end of cord.	90
Stage II	Intra-abdominal lymph node involvement, and/or tumor at point of spermatic cord resection, or involvement of scrotum.	29
Stage III	Distant metastases, or abdominal organ involvement.	9
	Total	128

Currently, nodes less than 5 cm are designated as A, and those larger
as B.

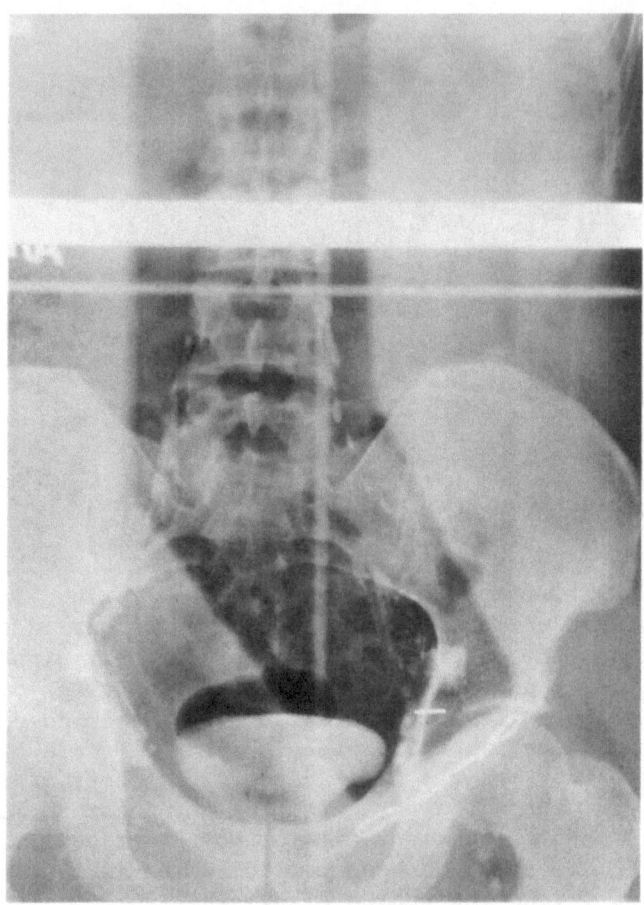

Fig. 2.   Typical irradiation field (Hockey Stick) used for Stage I
          disease. Note the contrast material in the lymph nodes, the
          metallic clip marking the transected end of the spermatic
          cord, and the lead marker showing the position of inguinal
          incision. The single large radiation field at a target skin
          distance of 160 cm obviates the need for matching adjacent
          fields. The horizontal white band is an artifact on the
          film produced by the cassette holder.

     Stage II (tumors less than 10 cm). In addition to the above,
these patients had the contralateral pelvic nodes treated and also
received prophylactic irradiation to the mediastinum and supraclavicular lymph nodes to a dose of 2500 to 3000 rads in 2-1/2 to 3 weeks
(Figure 3). Both supraclavicular regions were treated because lymphangiography had demonstrated that up to 10% of patients may drain
either to the right or bilaterally[4]. Prophylactic treatment was

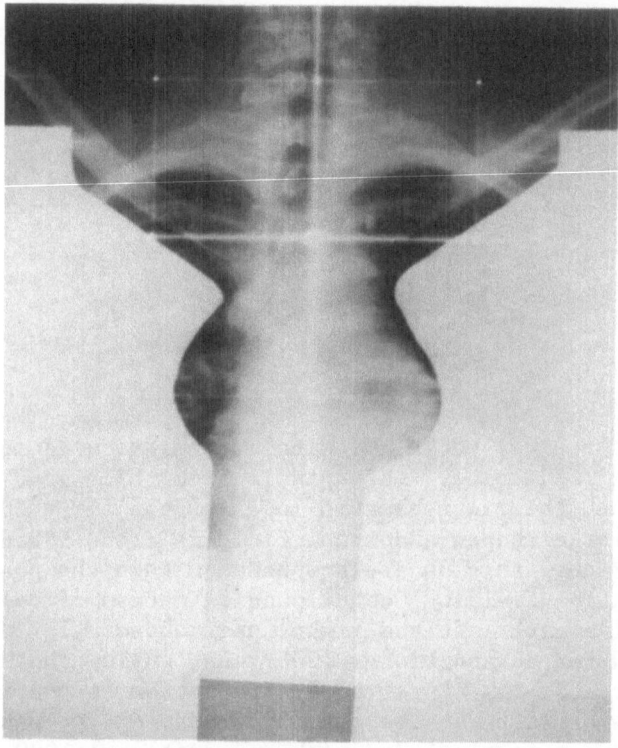

Fig. 3. Typical prophylastic irradiation field added for Stage II
         disease. The mediastinal, hilar and bilateral supra-
         clavicular lymph nodes were treated. Prophylaxis of the
         mediastinum is no longer used as routine practice in Stage
         II but it was carried out in this series.

limited to 3000 rads to prevent pulmonary complications. This
prophylactic treatment is no longer used (see Discussion).

    Since 1965 Stage II patients with bulky disease have been
treated with a whole abdominal technique (Figure 4)[5]. This is the
technique we also use for extensive abdominal non-Hodgkin's lymphoma.
When tumor localization required the inclusion of one kidney, the
contralateral kidney was shielded both anteriorly and posteriorly
throughout the treatment. These patients also received prophylactic
mediastinal and supraclavicular irradiation, as well as 1200 to 1500
rads to each lung. In view of recent advances in chemotherapy, the
prophylactic aspects of this approach are no longer used (see
Discussion).

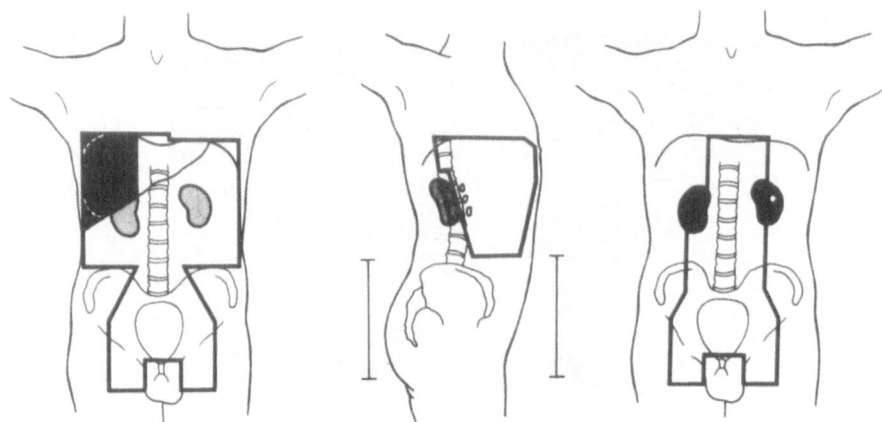

Fig. 4.   The three-way whole abdominal technique used for bulky
          Stage II disease. Left panel: 5 cm thick lead blocks
          protect the right hepatic lobe for the first 2000 rads.
          The large shaped abdominal field is cross-fired both AP-PA.
          The kidney shields are suspended within the posterior
          radiation fields. Center panel: In case of celiac node
          involvement, left and right upper abdominal fields are
          added for an additional 1000 rads, while concomitantly the
          lower abdominal inverted Y irradiation is continued by
          AP-PA cross-fire also for 1000 rads. Right panel: The
          treatment is concluded with an additional 1000-1500 rads
          to the entire para-aortic and iliac lymph node chains by
          AP-PA cross-fire with the kidneys shielded.

     Stage III.   In the past we have elected to treat Stage III cases
aggressively on an individual basis.   In cases with bulky para-
aortic, mediastinal, or supraclavicular disease, or with pulmonary
metastases, whole lung irradiation of 1500 rads over 3 to 4 weeks was
given, while lymph node bearing areas above the diaphragm received
treatment similar to that of the abdominal lymph nodes.   An additional 1500 to 2000 rads supplement was given to individual pulmonary metastases.   Eight of the 9 Stage III patients were treated
before 1974, when Einhorn introduced combination chemotherapy consisting of cisplatin, velban and bleomycin (PVB) for the treatment of
testicular neoplasms[6].   We currently favor reserving radiation
therapy only for those Stage III patients whose nodal disease in any
site is less than 5 cm and whose age, pulmonary and/or renal status
preclude the use of PVB, or for patients whose disease sites are
either unresectable or fail to respond to chemotherapy.

Treatment Results - January 1956-December 1980

     Actuarial survival for the entire group was 84% at 20 years
(Figure 5).   When actuarial survival is examined on the basis of

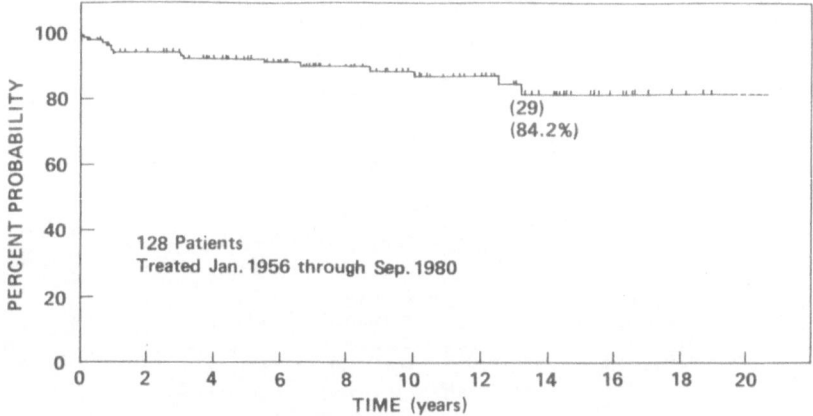

Fig. 5. Pure testicular germinoma, actuarial survival for all
stages.

Stage, the survival for Stage I is 89.5%; for Stage II, 81,0%; and
for Stage III, 31.1% (Figure 6). These survival rates compare favorably with other irradiation series of greater than 100 patients
(Table 2). Gehan analysis between Stages I and II demonstrated no
significant difference (P=0.66), but significant differences between
Stage I and III (P<0.0005) and between Stages II and III (P<0.0001)
were observed.

Of the 90 patients with Stage I disease, only one has died of
seminoma. He experienced a massive abdominal recurrence 5 months
after completion of therapy to the iliac and para-aortic lymph nodes.
Seven patients in the Stage I group have expired from intercurrent
causes: two from second malignancies, three from cardiac events, one
from suicide, and one from accidental death.

In reviewing the 29 Stage II patients, six had bulky para-aortic
disease. Only one died from seminoma. He was one of the first
patients treated in 1957, before our whole abdominal technique with
megavoltage irradiation had been thoroughly developed. He succumbed
from an intra-abdominal relapse with complicating sepsis. There were
four intercurrent deaths in the Stage II group: one from lung
disease; one from a cardiac event; one from a second malignancy; and
one from acute myelogenous leukemia 11 years after completion of
iliac, para-aortic, mediastinal, and supraclavicular irradiation.
Two Stage II patients did not receive prophylactic irradiation of the
mediastium and supraclavicular region. They remain without evidence
of disease.

Four of the 9 Stage III patients died of seminoma. One of these
four had metastatic disease containing embryonal elements. Another

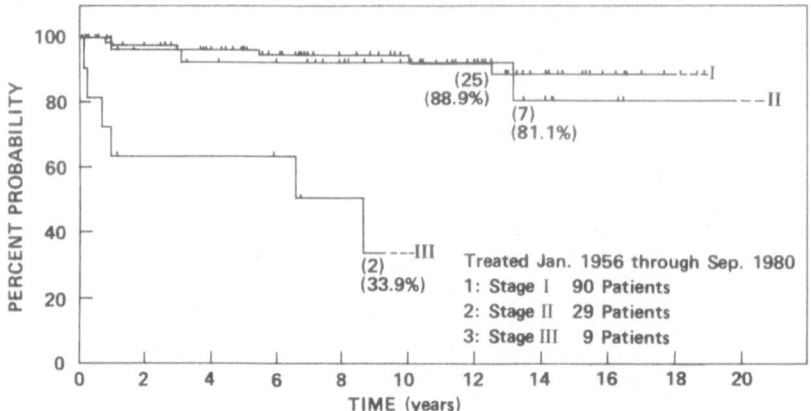

Fig. 6.   Pure testicular germinoma, actuarial survival by Stage,
          n = 128.

had intercurrent death from myocardial infarction.   One other patient
received 6 cycles of PVB following completion of radiotherapy, and
has been without evidence of recurrence for more than 5 years.

Complications

     Complications have been minimal and, in general, are related to
unusual treatment, such as a case of mild radiation pneumonitis when
a large mediastinal mass was treated.   Most occurred prior to 1962,
during the period when we were becoming more familiar with the use of
megavoltage irradiation.   There have been two cases of activation of
a duodenal ulcer.   In another patient, an inflammatory band produced
jejunal obstruction, which required surgical intervention.   Also, two
patients treated early in the series developed moderately severe
right lower quadrant subcutaneous fibrosis.   Another early patient
with Stage I disease had severe retroperitoneal fibrosis with
ureteral obstruction, and demyelinating disease, as well as gastrointestinal injury, about 10 years following treatment.   This fibrosis
occurred at the junction of adjacent fields and was apparently caused
by an overlap of the field margins.   With current techniques these
hazards no longer exist.

Treatment Failures and Recurrent Disease

     Treatment failure can be categorized either as local recurrence
of pure seminoma or the development of non-seminomatous metastases,
usually embryonal cell carcinoma[7].   Persistent or advancing disease
following treatment merits reexamination of the biochemical markers
and careful restaging, including biopsy, if possible.   Should patho-

Table 2. 5-Year Survival for Seminoma Patients Treated with
          Radiation Therapy on a Stage by Stage Basis (percentage)

	5-Year Survival		
	Stage I	Stage II	Stage III
SUH Bagshaw & Graham			
1956-1980	96.4	92.5	77.8
PHM Thomas et al.[10]			
1958-1976	94	74	32
MGH Dosoretz et al.[9]			
1950-1976	97	92	45
UCSF*Quivey et al.[20]			
1959-1973	100	90	75
RRTI*van der Werf-Messing[21]			
1950-1974	94.7	78	43

*Determinate survival; all others actuarial survival.

logic material reveal non-seminomatous elements, the patient should
receive combination chemotherapy.

    Testicular germinoma can recur as distant disease in previously
untreated areas, but it more frequently recurs adjacent to previously
treated fields as an anatomic miss, or it may recur if an insufficient dose has been delivered to the next echelon of potentially
involved lymph nodes. Eleven of the 16 patients evaluated in our
series who were treated elsewhere had recurrent disease and received
treatment at Stanford. One suffered brain metastases, was treated on
a palliative basis, and eventually succumbed to disseminated disease.
Only one other patient with recurrent disease, a patient who developed a second seminoma in the remaining testicle 26 years after
initial treatment of his first primary, died of seminoma. His iliac
and para-aortic regions were retreated, and he received additional
new treatment to his mediastinum and supraclavicular regions. He
later experienced two separate nodal relapses: one in the axilla,
and one in an inguinal lymph node which responded to additional
radiation. He died from visceral metastases 18 months later. The
remaining 9 of 11 patients are disease free from 3 to 19 years following radiotherapy to sites of relapse. Such sites included the
mediastinum, lung, supraclavicular and inguinal lymph nodes. Similar
success in treating recurrent disease has been reported by Fayos and
Kim[8], Dorosetz et al.[9] Thomas et al.[10] and Ball et al.[11].
Three of 11 patients also received adjuvant combination chemotherapy
consisting of either cisplatin, velban and bleomycin or cytoxan and
vincristine following third and fourth relapses involving bone. They
are now living 3, 4 and 5 years, respectively, disease free.

CHEMOTHERAPY

Until recently, the outlook for patients with visceral involvement, especially liver or bone, was bleak. Vugrin et al.[12] have
recently reported that of 9 patients with Stage III disease treated
with cisplatin and cyclophosphamide, five (56%) had complete remission with chemotherapy alone, and an additional three had a complete remission after a combination of chemotherapy, radiation and
surgery. Seven of these patients have remained in complete remission
with a minimum follow-up of 17 months. Einhorn in a private communication has recently described 31 patients, including five with extragonadal primaries, who were treated with combination chemotherapy
using regimens containing cisplatin, velban and bleomycin with or
without adriamycin. Twenty-one of 31 patients (68%) have achieved
complete remissions, and eight have had a partial response. Seventeen patients have remained free of disease from 18 to 69 months.

DISCUSSION

Clearly, the emergence of successful combination chemotherapy
alters the selection of patients to be treated with radiation therapy
as a primary modality. Advances in diagnostic radiology, e.g.,
lymphangiography and CT scanning, now enable the physician to stage
and substage each patient more accurately and to select the most
appropriate therapy. While lymphangiography may accurately assess
metastatic retroperitoneal disease in up to 90% of patients with
seminomas, CT scanning or ultrasonography are necessary as well[13].
Lymphangiography following direct cannulation of the testicular
lymphatics has demonstrated that the nodes identified through bipedal
lymphangiography are not the primary sites of lymphatic drainage of
the testes, and will show metastatic disease only after the primary
lymph nodes have become involved. The primary drainage from the left
testicle is the left lateral para-aortic nodes at the level of L1-2
just inferior to the left renal vein. For the right testicle, the
primary channels drain to lymph nodes at L1-3 between the right renal
vein and the aortic bifurcation, but they may also cross over and
drain directly into the left lateral para-aortic lymph nodes. These
sentinel lymph nodes lie lateral to the lumbar lymph nodes opacified
by bipedal lymphangiography[14]. While bipedal lymphangiography
offers the advantage of detecting architectural changes within non-
enlarged lymph nodes, it may miss involvement of the sentinel nodes.
CT scanning cannot detect disease less than 2 cm, yet it offers the
advantages of determining the bulk of the disease and the assessment
of lateral lumbar regions not visualized by bipedal lymphangiography,
as well as the detection of extralymphatic sites of involvement, i.e.
visceral disease. Thus, it appears that both lymphangiography and CT
are required to evaluate adenopathy in patients with seminoma. While
gray-scale ultrasound may be as accurate as CT scanning in assessing
retroperitoneal lymph nodes, we prefer CT scanning because it can

also be used for treatment planning, and its interpretation is more reproducible than that of ultrasound[15].

Patient selection for the choice of chemotherapy or radiotherapy can also be made on the basis of tumor marker. An elevated alpha-fetoprotein should prompt the physician to carefully reexamine the primary tumor for embryonal carcinoma, while an elevated β-HCG should lead to a reexamination for choriocarcinoma. Raghavan et al., have recently reported on 6 patients with pure seminoma and elevated alpha-proteins(16). Four of the 6 were treated with standard radiotherapy, and it was the authors' conclusion that radiation therapy was "of little value in three out of four patients" who subsequently expired after salvage chemotherapy, "and had no demonstrable effect as first line treatment for an abdominal mass in another patient". Javadpour reports that approximately 10% of pure seminomas treated at the National Cancer Institute had an elevated β-HCG which fell during treatment[17]. Persistently elevated levels after treatment indicate a need for restaging and treatment with combination chemotherapy.

Prophylactic irradiation of the mediastinum and supraclavicular lymph nodes has recently been shown to be unnecessary by Thomas et al., who demonstrated no survival benefit to patients treated prophy-lactically to the mediastinum[10]. Similar observations concerning the need for prophylactic treatment in the mediastinal and supra-clavicular regions have been made by Ball et al.[11] who reviewed the British experience, and Qian et al.[18] who reviewed the Beijing series. In the Princess Margaret Hospital series, isolated medias-tinal or supraclavicular recurrences were usually successfully sal-vaged with radiotherapy alone[10]. Patients with bulky abdominal disease frequently relapsed in multiple sites, and the authors argue that prophylactic treatment would compromise bone marrow tolerance if systemic chemotherapy were to be needed.

SUMMARY

Diagnostic and therapeutic technological advances have dramati-cally changed the treatment and outcome of patients with pure testic-ular germinomas. Patients, following inguinal orchiectomy, can now be selected to receive either radiotherapy or chemotherapy based on tumor markers, extent and volume of nodal disease, and stage. Those patients with an elevated alpha-fetoprotein should be treated by combination chemotherapy.

Patients with Stage I or IIA disease and elevated β-HCG can be successfully treated with radiation therapy, but will require careful monitoring of their β-HCG during and after treatment. Individuals with Stage I disease, normal markers, and staged with lymphangio-graphy and CT scanning, will experience survival rates greater than 95% when the regional lymph nodes are irradiated to 3000 rads. In

Stage IIA, survival rates of 75% to 90% have been achieved with
irradiation alone.

Prophylactic irradiation of the mediastinum and supraclavicular
lymph nodes is no longer indicated in patients with Stage II disease,
and such treatment may jeopardize salvage efforts with chemotherapy
in the event of systemic or widespread local relapse. Isolated nodal
relapses, however, often can be successfully treated with radiotherapy alone. Patients with Stage IIB, III, and IV disease should
receive chemotherapy in addition to local therapy, such as surgery of
irradiation to bulky disease. If age or pulmonary, cardiac, or renal
disease preclude the use of agents such as cisplatin, velban, bleomycin or adriamycin, radiation therapy may be successfully employed
even though whole abdominal or lung irradiation may be required. For
patients with Stage IV disease, radiotherapy may be useful in conjunction with systemic chemotherapy in the treatment of bony metastases or in whole brain treatment for brain metastases.

REFERENCES

1. J. D. Earle, M. A. Bagshaw, and H. S. Kaplan, Linear accelerator
    supervoltage radiation therapy: Testicular tumors, Radiology,
    9:1008 (1968).
2. J. D. Earle, M. A. Bagshaw, and H. S. Kaplan, Supervoltage
    radiation therapy of the testicular tumors, AJR, 117:653
    (1973).
3. L. D. Graham and M. A. Bagshaw, Treatment of testicular germi-
    nomas, in: "Urological Cancer", D. G. Skinner, ed., Grune and
    Stratton, New York, p.281 (1983).
4. R. G. Slawson, Radiation therapy for germinal tumors of the
    testis, Cancer, 42:2216 (1978).
5. R. H. Sagerman, G. E. Hanks and M. A. Bagshaw, Supervoltage
    radiation therapy. Use of the linear accelerator for
    treating ovarian adenocarcinoma, Calif.Med., 102(2):118
    (1965).
6. S. D. Williams and L. H. Einhorn, Advanced seminoma: Role of
    chemotherapy. In press.
7. D. E. Johnson, G. Appelt, M. L. Samuels, and M. Luna, Metastases
    from testicular carcinoma, Urology, 8:234 (1976).
8. J. V. Fayos and Y. H. Kim, Treatment of testicular tumors,
    Radiology, 128:471 (1978).
9. D. E. Dosoretz, W. U. Shipley, P. H. Blitzer, S. Gilbert, J.
    Prat, E. Parkhurst, and C. C. Wang, Megavoltage irradiation
    for pure testicular seminoma: Results and patterns of fail-
    ure, Cancer, 48:2184 (1981).
10. G. M. Thomas, W. D. Rider, A. J. Dembo, B. J. Cummings, M.
    Gospodarowicz, N. V. Hawkins, J. G. Herman, and C. W. Keen,
    Seminoma of the testis: Results of treatment and patterns of
    failure after radiation therapy, Int.J.Radiat.Oncol.Biol.
    Phys., 8:165 (1982).

11. D. Ball, A. Barrett, and M. J. Peckham, The management of metastatic seminoma testis, Cancer, 50:2289 (1982).

12. D. Vugrin, W. F. Whitmore, and M. Batata, Chemotherapy of disseminated seminoma with combination of cis-diamminedichloroplatinum (II) and cyclophosphamide, Cancer Clin.Trials, 4:423 (1981).

13. J. G. Maier, M. H. Sulak, and B. T. Mihemeyer, Seminoma of the testis: Analysis of treatment success and failure, AJR, 102:596 (1968).

14. J. R.T. Lee, B. L. McClennan, R. J. Stanley, and S. S. Sagel, Computed tomography in the staging of testicular neoplasms, Radiology, 130:387 (1979).

15. B. T. Burney and E. C. Klatte, Ultrasound and computed tomography of the abdomen in staging and management of testicular cancer, Radiology, 132:415 (1979).

16. D. Raghavan, A. L. Sullivan, M. J. Peckham, and A. M. Neville, Elevated serum alpha-fetoprotein and seminoma. Clinical evidence for a histologic continuum? Cancer, 50:982 (1982).

17. N. Javadpour, Management of seminoma based on tumor markers, Urol.Clin.North Am., 7:773 (1980).

18. T. Qian, Y. Hu, C. Chen, Y. Qi, D. Gu, and X. Gu, Radiation therapy of seminoma of the testis, Int.J.Radiat.Oncol.Biol. Phys., 7:717 (1981).

19. G. Boden and R. Gibb, Radiotherapy and testicular neoplasms, Lancet, 2:1195 (1951).

20. J. M. Quivey, K. K. Fu, K. A. Herzog, J. M. Weiss, and T. Phillips, Malignant tumors of the testis: Analysis of treatment results and sites and causes of failure, Cancer, 39:1247 (1977).

21. B. van der Werf-Messing, Radiotherapeutic treatment of testicular tumors, Int.J.Radiat.Oncol.Biol.Phys., 1:235 (1976).

IRRADIATION OF SEMINOMA

S. D. Fosså and J. F. Evensen

The Norwegian Radium Hospital
Oslo
Norway

INTRODUCTION

The treatment of seminoma is based on the experience that this testicular tumor is one of the most radiosensitive malignancies known. Total radiation doses of 30–40 Gy are sufficient to destroy all tumor tissue in the majority of patients. This high radio-sensitivity and the low frequency of irradiation side effects are the reasons why radiotherapy has been the standard treatment even in clinical stage I where only the minority of patients has microscopic retroperitoneal disease and where therapy is often purely prophy-lactic. In general, the disease also responds favorably to many of the known cytostatic drugs, especially the alkylating agents.

This report deals with the experience at The Norwegian Radium Hospital (NRH) with radiotherapy in patients with seminoma who were admitted between 1971 and 1977, inclusive. The review is based partly on the routinely computerized records of all hospitalized patients (stage I) and partly on a detailed review of patients' records (stage II, III, IV).

PATIENTS AND METHODS

The clinical staging is based on the system recommended by Peckham et al. [1]:

Stage I: No metastases.
Stage II: Retroperitoneal lymph node metastases.

         IIa: < 2 cm
         IIb: 2-5 cm
         IIc: > 5 cm

Stage III: Supradiaphragmatic lymph node metastases.
Stage IV: Extralymphatic metastases.

    Standard X-ray examinations included chest X-ray, bipedal
lymphography, and phlebography of the inferior vena cava. Human
choriogonadotropin in the urine and/or serum was determined in the
majority of patients.

    The WHO histological classification system[2] for testicular
tumors was used.

Stage I:

    There were 188 patients with stage I seminoma registered. Five
of these received no irradiation because of previous radiotherapy due
to cancer of the contra-lateral testicle (four cases) or advanced age
and decreased performance status (one case). In the other 183 cases,
radiotherapy consisted of a so called (anterior and posterior)
L-field (Figure 1). The irradiation field included the paraaortic
lymph nodes on both sides and the ipsilateral iliac lymph nodes. The
proximal border of the field was between the 10th and 11th thoracic
vertebrae. The reasons for irradiation of the external inguinal
region (applied by an anterior field only,) were:

1.   Previous operations in the inguinal region (herniotomy,
     orchiopexy).
2.   Tumor infiltration in the tunica vaginalis, the spermatic cord,
     epididymis or scrotal wall.
3.   Transscrotal orchiectomy.

    Usually a total dose of 40 Gy was applied, giving 2 Gy daily 5
times a week, by a Linear accelerator (8 MeV). After application of
20 Gy, urography was performed with the treatment field marked on the
patient in order to assess the amount of renal parenchyma within the
irradiation field. If more than one third of the kidney substance
was within the field, special lead blocks were applied for shielding
the renal tissue.

    The scrotum with the remaining testis was shielded with a 25 cm
thick lead block, placed as near the testis as possible. The gonadal
dose due to scattered irradiation was measured during the initial 3-4
exposures. These measurements formed the basis for calculation of
the total gonadal dose.

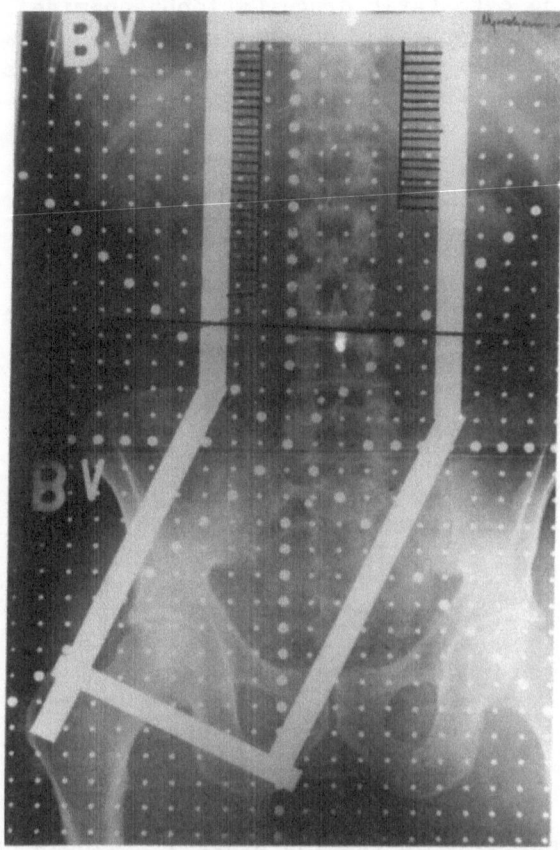

Fig. 1. Typical L-field (without external inguinal irradiation)
in patients with seminoma. ▤ Kidney shielding blocks.

Stage II:

There were 42 patients classified as stage II seminoma on
admission, of whom five were IIA, five were IIb and 32 were IIc.

As a rule these patients were treated with the L-field described
above. Four weeks after discontinuation of the infra-diaphragmatic
irradiation, prophylactic radiotherapy was started to the mediastinum
(anterior and posterior field) and the left supraclavicular fossa
(anterior field only), to a dose of 2 Gy x 20 over 4 weeks.

In stage II patients with extensive retroperitoneal masses,
large abdominal irradiation fields were used during the initial
course of the radiotherapy. After having received 20 Gy, the
patients had a 2-3 weeks treatment free interval. In the other cases

radiotherapy was preceded by 1-3 courses of chemotherapy, most often
with an alkylating cytostatic drug (Cyclophosphamide). The object of
this split course radiotherapy or pre-irradiation chemotherapy was to
allow reduction of the size of the irradiation field and thereby to
reduce irradiation of the neighboring organs (kidney, bowel).

Stage III:

For the 14 stage III seminoma patients the primary treatment was
the same as described for stage II. Depending on the initial extent
of the tumor mass, chemotherapy and/or split course technique was
used.

Stage IV:

Radiotherapy was used palliatively in stage IV patients (9
patients) whose treatment mainly consisted of combination chemotherapy.

RESULTS

Stage I:

The 5 year survival in the 188 stage I seminoma patients was
93%. Fourteen patients had died by January 1st, 1983, 12 of intercurrent disease with no evidence of their malignancy (NED) and 2 due
to seminoma. One of these two patients had not received any treatment due to his poor general medical condition and advanced age.
(78 years).

Six patients showed disease activity after L-field irradiation.
Relapses were observed after a mean of 11.5 months (range: 3-35).
Recurrent disease was observed in the mediastinum (2 patients), left
supraclavicular fossa (2 patients) and in the liver (1 patient). In
the 6th patient increasing HCG-levels were found without demonstrable
metastases. Second line treatment for this group consisted of radiotherapy (3 patients), chemotherapy (1 patient) or both (1 patient).
The 6th patient had liver metastases and did not receive any second
line treatment. Second line treatment was curative in all but that
one patient.

During the course of radiotherapy nausea and vomiting were
commonly observed. Most of the patients returned to work 4 weeks
after discontinuation of the irradiation. Persistent morbidity
(gastro/intestinal symptoms, and weakness was seen only rarely.

Figure 2 shows the gonadal doses in L-field irradiation, for those without or with inguinal irradiation. The gonadal dose tended to be greater in the group with external inguinal region irradiation.

Stage II:

All patients with stage IIa and stage IIb were cured by irradiation (Table 1). Seven of the 32 patients with stage IIc disease died of seminoma and 1 died of intercurrent disease with NED. Ten patients with stage IIc developed recurrences (retroperitoneum - 3, mediastinum/supraclavicular fossa - 2, lungs - 4, other locations - 1). Of the ten relapsing patients three were cured by second line

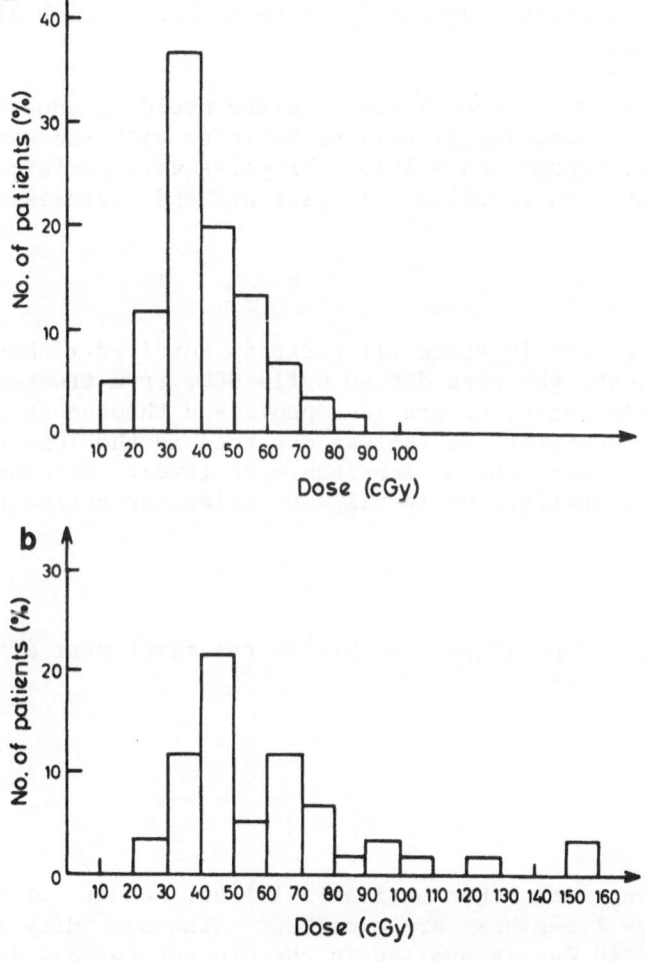

Fig. 2. Gonadal dose in patients with seminoma. (a) Without and (b) with external inguinal irradiation.

Table 1. Five Year Crude Survival in Patients with Seminoma Stage II
          (NRH: 1971-1977).

Substage	Total	No. of patients Dead	Survival(%)
IIa	5	0	100
IIb	5	0	100
IIc	32	8	75*
Total	42	8	81

* 1 patient dead of intercurrent disease with NED.

treatment which consisted primarily of chemotherapy with alkylating
agents and adriamycin.

Extensive retroperitoneal tumor masses could be observed for
several years in surviving irradiated patients with IIc seminoma
(Figure 3). Laparotomy and multiple biopsies were performed in some
of these patients and revealed only fibrosis and necrosis.

Stage III:

Only four of the 14 stage III patients survived (Table 2). One
of the ten patients who died did so while NED, from treatment-related
side effects (persistent severe leucopenia and thrombopenia). Among
the other nine recurrent disease was observed in the lung (three),
retroperitoneum (four) and mediastinum/neck (two). Second-line
chemotherapy was ineffective in all nine relapsing patients.

Stage IV:

All 9 stage IV patients died within the first year after
orchiectomy (Table 2).

DISCUSSION

Stage I:

In agreement with other series (3-10) our results of radiotherapy in stage I seminoma are excellent. The morbidity is minimal.
The total dose, 40 Gy, as applied in the present series, is un-
necessarily high for stage I disease and has now been reduced to 36
Gy.

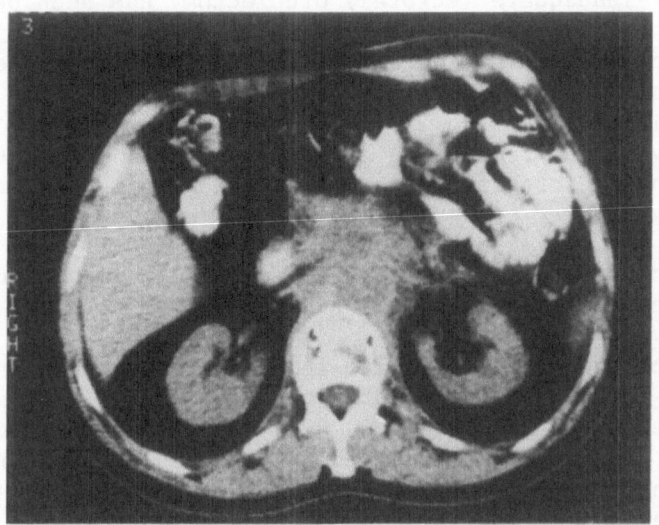

Fig. 3. Large retroperitoneal tumor mass in a patient with seminoma
         stage IIc 3 years after successful irradiation and
         chemotherapy. Laparotomy and biopsy showed only necrotic
         tissue and fibrosis.

Table 2. Five Year Crude Survival of Patients with Seminoma Stage
          III and IV (NRH: 1971-1977).

Stage	Total	No of patients Dead	Survival(%)
III	14	10*	29
IV	9	9	0
Total	23	19	17

* 1 patient dead (NED) of treatment-related side effects.

     The necessity for routine external inguinal irradiation in stage
I is questionable, even if the conditions given above are fulfilled.
The reasons for this are:

1.   Clinical experience indicates that external inguinal recurrences
     are extremely rare.
2.   The external inguinal region can easily be observed by clinical
     follow-up and recurrences can be detected at an early curable
     stage.
3.   In 11 patients fulfilling the above criteria for external
     inguinal irradiation no lymph nodes were demonstrated in the
     external or internal inguinal groups by testicular lympho-
     graphy[11].

4. Even if inadequate (scrotal) orchiectomy has been performed
    within a few weeks before the start of irradiation inguinal node
    involvement will not have occurred as lymph drainage to the
    external inguinal lymph nodes will not by that time have been
    established[12].
5. The advantage of omitting external inguinal irradiation is that
    the gonadal irradiation dose to the remaining testis is
    decreased.

Prophylactic irradiation of the external inguinal lymph nodes has now
therefore been discontinued at the NRH.

The technique for testicular shielding, described above, reduces
the gonadal irradiation to 1% of the total dose. Improvement of the
shielding technique may result in even lower testicular dosage[13],
but will be rather resource-demanding in a busy radiotherapy department. An ongoing study at the NRH deals with the clinical significance of the gonadal dose by a study of postradiation fertility
disturbances. The results of the study will indicate whether the
present rather simple technique of gonadal shielding during L-field
irradiation has to be changed in the future.

Stage II:

During recent years the necessity for prophylactic supradiaphragmatic irradiation in stage II seminoma has been discussed[10,14] with reference to the following issues:

1. Only a minority of patients with clinical stage II disease have
    microscopic seminomatous foci above the diaphragma and need this
    type of radiotherapy.
2. The regions of probable disease manifestation (neck,
    mediastinum) are easy to observe during follow-up examinations.
3. The disease is still curable when relapses occur in the lymph
    nodes above the diaphragma after primary infradiaphragmatic
    irradiation.
4. Bone marrow reserves are preserved if routine supradiaphragmatic
    irradiation s omitted. This may be of importance in patients
    who need intensive chemotherapy during follow-up.

The results of radiotherapy in stage IIa and stage IIb patients
are adequate and do not warrant a change of policy in the majority of
patients. However, attempts should be made to identify possible
sub-groups of high-risk patients, for example, by the level of serum
choriogonadotropin, histological sub-classification of seminoma or
small vessel invasion in the primary tumor. Radiotherapy, even if
combined with "conventional" chemotherapy does not cure all patients
with seminoma stage IIc. The results from other institutions are
even worse than those from the NRH[10,14]. Newer treatment

principles, mainly based on cisplatinum containing combination chemotherapy, have produced excellent results[14-17] and should be
generally adopted.

During the last 4 years the NRH's treatment policy for advanced
seminoma (stage IIc, II, IV) has been similar to that in non-seminoma
patients with comparable stages. So far 15 patients with stage IIc
have been treated with cisplatinum based chemotherapy followed by
radiotherapy (nine patients) or by retroperitoneal surgery (five
patients) (Figure 4). The remaining patient, a 68 year old man, died
of treatment complications after the first cycle of chemotherapy.
The others are alive with NED. In one of the operated patients small
foci of vital residual seminoma were demonstrated, but no tumor
tissue was found in the remaining cases.

Stage III and IV:

The poor results of stage III and stage IV seminoma warrant a
change of the treatment policy. Again, cisplatinum containing chemotherapy seems to improve the survival rates. Of 10 seminoma patients
with stage III and IV disease so treated (1978-1983) at the NRH, 6
are alive with NED, whereas 4 are dead (1 with NED because of postoperative complications) (Figure 5, Table 3).

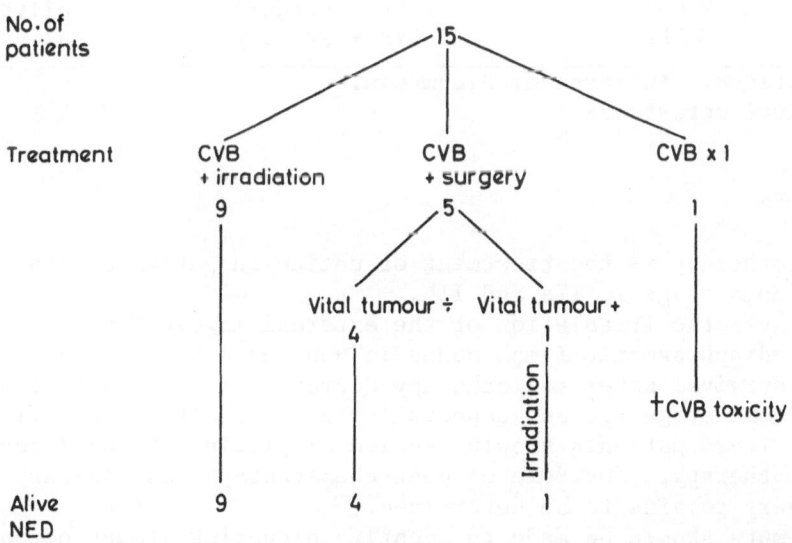

Fig. 4.   Treatment of seminoma stage IIc (NRH: 1978 - May 1983).

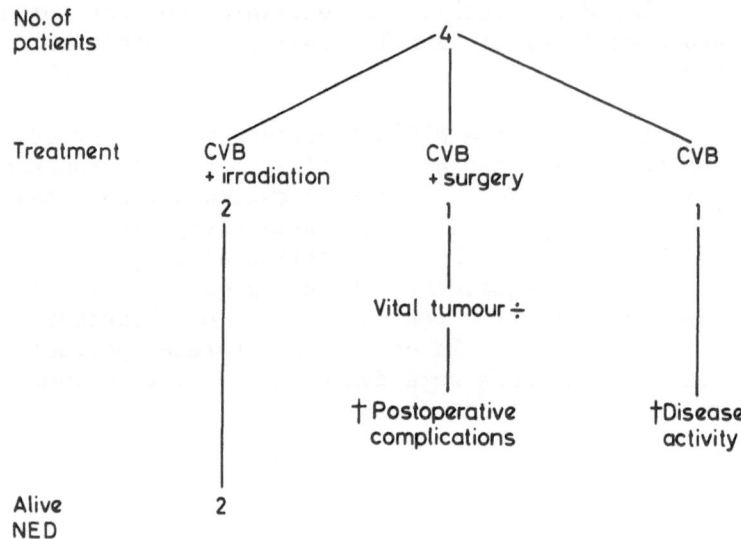

Fig. 5. Treatment of seminoma stage III (NRH: 1978 – May 1983).

Table 3. Treatment of Seminoma Stage IV (NRH: 1978 – May 1983).

Indent.	Sub-stage	Treatment	Status
K.R.	IV 00 (HCG ↑)	CVB*	alive NED
E.S.	IV CL_1	CVB + irradiation	alive NED
A.J.	IV CL_3	CVB	dead
M.E.	IV CL_3	CVB	dead
T.J.B.	IV 00°	CVB + surgery	alive NED
B.H.	IV CL_3	CVB + surgery	alive NED

* Cis-Platinum, Vinblastine, Bleomycin.
° Intestinal metastasis.

CONCLUSIONS

1.   Radiotherapy is the treatment of choice in patients with
     seminoma stage I, IIa and IIb.
2.   Prophylactic irradiation of the external inguinal and
     supradiaphragmatic lymph nodes is unnecessary.
3.   The survival after radiotherapy decreases in patients with
     seminoma stage IIc and especially in those with stage III and
     IV. These patients should receive cisplatinum based induction
     chemotherapy. The role of post-chemotherapy radiotherapy or
     surgery remains to be determined.
4.   Attempts should be made to identify high-risk groups of patients
     with seminoma in terms of serum tumor markers, sub-classifi-

cation of histology and extent or localization of tumor manifes-
tations.

REFERENCES

1. M. J. Peckham, T. J. McElwain, A. Barrett, and W. F. Hendry,
    Combined management of the malignant teratoma of the testis,
    Lancet, 2:267 (1979).
2. F. K. Mostofi and L. H. Sobin, Histological typing of testis
    tumors, in: "International Histological Classification of
    Tumors No. 16", World Health Organization, Geneva, (1977).
3. J. F. Doombos, D. H. Hussey, and D. E. Johnson, Radiotherapy for
    pure seminoma of the testis, Radiology, 116:401-404 (1975).
4. B. van der Werf-Messing, Radiotherapeutic treatment of
    testicular tumors, Int.J.Rad.Oncol.Biol.Phys., 1:235-248
    (1976).
5. J. M. Quivey, K. K. Fu, K. A. Herzog, J. M. Weiss, and T. L.
    Phillips, Malignant tumors of the testis: Analysis of the
    treatment results and sites and causes of failure, Cancer,
    39:1247-1253, (1977).
6. M. A. Batata and A. Unal, The role of radiation therapy in
    relation to stage and histology of testicular cancer,
    Sem.Oncol., 6:69-73 (1979).
7. Q. Tu-Nan, H. Yu-Hua, C. Chih-Xian, Q. Yu-Quin, G. Da-zhong, and
    G. Xian-Zhi, Radiation therapy of seminoma of the testis,
    Int.J.Rad.Oncol.Biol.Phys., 7:717-720, (1981).
8. G. E. Hanks, D. F. Herring, and S. Kramer, Patterns of care
    outcome studies results of the national practice in seminoma
    of the testis, Int.J.Rad.Oncol.Biol.Phys., 7:1413-1417
    (1981).
9. M. J. Peckham, Seminoma testis, in: The Management of Testicular
    Tumors, M. Peckham, ed., Edvard Arnold (Publ.) Ltd., 134-151
    (1981).
10. G. M. Thomas, W. D. Rider, J. Dembo, B. J. Cummings, M.
    Gospodarowicz, N. V. Hawkins, J. G. Herman, and C. W. Keen,
    Seminoma of the testis: Results of treatment and patterns of
    failure after radiation therapy, Int.J.Rad.Oncol.Biol.Phys.,
    8:165-174 (1982).
11. H. H. Lien, A. Kolbenstvedt, A. Miller, and S. J. Bakke,
    Antegrade testicular vein phlebography and funicular
    lymphography in testicular tumors, Act.Rad.Diagn., 21:603-613
    (1980).
12. D. E. Johnson and R. J. Babaian, The case for conservative
    surgical management of the ilioinguinal region after
    inadequate orchiectomy, J.Urol., 123:44-46 (1980).
13. H. Kubo and W. U. Shipley, Reduction of the scatter dose to the
    testicle outside the radiation treatment fields, Int.J.Rad.
    Oncol.Biol.Phys., 8:1741-1745 (1982).

14. D. Ball, A. Barrett, and M. J. Peckham, The management of
     metastatic seminoma, Cancer, 50:2289-2294 (1982).
15. R. Wajsman, S. A. Beckley, and J. E. Pontes, Changing concepts
     in the treatment of advanced seminomatous tumors, J.Urol.,
     129:303-306 (1983).
16. H. Cortés Funes, A. T. Van Oosteron, S. D. Williams, C. P.
     Vendrik, and W. W. ten Bokkel Huinink, The treatment of
     advanced seminomas with cisplatin, velban and bleomycin, 13th
     Int.Cancer Congr., (Proceedings), Seattle, Washington, USA,
     Sept. 8-15:180 (1982).
17. W. L. Mendenhall, S. D. Williams, L. H. Einhorn, and J. P.
     Donohue, Disseminated seminoma: Re-evaluation of treatment
     protocols, J.Ruol., 126:493-496 (1980).

TUMOR OF THE TESTIS - NEW DRUGS

AND DRUG COMBINATIONS

G. Stoter

Department of Oncology
Free University Hospital
Amsterdam, The Netherlands

The cure rate for patients with disseminated testicular non-
seminomatous tumors is 70%[1-3]. The 30% of patients with a low
probability of cure are characterized by a large tumor burden, high
levels of serum markers and previous treatment[1-5]. The first two
risk factors decrease the potential for chemotherapy to achieve
complete malignant cell kill, whereas previous radiotherapy and
chemotherapy increase the risk of lethal toxicity when modern induction chemotherapy is given[3].

Most investigators today acknowledge that the patients who are
candidates for chemotherapy should be divided in two categories:

I) Patients with low tumor burden and low marker levels.
II) Patients with high tumor burden and high marker levels.

The first category has a 90% chance for complete response,
whereas the second group achieves complete response in only 40-60% of
the cases[1-5]. In patients with good prognostic factors (category
I) the main good is to minimize the side effects of the treatment,
which parallels the situation in Hodgkin's disease. However, in
patients with poor prognostic factors (category II) the treatment
should be intensified.

Intensified treatment can be achieved by, 1) the addition of
other active agents to existing combinations, 2) the replacement of
agents by more effective new drugs, 3) the use of alternating, non
cross-resistant, combinations of drugs, and 4) combined modality
treatment, including surgery and radiotherapy.

Addition of New Agents

Several investigators have tried the addition of other drugs to
the core of cisplatin, vinblastine and bleomycin (PVB). The VAB
regimens at MSKCC, which include actinomycin-D, chlorambucil, cyclophosphamide and adriamycin in different schedules and dosages, have
yielded no superior results over the PVB regimen used elsewhere[3,6-10]. At Indiana University the addition of adriamycin to
the PVB regimen did not improve the results[11]. In a German multicenter trial, iphosphamide was added to the PVB combination in a
randomized fashion. The complete response rate was 63% for PVB and
65% for PVB plus iphosphamide. More severe toxicity occurred in
cases treated by the four drug combination[12]. In general, the
addition of further drugs to an already active 2- to 3-drugs combination does not seem to improve the therapeutic results[13].

Replacement with New Drugs

VP-16-213 has been shown to be very effective in the treatment
of non-seminomas of the testis. In a total of 76 patients with
tumors resistant to first-line chemotherapy, a response rate of 40%
was achieved[14-16]. At Indiana University 51 patients who developed
relapse after PVB were treated with PVB + VP-16-213. A response rate
of 82% was obtained including 40% with complete responses[14,17].
Although VP-16-213 has a myelosuppressive effect similar to that of
vinblastine, it has less mucosal toxicity and no neurological side
effects, such as paralytic ileus. For these reasons VP-16-213 as
first line therapy in place of vinblastine has potential. The Southeastern Cancer Study Group is randomizing patients between PVB and
PEB (E=Etoposide, B=VP-16-213). At the Royal Marsden Hospital a
phase II study of first-line PEB chemotherapy is ongoing. The preliminary results indicate that PEB is as effective as PVB but less
toxic.

Alternating Drugs

The high response rate with PVB + VP-16-213 in patients relapsing after PVB suggests that these combinations are not cross-
resistant. This would raise the possibility of using a sequence of
two non cross-resistant combinations in the hope of a higher complete
response rate in patients with negative prognostic characteristics.
A similar approach is attempted in Hodgkin's disease with the sequence of MOPP and ABVD[18]. For this reason the EORTC (European
Organisation for the Treatment of Cancer) is now doing a randomized
study of PEB alone versus the alternating regimen of PVB and PEB for
a total of 4 cycles.

Combined Modalities

Multimodality treatment with chemotherapy and/or surgery has
been discussed elsewhere in this book. It is sufficient to mention
here that debulking surgery after induction chemotherapy will render
an additional 15% of the patients free of tumor. There are few data
on combined chemotherapy and radiotherapy. At the Royal Marsden
Hospital, London the combination has been abandoned and they now use
chemotherapy alone.

As mentioned before, the potential for cure in patients with low
lumor burden is so high, that the main goal for clinical research is
to decrease the toxicity of the treatment. The most important side
effects of PVB combination chemotherapy are myelosuppression, renal
damage and pulmonary fibrosis[1,3]. Einhorn and coworkers[19] demonstrated that low dose vinblastine (0.3 instead of 0.4 mg/kg/cycle) is
as effective and less toxic with regard to granulocytopenic fever and
sepsis[29]. On the other hand, two studies have demonstrated that
reduction of the dose of cisplatin is accompanied by a decrease of
the complete remission rate and an increase of the relapse
rate[20,21]. Bleomycin causes clinical signs of pulmonary fibrosis
in 5-10% of the cases and a toxic death rate of 1-5%[22,23]. In a
recent EORTC study of PVB, 56 out of 126 patients (44%) had reductions of more than 20% in their vital capacity (FVC) and 65 of 207
patients (27%) had fibrotic changes on chest X-ray. These data
suggest that bleomycin should be omitted from chemotherapy regimens
in "good risk" patients with a low tumor burden and low levels of
tumor markers. At present the EORTC is randomizing such "good risk"
patients to receive either 4 cycles of cisplatin, VP-16-213, and
bleomycin (PEB) or cisplatin and VP-16-213 without bleomycin (EP).
At MSKCC a randomized study of VAB-6 versus EP is ongoing.

An intensive search of platinum analogs is ongoing to identify
agents which are less toxic, especially with regard to nephrotoxicity, nausea and vomiting.

A variety of drugs, such as TNO-6, CBDCA(JM-8), JM-40 and CHIP
are under investigation[24-29]. CBDCA may be less nephrotoxic than
cisplatin and has shown anti-tumor activity in ovarian and breast
cancer[26,28]. This drug does not seem to be cross-resistant with
cisplatin[26].

In conclusion, in future trials patients should be divided into
"bad risk" and "good risk" categories. In "bad risk" patients the
treatment should be intensified. Ways to do this include the introduction of new active agents (VP-16-213, cisplatin analogs) and
possibly the use of alternating non cross-resistant combinations
(PVB/PEB). In "good risk" patients the toxicity of the treatment
should be ameliorated by the omission of toxic drugs (bleomycin) or
the reduction of the drug dosage provided that this does not decrease

the therapeutic effect (vinblastine). Also, the introduction of new active agents, like CBDCA, which has less toxicity than the existing drugs, is an important goal to improve further the results of chemotherapy.

REFERENCES

1. L. H. Einhorn, S. D. Williams, I. Mandelbaum, and J. P. Donohue, Surgical resection in disseminated testicular cancer following chemotherapeutic cytoreduction, Cancer, 48:904 (1981).

2. D. Vugrin, W. F. Whitmore, P. C. Sogani, M. Bains, H. W. Herr, and R. B. Golbey, Combined chemotherapy and surgery in the treatment of advanced germ-cell tumors, Cancer, 47:2228 (1981).

3. G. Stoter, C. P. J. Vendrik, A. Struyvenberg, D. Th. Sleyfer, R. Vriesendorp, H. Schraffordt Koops, A. T. van Oosterom, W. W. ten Bokkel Huinink, and H. M. Pinedo, The 5-year survival of patients with disseminated non-seminomatous testicular cancer treated with cisplatin, vinblastine and bleomcyin, Cancer, (in press).

4. E. S. Newlands, R. H. J. Begent, G. J. S. Rustin, D. Parker, and K. D. Bagshaw, Further advancement on the management of malignant teratomas of the testis and other sites, The Lancet, 1:948 (1983)

5. M. J. Peckham, A. Barrett, T. J. McElwain, W. F. Hendry, and D. Raghavan, Non-seminoma germ cell tumors (malignant teratoma) of the testis, Brit.J.Urol., 53:162 (1981).

6. T. F. Reynolds, D. Vugrin, E. Cvitkovic, and E. Cheng, VAB-3 combination chemotherapy of metastatic testicular cancer, Cancer, 48:888 (1981).

7. D. Vugrin, E. Cvitkovic, W. F. Whitmore, E. Cheng, and R. B. Golbey, VAB-4 combination chemotherapy in the treatment of metastastic testis tumors, Cancer, 47:833 (1981).

8. D. Vugrin, H. W. Herr, W. F. Whitmore, E. C. Sogani, and R. B. Golbey, VAB-6 combination chemotherapy in disseminated cancer of the testis, Ann.Intern.Med., 95:59 (1981).

9. L. H. Einhorn and J. P. Donohue, Cis-diamminedichloroplatinum, vinblastine, and bleomycin combination chemotherapy in disseminated testicular cancer, Ann.Intern.Med., 87:293 (1981).

10. L. H. Einhorn and S. D. Williams, The management of disseminated testicular cancer, in: "Testicular tumors, management and treatment", L. H. Einhorn, ed., Masson Publishing USA, New York, p.119 (1980).

11. L. H. Einhorn, S. D. Williams, T. Troner, R. Birch, and F. A. Greco, The role of maintenance therapy in disseminated testicular cancer, N.Eng.J.Med., 305:727 (1981).

12. H. J. Schmoll, V. Diehl, and J. Hartlapp, BUN/VBL/DDP vs BLM/
     VBL/DDP/IPP in disseminated testicular cancer: a prospective
     randomized trial, Abstr.UICC Conf.Clin.Oncol.Lausanne, 128
     (1981).

13. P. Alberto, W. Berchtold, and R. Sonntag, Chemotherapy of small
     cell carcinoma of the lung: comparison of a cyclic altern-
     ative combination with simultaneous combinations of 4 and 7
     agents, Europ.J.Cancer, 17:1027 (1981).

14. R. A. Dhafir, L. H. Einhorn, S. Williams, P. Rosenbaum, and
     B. Issell, The effect of etoposide (VP-16), used alone or in
     combination on the survival of patients with refractory
     testicular cancer, Proc.Amer.Soc.Clin.Oncol., 22:464 (1981).

15. B. M. Fitzharris, S. B. Kaye, and S. Saverymuttu, VP-16-213 as a
     single agent in advanced testicular tumors, Europ.J.Cancer,
     16:1193 (1980).

16. E. S. Newlands and K. D. Bagshaw, Anti-tumor activity of the
     epipodophyllin derivative VP-16-213 (etoposide: NSC-141540)
     in gestational choriocarcinoma, Europ.J.Cancer, 16:401
     (1980).

17. S. D. Williams, L. H. Einhorn, and F. A. Greco, VP-16-213
     salvage therapy for refractory germinal neoplasm, Cancer,
     46:2154 (1980).

18. G. Bonadonna, A. Santoro, V. Bonfante, and P. Valagussa, Role
     of ABVD chemotherapy in Hodgkin's disease, UICC-Conf.Clin.
     Lausanne, abstr.09-0192:28 (1981).

19. L. H. Einhorn and S. D. Williams, Chemotherapy of disseminated
     testicular cancer: a random prospective study, Cancer,
     46:1339 (1980).

20. G. J. Bosl, R. Kwong, P. A. Lange, E. J. Fraley, and B. J.
     Kennedy, Vinblastine, intermittent bleomycin and single-dose
     cis-dichlorodiammineplatinum in the management of stage III
     testicular cancer, Cancer Treat Rep., 64:331 (1980).

21. K. Samson, R. L. Stephens, and R. C. Kludo, Positive dose-
     response of high (H) versus low (L) dose cis-platinum (DDP),
     vinblastine, velbe and bleomycin (Bleo) in disseminated germ
     cell neoplastic of the testis, Proc.Am.Soc.Clin.Oncol., 22:
     470 (1981).

22. R. H. Blum, S. K. Carter, and K. Agre, A clinical review of
     bleomycin - a new anti-neoplastic agent, Cancer, 31:903
     (1973).

23. M. L. Samuels, D. E. Johnson, P. Y. Holoye, and V. J. Lanzotti,
     Large-dose bleomycin therapy and pulmonary toxicity. A
     possible role of prior radiotherapy, J.Amer.Med.Assoc., 235:
     1117 (1976).

24. A. Boven, W. J. F van der Vijgh, I. Klein, H. Schluper, R. Nauta
     and H. M. Pinedo, Comparative activity and distribution
     studies of 5 platinum analogs in 2 human ovarian carcinoma
     lines grown in nude mice, Proc.Amer.Assoc.Cancer Res., 24:
     293 (1983).

25. S. Kaplan, R. Joss, C. Sessa, A. Goldhirsch, M. Cattanes, and
     F. Cavalli, Phase I trials of cis-diammine-1,1-cyclobutane
     dicarboxylate platinum II (CBDCA) in solid tumors, Proc.Amer.
     Assoc.Cancer Res., 24:132 (1983).

26. B. D. Evans, A. H. Calvert, S. J. Harland, K. Shanti Raju, A.
     Jones, and E. Wiltshaw, Phase II and early phase III trials
     with JM-8 (Cis-diammine-1,1-cyclobutane dicarboxylate
     platinum II) in advanced ovarian carcinoma, Proc.Amer.Assoc.
     Cancer Res., 24:154 (1983).

27. H. Pinedo, W. ten Bokkel Huinink, H. Gall, J. McVie, G.
     Simonetti, W. van der Vijgh, L. Farber, and J. Vermorken,
     Toxicity of cis-1,1-di(aminomethyl)-cyclohexane platinum (Pt)
     II sulphate (TNO-6) in relation to method of administration,
     Proc.Amer.Soc.Clin.Oncol., 2:33 (1983).

28. G. A. Curt, J. J. Grygiel, R. Weiss, B. Corden, R. Ozols,
     D. Tell, J. Collins, and C. E. Myers, A phase I and pharma-
     cokinetic study of CBDCA (NSC 241240) (1983).

29. M. Lassus, T. Ohnuma, S. Leyvraz, and J. F. Holland, A phase I
     study of CBDCA (carboplatin), Proc.Amer.Socl.Clin.Oncol.,
     2:37 (1983).

TUMORS OF THE TESTIS - INDUCTION AND MAINTENANCE CHEMOTHERAPY

FOR DISSEMINATED NON-SEMINOMAS: THE EORTC EXPERIENCE

G. Stoter[1], W. W. ten Bokkel Huinink[2], C. P. J. Vendrik[3],
A. T. van Oosterom[4], W. G. Jones[5], S. B. Kaye[6], and
D. Th. Sleyfer[7], for the EORTC Genito-Urinary Group

[1]Free University Hospital, Amsterdam, The Netherlands
[2]Netherlands Cancer Institute, Amsterdam, The Netherlands
[3]University Hospital, Utrecht, The Netherlands, [4]University
Hospital, Leiden, The Netherlands, [5]Cookridge Hospital, Leeds
U.K., [6]Gartnavel Hosptial, Glasgow, U.K., [7]University Hospital
Groningen, The Netherlands

229 patients were entered on a randomized study of PVB induction
chemotherapy which consisted of cisplatin (P) 20 mg/kg i.v. days 1-2,
q 3 weeks, vinblastine (v) 0.15 mg/kg or 0.20 mg/kg i.v. days 1-2 q 3
weeks and bleomycin (B) 30 mg i.v. on day 2, then weekly for 12
weeks. After 4 cycles of PVB the response to treatment was assessed
and complete responders were randomized to receive maintenance chemotherapy for the duration of one year with cisplatin 50 mg/m^2 i.v.
every 6 weeks and vinblastine 0.20 mg/kg i.v. every 3 weeks.

Of the 228 patients entered, 25 had been previously treated
leaving 203 patients for randomization. Of these, 69 patients are
not evaluable for response because of insufficient data (24
patients), insufficient time of follow-up (44 patients), and non-
cancer related early death (1 patient). Thus 134 patients remain for
evaluation of response and of these, 64 received high dose PVB (0.4
mg V/kg/cycle) and 70 received low dose PVB (0.3 mg V/kg/cycle). The
25 previously treated patients were not randomized but assigned to
low dose PVB. Of the randomized patients 45 of 64 in the high dose
PVB arm achieved complete response CR (71%), 13 achieved partial
response PR (20%) and 6 showed progressive disease (P) (9%). Of the
randomized patients in the low dose PVB arm 50 of 70 achieved CR
(71%), 16 achieved PR (23%) and 4 had progression (6%). Five of the
patients with complete response in the high dose group had a subsequent relapse, but none of those in the low dose PVB group did so.
This difference is not statistically different at this time.

We also analysed the response to treatment according to the extent of the disease. Using Peckham's staging system[1] patients were classified "High Volume Metastases" (HVM) when they had stages IIC, $IIIM_3$, $IIIN_3$ or IVL_3. Of the patients with HVM 24 or 41 in the high dose PVB group achieved CR (59%) as compared to 23 of 38 in the low dose PVB group (61%). Of the patients with "Low Volume Metastases" (LVM) 21 of 24 in the high dose PVB arm achieved CR (88%) as compared to 27 or 31 (87%) in the low dose PVB group. Notice that patients with HVM have a 30% less chance of achieving CR than patients with LVM. There was no therapeutic difference between high dose and low dose PVB treatment.

A comparison was made of the toxicity produced by high dose and low dose PVB therapy. There was no significant difference in the incidence of non-hematological side effects, except for a higher incidence of mucositis in the high dose PVB group (55% versus 36%). However, 30% of the patients on high dose PVB developed leukocytopenia below $1000/mm^3$ during the first treatment cycle as compared to only 13% (p=0.001) on the low dose PVB arm. Granulocytopenic febrile episodes occurred in 55% of the high dose PVB treatment cycles and in only 30% of the low dose PVB. There was no difference in the lowest counts of platelets between the two arms.

Of the 95 complete responders 68 were randomized to maintenance chemotherapy (37 patients) or no maintenance therapy (31 patients). One patient (3%) on the maintenance arm, and 2 (6%) on the control arm relapsed during a follow-up period of 10-40 months. This difference is not statistically significant. Although it is too early to draw a definitive conclusion from this study of maintenance chemotherapy, in light of results from other studies[2,3], it is expected that maintenance chemotherapy with cisplatin and vinblastine will probably not reduce the relapse rate in well-documented complete responders. Our data confirm the results of Einhorn[2] showing that the low dose vinblastine is as effective as the high dose and is less toxic. Therefore low dose PVB should be the standard regimen.

Other studies of maintenance chemotherapy have failed to demonstrate a benefit for the patients in whom a complete response was well documented[2,3]. We believe that maintenance chemotherapy should be omitted in testicular cancer as is now done in Hodgkin's disease.

REFERENCES

1.   M. J. Peckham, A. Barrett, T. J. McElwain, W. F. Hendry, and D. Raghavan, Non-seminoma germ cell tumors (malignant teratoma) or the testis, Brit.J.Urol., 53:162 (1981).

2.   L. H. Einhorn, S. D. Williams, M. Troner, R. Birch, and F. A.
      Greco, The role of maintenance therapy in disseminated
      testicular cancer, N.Eng.J.Med., 305:727 (1981).
3.   D. Vugrin, H. W. Herr, W. F. Whitmore, P. C. Sogani, and R. B.
      Golbey, VAB-6 without maintenance: progress in chemotherapy
      of testicular germ cell tumors, Proc.Amer.Soc.Clin.Oncol.,
      22:474 (1981).

MANAGEMENT OF BULKY ADVANCED DISSEMINATED

TESTICULAR CANCER

N. Javadpour

The Surgery Branch, National Cancer Institute
National Institutes of Health
Bethesda, Maryland, USA

Recent advances in the chemotherapy of advanced testicular cancer and the availability of the specific serum tumor markers alfa-fetoprotein (AFP) and human chorionic gonadotropin (HCG) have made a dramatic improvement in the management of patients with non-seminomatous testicular cancer. Cis-platinum containing combination chemotherapy regimens have produced improved overall response rates in metastatic disease[1-15]. Over two-thirds of patients with Stage III disease can be expected to achieve a complete response to chemotherapy. The relapse rate in these patients is rare, with over 90% of the relapses occurring in the first 2 years. Since the median follow-up in a number of reported series is greater than 2 years, it is likely that the majority of patients who achieve a complete response to chemotherapy are cured of their disease.

However, within the same group of patients with disseminated disease, two distinct subsets of patients can be identified with a markedly different prognosis. Patients with minimal disseminated disease have a complete response rate of 90 to 95% to chemotherapy. In contrast, in patients with advanced, massive, bulky stage III disease, the complete response rate to chemotherapy is only 43 to 50% (Table 1). Similar correlations between tumor volume and response to chemotherapy have been reported for other malignancies, such as oat cell carcinoma of the lung, ovarian carcinoma, Burkitt's lymphoma, rhabdomyosarcoma, neuroblastoma, and some other tumors. These clinical observations, as well as similar results in experimental systems, have frequently been used to support a role for cytoreductive surgery in patients with bulky tumors. However, there has been no previously reported prospective randomized trial in which the role of cytoreductive surgery in any human malignancy has been critically analyzed.

Table 1. Extent of Disease

	Number	Complete Remission (%)	NED + Surgery (%)
Minimal plum.	14	13 (93)	0
Advanced pulm.	20	10 (50)	3(15)
Minimal plum & abd.	13	13(100)	0
Advanced abd.	23	10 (43)	8(35)
Elevated markers only	4	4(100)	0

Einhorn, LH and Williams, SD Cancer 46:1339,80

Advanced testicular cancer is a disease state well suited for the evaluation of the efficacy of cytoreductive surgery primarily because of the availability of effective chemotherapy for patients with a small volume of residual disease. Patients with bulky disease and a poor prognosis could hypothetically be converted to a good prognosis group by cytoreductive surgery prior to the administration of chemotherapy. Furthermore, the following characteristics make testicular cancer particularly amenable to such a clinical trial:

1. testicular cancer primarily metastasizes to sites which surgically accessible, e.g., the retroperitoneum and the lungs:
2. serum levels of AFP and HCG frequently correlate with the volume of tumor and consequently can be used to estimate the actual volume of tumor resected by cytoreductive surgery; and
3. testicular carcinoma is a disease of otherwise healthy young men who can tolerate extensive surgery with acceptable morbidity. Accordingly, in 1976 the Surgery and Medicine Branches of the National Cancer Institute (NCI) started a prospective randomized trial of cytoreductive surgery followed by chemotherapy versus chemotherapy alone in patients with bulky stage III testicular carcinoma (Figure 1).

Results of the trial are summarized in Table 2. There were no statistically significant differences in overall response rate (75% versus 84%) or in complete response rates (50% versus 37%) between the group randomized to cytoreductive surgery and the group treated with chemotherapy alone. The complete responders in the chemotherapy group include patients who underwent complete resection of mature teratoma following chemotherapy. When all 39 patients are considered, there is a suggestion that those patients treated with chemotherapy as initial treatment have a survival advantage (p = 0.055). From this result one can conclude that it is unlikely (p<0.028) that a true beneficial effect of cytoreductive surgery was missed due to the relatively small number of patients in each treatment arm.

**NCI RANDOMIZED CLINICAL TRIAL FOR BULKY
STAGE III NON-SEMINOMATOUS TESTICULAR TUMOR**

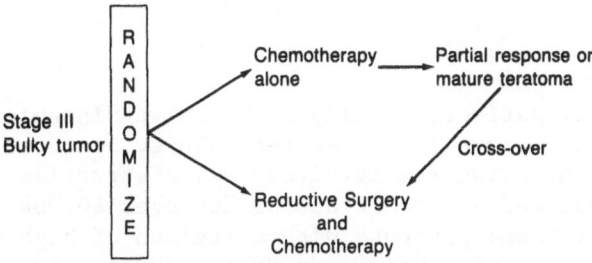

Fig. 1. Previous NCI protocol.

The data in this trial suggest that cytoreductive surgery prior to chemotherapy is not indicated in patients with advanced bulky testicular cancer. However, surgery may still have an important role in the management of patients with advanced disease after intensive chemotherapy.

Furthermore, it is apparent that there is a subset of patients with poor prognostic features, including massive, bulky, disseminated tumors.

DISSEMINATED MINIMAL DISEASE

After treating these patients with three to four cycles of platinum, vinblastine, and bleomycin (PVB), the patients are clinically restaged. Those without any evidence of disease will be followed, and those with abdominal pulmonary, and/or neck masses will undergo resection of the masses. The latter group of patients are followed without any further treatment. However, patients who have

Table 2.

Results	Cytoreductive surgery followed by chemotherapy	Chemotherapy as initial treatment
Patients fully restaged	19/20	19/19
Number with response:		
Complete response	10/20 (50%)	7/19 (37%)*
Partial response	5/20 (25%)	9/19 (47%)
No response	5/20 (25%)	3/19 (16%)

* This includes three patients who had resection of mature teratomas
  following chemotherapy.

viable tumor in their surgically resected specimens are treated with
further chemotherapy.

MASSIVE BULKY TUMOR

This subset of patients usually has massive disseminated disease
with one of the poor prognostic features. These features include
obstructive uropathy, visceral involvements, obstruction of the
inferior vena cava, and serum AFP and/or HCG over 10,000 ng per ml.
Currently we treat these patients with a regimen of high dose cis-
platinum, vinblastine, bleomycin, and VP-16 (PVBV). If the disease
disappears, we follow the patients for recurrence. If patients have
resectable masses with normal serum markers, these masses are resected. The presence of malignant elements will call for further
therapy. The rational for this protocol is as follows:

1. VP-16 is an active agent in relapsed patients with testicular
    cancer and has synergistic activity with platinum in L1210
    leukemia. The major toxicity of PVB is due to the myelosup-
    pressive effects of vinblastine. By lowering the dose of vin-
    blastine and adding VP-16, which also has myelosuppression as
    its major toxicity, the overall hemotologic toxicity of PVBV is
    still acceptable.
2. Testicular cancer has a steep dose-response curve to platinum.
    Treatment of patients who had relapsed on a low dose platinum
    regimen with a high dose of platinum and diuresis was associated
    with a response rate of 55%. Animal toxicity studies suggested
    that renal toxicity due to platinum could be further minimized
    by maintenance of a brisk chloride diuresis[12].

Accordingly, patients in the pilot protocol of PVBV received 40 mg
per M^2 of platinum for 5 days. A chloride diuresis was maintained
with 5 liters of normal saline hydration on every day of platinum
administration. The schema of the current protocol is shown in
Figure 2. This protocol is ongoing, and we have treated 19 patients
with multiple poor prognostic features with 21 day cycles of PVBV:
cis-platinum (P) (40 mg per M^2 IV days 1 to 5), vinblastine (V) (0.2
mg per kg IV day 1), bleomycin (B) (30 units IV every week), and
VP-16 (V) (100 mg per M^2 per IV days 1 to 5). Twice the dose of P
(in 250 ml of 3% saline) used in standard regimens was administered
over 30 minutes with continuous hydration (250 ml per hour normal
saline) for 5 days. This is based on experimental studies by
Litterst[10] which demonstrated that hypertonic saline can decrease
the renal toxicity of Platinum. Thirteen of 15 (80%) previously
untreated patients achieved a complete response. There was one death
due to sepsis, and another patient died from a gastrointestinal
mucosal slough. There has been no renal toxicity other than transient hypomagnesemia with 60 cycles of high dose platinum even when
used in patients with tumor-related obstructive uropathy. These

**A RANDOMIZED CLINICAL TRIAL
FOR DISSEMINATED BULKY TESTICULAR CANCER**

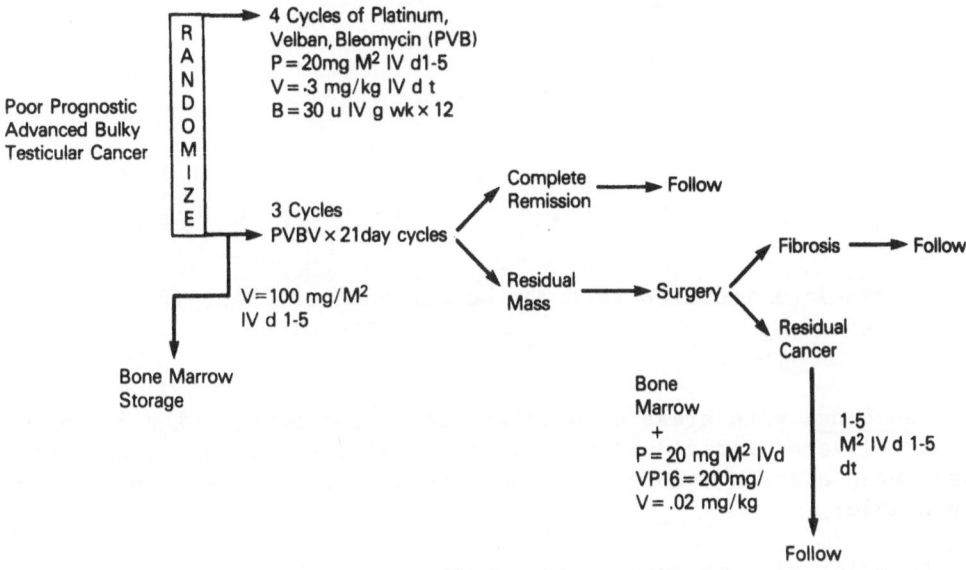

Fig. 2.   Current NCI protocol.

results indicate that PVBV is an effective regimen for poor prognosis
nonseminomatous testicular carcinoma (NSTC) patients and that high
dose platinum can be safely administered in hypertonic saline. A
prospective randomized trial of PVB is currently in progress at the
NCI for poor prognosis NSTC patients.

Generally, after administration of three to four cycles of
intensive chemotherapy containing cis-platinum, if there is still a
residual mass, we advocate resection to eliminate possible tumor.
The histopathologic examination of these tumors will determine the
need for additional therapy.

SURGICAL TECHNIQUE

Transabdominal Approach

Patients undergoing cytoreductive surgery receive a midline
vertical incision from the xiphoid process to the pubis (Figure 3).
On entering the peritoneal cavity, a thorough exploration is performed. The kidneys and ureters are identified and dissected free
and usually represent the lateral margins of the dissection. Control
of the vena cava and aorta is gained distally at a convenient site,
and control proximally is achieved at a level just beneath the renal
vessels. The mass of retroperitoneal tumor and vena cava are re-

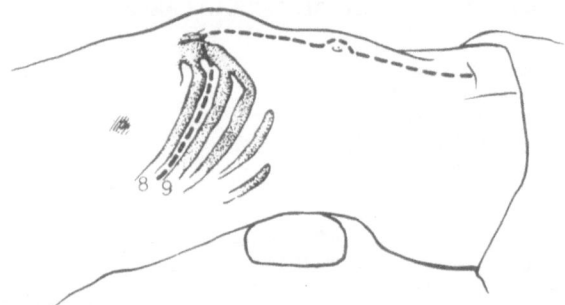

Fig. 3.   Mid-line incision that may be extended to chest if
          necessary.

sected en bloc, with great care taken around the aorta. The lumbar
veins are ligated individually. By resecting the inferior vena cava
we have been able to remove much tumor that would otherwise have been
inaccessible.

     The major risk of surgery is uncontrollable hemorrhage from
injury either to the inferior vena cava or to the aorta which is
encased in tumor. En bloc resection of the vena cava eliminates the
IVC as a source of hemorrhage and, by allowing greater access around
the aorta, diminishes the likelihood of injury to this structure.
Intraluminal clot or tumor is frequently documented on preoperative
inferior vena-cavography. Most patients experience varying degrees
of postoperative leg edema, which has been well controlled with
conservative measures. We attribute this lack of significant postoperative edema to the gradual development of extensive collateral
vessels forming as a result of the tumor encroaching on the IVC and
aorta (Figure 4).

Thoracoabdominal Approach

     In patients with large masses involving the crura of the dia-
phragm, extending into the thoracic cavity, we advocate a thoracoabdominal approach. The position of the patient is extremely important. For a left thoracoabdominal approach the patient is positioned
with flat pelvis, and the chest is rotated about 30° and the kidney
rest elevated (Figure 5). The incision should be made in such a
manner that the pedicle of the kidney will be in the center of the
operative field. We prefer an incision started as paramedian and
continued over the tenth rib, which is also resected. The advantages
of this incision over the higher incisions are that there is no need
to close the ribs, and no need to convert the incision into a T-shape
as advocated by some surgeons. After the pleura, the retroperitoneal
structures, including renal pedicle, aorta, and IVC, are exposed

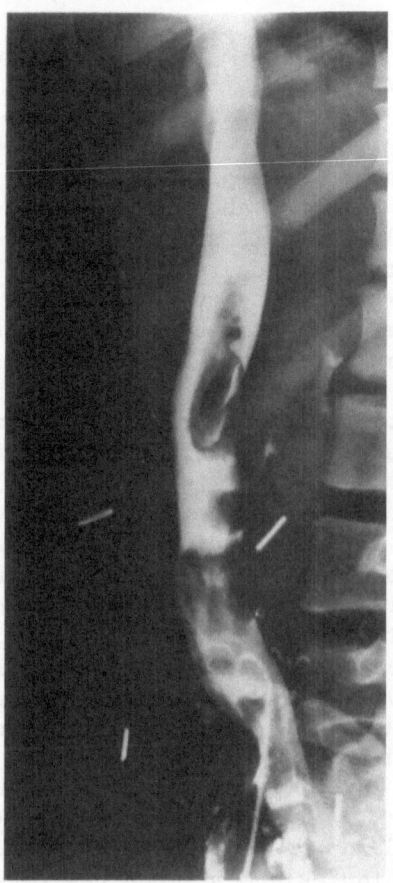

Fig. 4. Inferior venacavogram of a patient with retro-peritoneal testicular tumor causing displacement and filling defects.

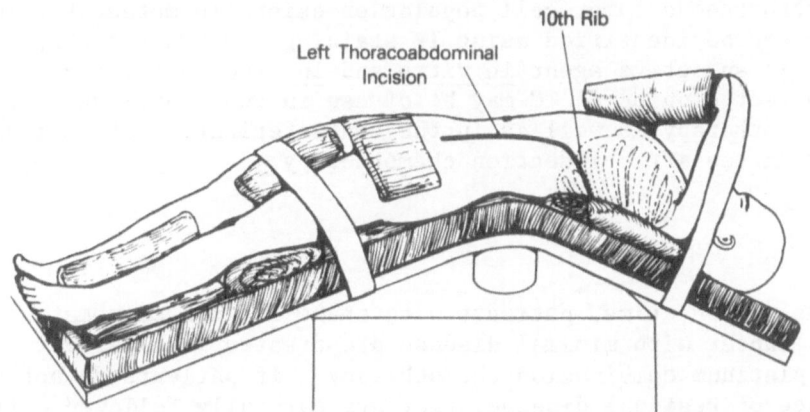

Fig. 5. Position of a patient on operating table for a left thoraco-abdominal incision.

(Figure 6). The tumor is dissected en bloc, and this may require
resection of a part of colon, small intestine, and IVC. If the tumor
is very large, the peritoneum is opened, and the intestines are
packed to the opposite side before the tumor is resected by this
intraperitoneal approach (Figure 7 and Figure 8). The boundary of
the tumor is mapped by titanium clips. We avoid the conventional
metal clips, since they interfere with future computed tomograms.
The closure of the wound is accomplished layer by layer, and a number
30 to 36 chest tube is inserted and kept for 24 hours. Patients are
ambulated early and tolerate this approach very well with minimal
postoperative discomfort. We generally perform clonogenic assay and
chemosensitivity on surgically removed specimens. Persistent residual colony that is removed surgically appears to eliminate several
limitations and problems inherent in clonogenic assays of other
tumors. To investigate this problem we have studied 16 testicular
cancer (TC) specimens obtained surgically at diagnosis or following
chemotherapy which were cloned in soft agar. Seven (41%) of the
specimens cultured formed colonies, with a mean cloning efficiency of
0.021% Colony formation was observed with all the common histologic
subtypes of TC. Three of four specimens demonstrated a decrease in
colony formation to less than 50% of controls after a 1-hour exposure
to VP-16 to 300 microgram per ml (Table 3). Two of these patients
had a response to treatment with a VP-16 based salvage regimen. We
have utilized monoclonal antibodies to cellular AFP and HCG to demonstrate the authenticity of cloned TC cells by immunoperoxidase (IP)
staining of tumor colonies. IP staining of the colonies for AFP and
HCG was correlated with the serum levels of these tumor markers. In
two instances the same markers were elevated in the serum as detected
within tumor colonies. In two cases, the markers(s) elevated in the
serum was not expressed in the colonies. In one case a residual
teratoma mass formed colonies and stained positive for AFP. The
patient subsequently developed an elevated serum AFP. These studies
demonstrate that

1.   TC can be cloned directly in soft agar;
2.   a heterogeneous tumor cell population exists in metastatic TC
     which can be identified using IP staining for HCG and AFP;
3.   VP-16 is an active agent in vitro and in vivo in TC; and
4.   the direct cloning of TC may be of use in the individualization
     of chemotherapy as well as in the characterization of residual
     tumor masses after induction chemotherapy.

SUMMARY

     At the present time, patients with stage III nonseminomatous
testicular cancer with minimal disease are treated with three to four
cycles of platinum combination chemotherapy. If patients do not show
any evidence of residual disease, they are carefully followed. If
they demonstrate evidence of disease in terms of residual tumor,

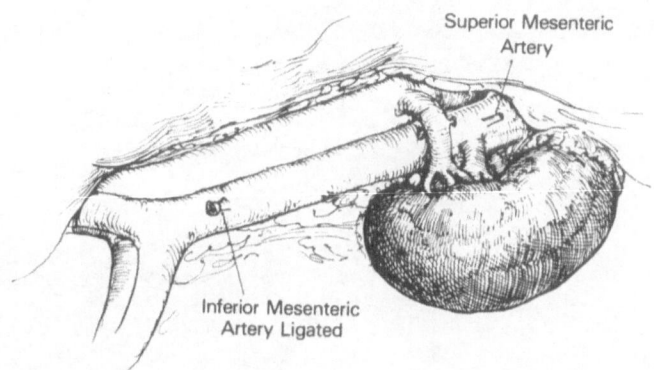

Fig. 6. Aorta and IVC exposed.

Table 3. PVB-VP-16 (n=15) vs PVB (n=7) (2 to 1 Randomization)

Status	PVB-VP-16	PVB
CR	87%	70%
Alive & NED	80%	57%

fibrous and/or necrotic tumor, the mass is surgically removed. In our experience with the conventional chemotherapeutic regimens, one-fourth of these masses have morphologically recognizable tumor, and the only reliable means for diagnosis and therapy is to remove them surgically. If the mass contains tumor, further chemotherapy is considered for these patients.

Patients with massive, bulky, disseminated tumors or patients with bulky extragonadal tumor or those with visceral metastases are treated with three to four cycles of intensive chemotherapy (PVBV) containing VP-16, and, if there is no evidence of disease, they are followed carefully. The residual disease is resected; if it contains tumor, ablative therapy is pursued. The efficacy of ablative therapy in disseminated bulky testicular cancer is being tested in a randomized study. The preliminary results are encouraging. It is important to consider disseminated testicular tumor as minimal, moderate, and maximal disease with presence of poor prognostic features. The first group has a cure rate of more than 95%. However, patients with poor prognostic features are reported to have lower survival rates. The new protocol of high dose platinum in hypertonic saline (40 mg per M^2 x 5 days) and the utilization of VP-16 as a primary agent with vinblastine, bleomycin and autologous bone marrow, if necessary, have

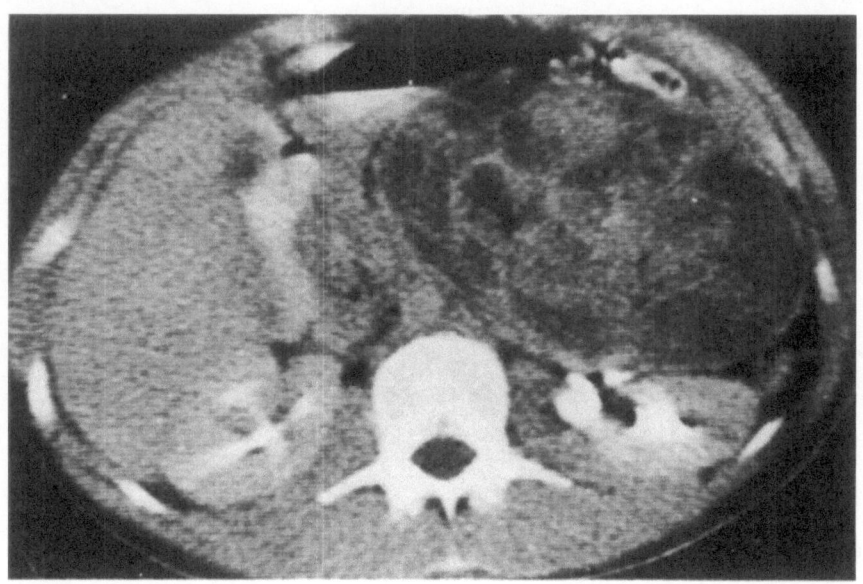

Fig. 7.   An abdominal C.T.   Note a left upper quadrant mass with
cystic components.

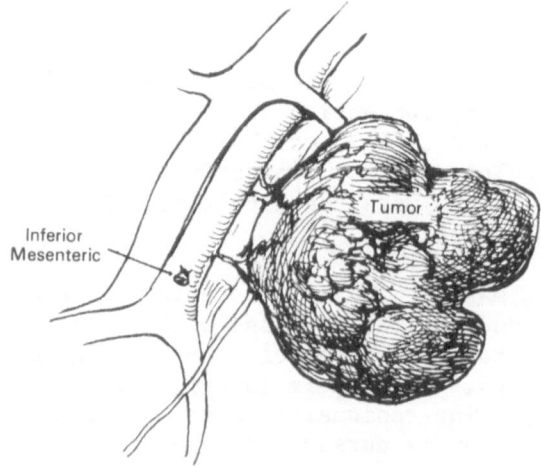

Fig. 8.   A large tumor being dissected from aorta.

made a dramatic improvement in the survival of patients with poor
prognostic features, with an 80% or better complete response in this
program (Table 4).

Table 4.   Testicular Carcinoma Cloning in Soft Agar.

Histologic type	Colony Growth	> 25 colonies per plate	Drug testing
Choriocarcinoma	1/1	1/1	0/1
Embryonal or Mixed	5/7	5/5	3/5
Seminoma	2/3	2/2	1/2
Total	8/11	8/11	4/11

REFERENCES

1. T. Anderson, T. A. Waldmann, N. Javadpour, et al., Testicular germ cell neoplasms. Recent advances in diagnosis and therapy, Ann.Intern.Med., 90:373-385 (1979).

2. R. B. Bracken, D. E. Johnson, O. H. Frazier, et al., The role of surgery following chemotherapy in stage III germ cell neoplasms, J.Uzol., 129:39 (1983).

3. L. H. Einhorn and S. D. Williams, Chemotherapy of disseminated testicular cancer. A random prospective study. Cancer 46:- 1339-1344 (1980).

4. E. E. Fraley, P. H. Lange, and B. J. Kennedy, Germ cell testicular cancer in adults, N.Engl.J.Med., 301:1370-1377 (1979).

5. E. Gehan, A generalized Wilcoxon test for comparing arbitrarily single censored samples, Biometrika 52:203-223 (1965).

6. N. Javadpour, K. R. McIntire, T. S. Waldmann, et al., The role of radioimmunoassay of serum alphafetoprotein and human chorionic gonadotropin in the intensive chemotherapy and surgery of metastatic testicular tumors, J.Urol., 119:759-762 (1978).

7. N. Javadpour, R. F. Ozols, T. Anderson, A. B. Barlock, R. Wesely, and R. C. Young, A randomized trial of cytoreductive surgery followed by chemotherapy versus chemotherapy alone in bulky stage III testicular cancer with poor prognostic feature, Cancer 48:2004-2020 (1982).

8. N. Javadpour, R. F. Ozols, and R. C. Young, Prospective randomized trial of cytoreductive surgery followed by chemotherapy versus chemotherapy alone in bulky disseminated testicular cancer, in: "Proceedings of the American Urological Association," (1982).

9. E. Kaplan, and P. Meir, Non-parametric estimation from incomplete observations, J.Am.Stat.Assoc., 53:457-481 (1958).

10. C. L. Litterst, Alterations in the toxicity of cis-dichlorodiammineplatinum and in tissue localization of platinum as a function of NaCl concentration in the vehicle of administration, Toxicol.Appl.Pharmacol., 61:99 (1981).

11. C. E. Merrin, H. Takita, Wever, et al., Combination of radical
     surgery and multiple sequential chemotherapy for the treat-
     ment of advanced carcinoma of the testes (stage III). Cancer
     37:20-29 (1976).
12. F. R. Ozols, N. Javadpour, G. L. Messerschmidt, and R. C. Young,
     Poor prognosis nonseminomatous testicular cancer: An effect-
     ive high dose cis-platinum regimen without increased renal
     toxicity, ASCO, (1982).
13. M. L. Samuels, P. Y. Holoye, and D. E. Johnson, Bleomycin
     combination chemotherapy in the management of testicular
     neoplasia, Cancer 30:318-326 (1975).
14. D. Vugrin, E. Cvitkovic, E. F. Whitmore, Jr. et al., VAB-4
     combination chemotherapy in the treatment of metastatic
     testis tumor, Cancer 47:833-389 (1981).
15. D. Vugrin, W. E. Whitmore, H. Herr, et al., Indication for
     retroperitoneal lymph node dissection after chemotherapy for
     advanced nonseminomatous testicular cancer, ASCO (1982).

EFFECT OF COMBINATION CHEMOTHERAPY RELATED TO THE
SIZE OF RETROPERITONEAL LYMPH NODE METASTASES IN NON
SEMINOMATOUS GERM CELL TUMORS: A PRELIMINARY REPORT

S. Guazzieri, G. Ferro, A. Lembo, C. Vergani,
P. Sperandio, and F. Pagano

Institute of Urology and Division of Medical Oncology
University of Padua, Italy

Tyrrel and Peckham[1] demonstrated that radiation therapy cannot achieve an acceptable cure rate for patients with metastatic lymph nodes larger than 2 cm in diameter but the combination of intensive chemotherapy with planned surgery had dramatically changed the prognosis of these patients[2-5].

Surgery alone does not achieve a complete control of bulky masses either because of the amount of extranodal invasion or because of the incidence of the extraregional micrometastases[6-7].

The strength of the PVB regimen has been clearly demonstrated[4,5,8] but its effectiveness in relation to the size of the retroperitoneal lymph node metastases had not been well evaluated.

MATERIALS AND METHODS

From 1979 through July 1982 12 consecutive patients with retroperitoneal metastases larger than 2 cm on CAT scan were reviewed, and the largest transverse diameter was determined.

For staging purpose, funicular lymphangiograms were performed only in patients who had undergone orchiectomy in our Institute, but all had a CAT scan of the chest and abdomen as well as an IVP.

Patients with non bulky metastases also had a bipedal lymphangiogram for more exact assessment of the extension of disease.

In selected cases hepatic sonography brain scans, venocavograms and aortograms were carried out. CAT scan of the whole chest and abdomen were repeated at the end of the fourth course of chemotherapy and at the end of treatment.

According to the classification of the Tumor Institute of Milan (Table 1) four patients were placed in stage II B, one patient in stage II C, two patients in stage II D and five patients in stage III.

All patients were treated with from four to six courses of multiple intensive chemotherapy as in the Einhorn schedule.[5] Four courses were given either to patients with side effects requiring cessation of treatment or to those without evidence of disease after the fourth course.

A radical transperitoneal lymphadenectomy as described by Donohoue[9] was performed in all patients from 21 to 30 days after the completion of chemotherapy. No patients in stages lower than stage III had residual disease in the chest at the time of surgery.

Table 1. Staging System in Use in our Institution.

Stage I	:	No evidence of metastases
Ia	:	Tumor is limited to the testis and epididymis
Ib	:	Tumor is invading spermatic cord
Ic	:	Tumor is invading scrotal wall
Stage II	:	Metastases to the subdiaphragmatic nodes only
IIa	:	Nodal metastases with a maximum diameter < 2 cm. and less than 5 in number
IIb	:	Nodal metastases with a diameter > 2 cm (but < 5 cm) or < 2 cm but more than 5 in number
IIc	:	Retroperitoneal metastases with a diameter > 5 cm or with involvement of the retroperitoneal vessels
IId	:	Palpable abdominal mass
Stage III	:	Distant metastases are present
IIIa	:	Metastases to the mediastinal or supraclavicular nodes only
IIIb	:	Hematogenous metastases to the lung only
		- minimal pulmonary disease: less than 5 nodules per lung and none > 2 cm.
		- Advanced pulmonary disease: 5 or more nodules in both the lungs or at least 1 nodule larger than 2 cm, or pleural involvement
IIIc	:	Hematogenous metastases in multiple organ sites

RESULTS

In the group of seven patients with retroperitoneal nodes smaller than 10 cm in diameter the preoperative chemotherapy was curative in 6 (86%) and because of the absence of viable tumor in all the removed tissue we were able to stop all treatment (Figure 1).

The seventh patient had six additional second line chemotherapy courses based on ADM and VP-16 but died with hepatic and pulmonary metastases seven months after retroperitoneal node dissection. This patient had a slight elevation of AFP before surgery.

The other five patients, in whom retroperitoneal metastases were larger than 10 cm in diameter, had viable tumor in pathologic specimens in four cases out of five (80%). Furthermore, three cases out of the five with viable tumor in the retroperitoneal area had negative serum markers at the time of surgery.

The pathology of the metastases relative to the type of primary tumor is reported in Table 2.

All tumor positive cases were treated after surgery with another six courses of chemotherapy as reported before.

All patients are alive and in complete remission after 4-34 months (mean 21) months from the completion of therapy.

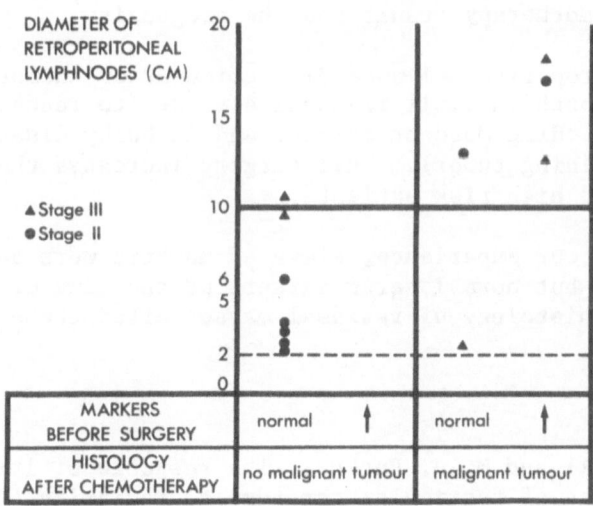

Fig. 1. Relation between the size of retroperitoneal nodes before chemotherapy and the post-therapy pathology.

Table 2. Pathology Post-chemotherapy Compared to Stage at
          Presentation. The pathology is worse in highest stages.

Initial Stage	No. of Patients	Inflammatory Reaction	Fibrosis and Necrosis	Necrosis	Malignant Teratoma	Embryonal Carcinoma
II° B	4	3	1	0	0	
II° D	2				1	1
III° A	1			1		
III° B	4			1	1	2
III° C	1			1		
Total	12	3	1	3	2	3

CONCLUSIONS

It is well known that the non seminomatous testicular tumors are
more responsive to chemotherapy than to radiation therapy and that
the latter cannot cure retroperitoneal metastases larger than 2 cm in
diameter.

In our experience, however limited, combined PVB chemotherapy
was able to cure metastases up to 10 cm in diameter, while patients
with larger abdominal masses had persistent tumor in 80% of cases.
We conclude that a correlation exists between the size of the metastases, the chemotherapy result and the prognosis.

Radical retroperitoneal node dissection after chemotherapy is
still mandatory both in small residual disease (to recognize unre-
sponsive cases needing more treatment) and in bulky disease (to
destroy the remaining tumor). This surgery increases the chance for
recovery in these high risk patients.

Finally, in our experience, elevated markers were an ominous
prognostic sign, but normal serum markers at the time of surgery did
not predict the histology of residual masses after chemotherapy.

REFERENCES

1. C. J. Tyrrell and M. J. Peckham, The response of lymph node
    metastases of testicular teratoma to radiation therapy,
    Brit.J.Urol., 48:363 (1976).
2. T. Anderson, N. Javadpour, R. Schilsky, A. Barlock and R. C.
    Young, Chemotherapy for testicular cancer: current status of
    the National Cancer Institute combined modality trial, Cancer
    Treat.Rep., 63:1687 (1979).

3. W. D. DeWys, C. Begg, R. Slayton, R. G. Hahn, and J. Brodsky, Chemotherapy for advanced germinal cell neoplasma: preliminary report of an E.C.O.G.S., Cancer Treat.Rep., 63:1675 (1979).

4. L. H. Einhorn, Platinum combination chemotherapy in disseminated testicular cancer, in: "Cancer of the Genito-urinary tract," New York, Raven Press, (1979).

5. L. H. Einhorn, Combination chemotherapy with cis-dichlorodiammineplatinum (II), Cancer Treat.Rep., 63:1659 (1979).

6. D. H. Hussey, K. H. Luk and D. E. Johnson, The role of radiation therapy in the treatment of germinal cell tumors of the testis other than pure seminoma, Radiology, 123:175 (1977).

7. E. M. Jacobs and F. M. Muggia, Testicular cancer, risk factors and the role of adjuvant chemotherapy, Cancer, 45:1782 (1980).

8. M. B. Garnick, G. P. Canellos, G. P. Richie and J. J. Stark, Sequential combination chemotherapy and surgery for disseminated testicular cancer, Cancer Treat.Rep., 63:1681 (1979).

9. J. P. Donohue, Retroperitoneal lymphadenectomy, the anterior approach including bilateral suprarenal hilar dissection, Urol.Clin.N.Am., 4:500 (1977).

ROUND TABLE:

CHEMOTHERAPY OF TESTIS TUMORS

G. Stoter

Department of Oncology
Free University Hospital
Amsterdam, The Netherlands

The following questions were discussed:

1) How can we improve the treatment results in patients with a poor prognosis.
2) What are the poor risk factors.
3) How can we reduce the toxicity of chemotherapy without loss of therapeutic effectiveness.
4) What is the chemotherapy of choice in seminoma as compared to non-seminoma.
5) What is the role of adjuvant chemotherapy following retroperitoneal lymph node dissection (RPLNO) in stage II non-seminoma.

In short introductions by Peckham, Stoter and Torti the poor risk factors were defined as the extent of the disease (bulk) and tumor marker levels. Torti made an interesting observation from a multivariate analysis at Stanford University, U.S.A. that the level of β-HCG should decrease to 1/200 of its original level between day 1 and 22 of the treatment. For example, a patient with a serum concentration of β-HCG of 200,000 ng/ml at the start of chemotherapy must have a reduction to 1000 ng/ml at the start of his second treatment cycle (day 22), otherwise he has a poor prognosis. This does not apply to AFP, and Torti speculated that it might also not be true when cisplatin is given over 5 days instead of 1 day as is the practice at Stanford.

The problem of how to improve the treatment results in patients with a poor prognosis was addressed by several participants. Pizzocaro (Int-Milano) presented data to show that with the use of PVB patients with extra-pulmonary hematogenous metastases and those with very bulky disease (abdominal metastases >10 cm, lung metastases

237

>5cm) achieved complete responses in only 33% and 47%, respectively.
These disappointing results led him to replace vinblastine by VP-16
in a dose of 100 mg/m^2 i.v. on days 1-5 in combination with cisplatin
20 mg/m^2 i.v. days 1-5, and bleomycin 100 mg/m^2 i.v. on day 2. With
this combination he achieved durable complete responses in 13 of 14
patients (92%) with very bulky disease. Peckham briefly summarized
the treatment strategy for non-seminomatous tumors at the Royal
Marsden Hospital, U.K. Patients with stage I disease go into a
surveillance study, after orchiectomy. Those who show progressive
disease (19%) and all patients with stages IM-IV (IM = markers persistently elevated after orchiectomy without other signs of metastasis), will receive induction chemotherapy with PEB, which is
cisplatin (P) 20 mg/m^2 i.v. days 1-5, VP-16 (E = Etoposid®) 120 mg/m^2
i.v. days 1-3, and bleomycin (B) 30 mg i.v. on day 2. Earlier in the
morning session Peckham had shown that PEB was as effective as PVB
and less toxic when the 3-day course of VP-16 was used, as compared
to PVB, but he had been unable to show improved results in patients
with bulky disease, in contrast to Pizzocaro. Javadpour referred to
studies at the National Cancer Institute, Bethesda, U.S.A., and by
EORTC aiming for improved results in poor risk patients. These
studies implement the use of high dose cisplatin 40 mg/m^2 days 1-5
with hypertonic saline to prevent renal toxicity, and the use of
alternating non cross-resistant combinations such as PVB and PEB.
Peckham mentioned the new drugs under investigation at the Royal
Marsden with which they hope to see more effect and less toxicity.
JM-8 (CBDCA, a carboplatinum compound) has been shown to be less
nephrotoxic and to cause less vomiting than cisplatin. The drug has
activity in ovarian and breast cancer and at the Royal Marsden testicular cancer patients with bulky disease are now primarily treated
with JM-8 300 mg/m^2 day 1, VP-16 120 mg/m^2 days 1-3, and bleomycin 30
mg day 2. This is the most recent contribution of the Royal Marsden
aimed at an increase of therapeutic results and a decrease of the
toxicity.

Ways to decrease the toxicity were also indicated by Javadpour
(hypertonic saline with cisplatin to reduce nephrotoxicity) and by
Stoter who mentioned studies at the Memorial Sloan Kettering
Hospital, New York and by EORTC in which bleomycin has been omitted
in good risk patients. The results are awaited.

The choice of chemotherapy in seminoma was briefly discussed.
There was general agreement that stage IIC-IV seminoma patients
should receive chemotherapy primarily. Stoter warned that monochemo-
therapy with alkylating agents is obsolete because of the poor results. Javadpour and Torti stated that the chemotherapy of seminoma
should be the same as of non-seminoma. Peckham added that he wished
to consider PEB as the standard regimen instead of PVB, because of
the more severe toxicity of the latter combination. A word of
caution was spoken with regard to the use of bleomycin in patients
who had previously received radiotherapy to the lung, because of the
increased risk of lung fibrosis.

Soloway (University of Tennessee) introduced the subject of adjuvant chemotherapy. He is in favor of RPLND in stage I patients, because in the U.S.A. it will often be impossible to do a close observation because of geographic and other reasons, whereas a RPLND is an operation with low morbidity and no mortality. With regard to resectable stage II disease he strongly supported the present U.S.A. Intergroup Study where patients with resectable stage II are randomized to 2 cycles of PVB or VAB=6 adjuvant chemotherapy or control. Javadpour presented preliminary results of that study in which 88 patients had received adjuvant chemotherapy; none relapsed. In the no treatment arm, 34 of 87 (39%) patients relapsed and 3 died; only 1 due to the cancer. Scardino (Dept. Urology, Baylor College of Medicine, Houston, U.S.A.) asked what the relative proportion of patients with stage II A, B, and C was in both arms, but these data were not available. Scardino also suggested that a future study should be aimed at the question whether it is possible to give less agressive adjuvant chemotherapy to good risk patients. Torti stated that he also supports the Intergroup Study, but for those patients who refuse to be randomized he follows the close observation procedure after RPLND with chemotherapy for relapse. Stoter pointed out that the situation in Europe is different as not many centers perform RPLND. For those who do he would prefer the approach by Pizzocaro who follows the patients closely after RPLND with chemotherapy for relapse. Javadpour added that in patients with stage II A he is definitely in favor of close observation and chemotherapy only for relapsing patients.

At the end of the session there were two more questions from the floor. Altwein (Dept. Urology, Bundeswehrkrankenhaus, Ulm, West Germany) asked Peckham whether he could say how many of the 61 patients in his PEB study (reported earlier in the morning) had abdominal metastases larger than 10 cm. Peckham promised to look this up and to send the information to Altwein. Finally, de Voogt (Dept. Urology, Free University, Amsterdam) drew attention to the heavy psychosocial stress that patients undergo as a result of the chemotherapy. Torti confirmed this and pointed at the particular problems during the first half year after the chemotherapy where the emotional burden is often the most intensive. Gradually, he said, the symptoms will subside when the patients find care and concern in their environment. Peckham spoke of a concentration camp-like syndrome, where emotional experiences tend to come back to the mind of the subject at unexpected moments. Stoter added that for the chemotherapy and the fear of death are not the only severe burdens for the patients, since society can also be cruel to them, mostly hidden under the mask of concern. This holds especially for long-term disease-free survivors who want reintegration in the community.

TESTICULAR FUNCTION AND HORMONE STATUS IN PATIENTS

WITH MALIGNANT GERM TUMORS BEFORE AND AFTER TREATMENT

S. D. Fosså

The Norwegian Radium Hospital
Oslo
Norway

INTRODUCTION

Testicular cancer is predominantly a malignancy of young men. The majority of these patients can be cured today, so that maintenance of fertility has become an important issue. It is, therefore, of interest to evaluate to what extent the various treatment regimes may influence fertility and hormone status in these patients.

The present report deals with the preliminary results of fertility evaluation in patients with testicular cancer before and after treatment.

PATIENTS AND METHODS

Fertility and hormone status were evaluated in 187 patients who were treated for testicular cancer at The Norwegian Radium Hospital between 1977 and 1982. Only those patients were included in whom treatment had resulted in complete remission. Post-treatment analysis was therefore done on disease-free patients. Table 1 gives details of tumor histology and stage-distribution (according to The Royal Marsden classification system)[1].

Treatment

Up till 1979 the primary treatment of seminomatous and non-seminomatous testicular cancer, stage I, II, and III, most often consisted of high-voltage irradiation (seminoma: 40 Gy/4 weeks; non-seminoma: 50 Gy/5 weeks). Patients with stage II and stage III

Table 1.   Histology and Stage in 187
           Patients with Testicular Cancer
           Evaluated for Fertility and Serum
           Hormone Status.

Stage	No. of patients Seminoma	Non-seminoma	Total
I	72	40	112
II	14	37	51
III	1	2	3
IV	2	19	21
Total	89	88	187

also received irradiation to the supraclavicular and mediastinal
lymph nodes. If the patient had undergone previous inguinal surgery
or if there was tumor involvement of the spermatic cord, tunica
vaginalis and epididymis the inguinal region was irradiated as well.
Patients with distant metastases or with recurrent tumor in
irradiated regions were treated with combination chemotherapy consisting of Vincristine, Adriamycin, Cyclophosphamide, Actinomycin C
and Medroxy-Progesterone acetate (VACAM)[2].

After 1978/1979 irradiation was replaced by retroperitoneal
surgery in patients with non-seminoma, clinical stages 1 and IIa[2].
Patients with surgical stage I did not receive additional treatment
after unilateral, retroperitoneal lymph node resection, whereas stage
II patients received adjuvant chemotherapy with Cis-Platinum, Vinblastine and Bleomycin (CVB) (3-4 cycles)[3].

During the last 5 years patients with non-seminoma, stage IIb or
more and patients with seminoma stage IIc or more, have been treated
with 4 induction cycles of CVB, and non-seminoma patients also received maintenance chemotherapy (CVB + Cyclophosphamide + Adriamycin
+ Vincristine) and consolidation chemotherapy (CCNU + Vinblastine)[4]. Surgery and radiotherapy were initially used as second and
third line treatment in non-seminoma patients. During a later period
surgery replaced routine consolidation and maintenance chemotherapy
in the majority of patients[5].

Initially, in patients with seminoma (stage IIC and IV) radiotherapy was used as consolidation treatment after CVB as often as
possible. Later on, post-chemotherapy surgery (after CVB) was used
more frequently instead of consolidation treatment with radiotherapy.

Table 2 shows the different treatment modalities used for the
187 patients.

<u>Evaluation of hormone status and fertility</u>. Serum lutenizing hormone (LH), follicle stimulating hormone (FSH) and Testosterone levels were determined by previously published methods [6]. Normal ranges were LH: 2-12 U/1; FSH: 2,8-12 U/1; Testosterone: 12-40 nmol/1.

Sperm analysis was done on semen obtained by masturbation after at least 3 days of sexual abstinence. The ejaculate was examined for the number of sperm cells per ml, the motility index (percentage of motile sperm x degree of motility, grades 0-4), and the percentage of morphologically abnormal sperm cells. The results were recorded either as <u>azoospermia</u>, <u>highly impaired sperm count</u> (< 10 million sperm cells/ml and/or motility index < 20, and/or 80% amorphous sperm cells), or <u>normospermic</u>.

Fatherhood was determined either by the patients report or by consulting the Norwegian Birth Registry, which identifies the parents of all children born in Norway after 1966.

RESULTS

Patients After Orchiectomy and Before Further Treatment

Sperm and hormone analysis were done in 59 orchiectomized patients before further treatment. There were 20 who were normospermic, 30 had highly impaired sperm counts and 9 patients were azoospermic (Table 3). Serum LH was increased above the normal range in 12 patients, most often due to β-HCG production in metastatic tumors. Only 6 patients, all of whom were either azoospermic or had impaired sperm counts, showed slightly increased FSH-values. In

Table 2. Treatment in 187 Patients with
Testicular Cancer Evaluated for
Fertility and Serum Hormone Status

Treatment	No. of patients
RLND* only	28
unilateral	27
bilateral	1
Irradiation only	84
CVB ± RLND	38
CVB + other cytostatics	10
CVB + irradiation	15
CVB + irradiation + other cytostatics	12
Total	187

* Retroperitoneal lymph node dissection.

Table 3.   Sperm Analysis and Serum Hormone Status in 59 Patients
           after Orchiectomy for Testicular Cancer and Before Further
           Treatment.

a) "Normospermic" patients (20)

Analysis	Mean	No of levels above* and below** the normal range
LH [U/1]	9.8	2
FSH [U/1]	5,2	0
Testosterone [mmol/1]	16,3	6

b) Patients with impaired spermatogenesis (30)

Analysis	Mean	No of levels above* and below** the normal range
LH [U/1]	14,6	6
FSH [U/1]	6,3	4
Testosterone [mmol/1]	19,7	3

c) Azoospermic patients (9)

Analysis	Mean	No of levels above* and below** the normal range
LH [U/1]	25,8	4
FSH [U/1]	8,6	2
Testosterone [mmol/1]	14,6	4

d) All patients (59)

Analysis	Mean	No of levels above* and below** the normal range
LH [U/1]	14,7	12
FSH [U/1]	6,3	6
Testosterone [mmol/1]	17,8	13

 * Serum LH/FSH
** Serum Testosterone

general, the serum testosterone values were in the lower levels of
the normal range, but testosterone values were subnormal in 13 cases.

After Treatment (all types) (Figure 1)

     Elevations of mean FSH values after treatment were observed most
often and were most pronounced in azoospermic patients, and to a
lesser degree, in patients with highly impaired spermatogenesis. In
normospermic cases and in those who achieved fatherhood during the

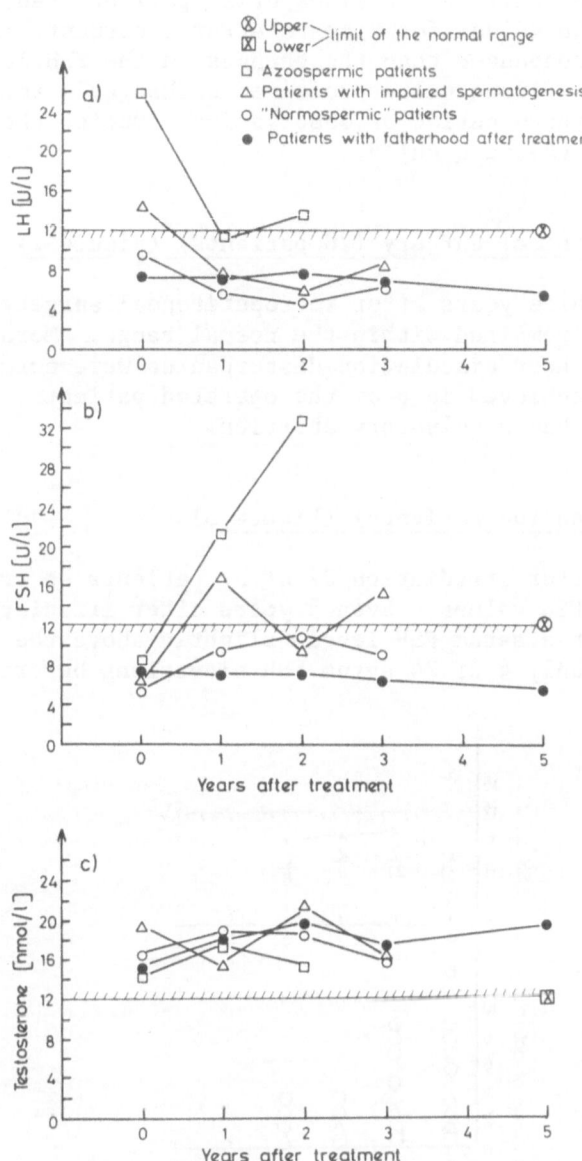

Fig. 1.   Mean serum hormone levels in orchiectomized patients with
          testicular cancer before and after further treatment.
          a) LH,   b) FSH,   c) Testosterone.

observation period the mean FSH values remained within the normal
range, when evaluated 1, 2, 3 and 5 years after discontinuation of
treatment.

In addition, half of the azoospermic patients had slightly elevated serum LH values for 2 years after treatment, but this tendency was less pronounced than the changes in the FSH levels. The mean serum testosterone values remained unchanged after treatment during the 1-5 years period of observation. During the observation period 15 men fathered a child.

After Retroperitoneal Surgery (28 patients) (Figure 2)

One, two and three years after retroperitoneal surgery alone the mean FSH values remained within the normal range. More than 50% of the patients without ejaculation disturbances were normospermic. Fatherhood was achieved in 6 of the operated patients. The wife of the 7th patient had a voluntary abortion.

After irradiation (84 patients) (Figure 3)

One year after irradiation 27 of 37 patients so treated had elevated serum FSH values. Even 5 years after irradiation 10 of 26 irradiated men had serum FSH levels slightly above the normal range as compared to only 4 of 24 serum FSH elevations before radiotherapy.

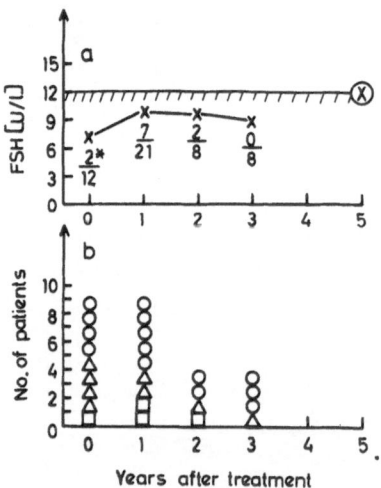

Ⓧ Upper limit of the normal range
☐ Azoospermic patients
△ Patients with impaired sperm analysis
◯ "Normospermic" patients
✳ No. of patients with elevated/evaluated FSH levels

Fig. 2. Mean serum FSH levels (a) and sperm analysis (b) in ochiectomized patients with testicular cancer before and after retroperitoneal surgery.

Nevertheless, there was a clear tendency towards a decrease in the serum FSH values after the first post-treatment year.

Only 3 of 23 men were normospermic one year after irradiation. The gonadal dose, which was measured in 69 patients (range: 40-160 cGy; mean 61 cGy) could not be correlated with the serum FSH elevations or the degree of disturbances of spermatogenesis. Fatherhood was achieved by 6 irradiated patients.

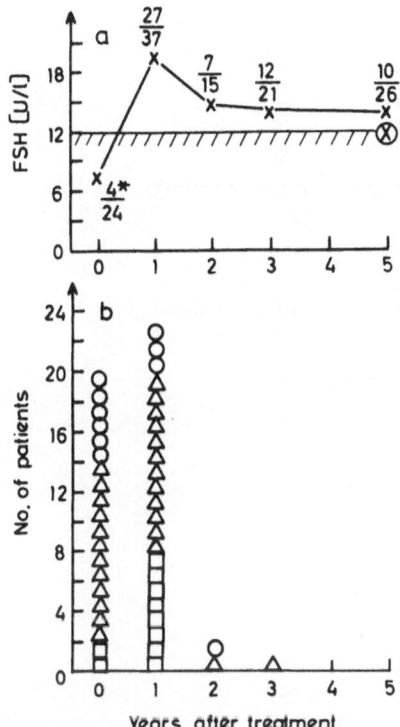

Mean serum FSH levels (a) and sperm analysis (b) in orchiectomized patients with testicular cancer before and after irradiation.

⊗ Upper limit of the normal range
☐ Azoospermic patients
△ Patients with impaired sperm analysis
○ "Normospermic" patients
* No. of patients with elevated/evaluated FSH levels

Fig. 3. Mean serum FSH levels (a) and spermatogenesis in orchiectomized patients with testicular cancer before and after irradiation.

After CVB +/- retroperitoneal surgery (38 patients) (Figure 4)

One year after discontinuation of CVB chemotherapy the serum FSH
value was elevated in 24 of 31 patients. There was a gradual decrease of the mean serum FSH values after the first year. One year
after CVB treatment, signs of spermatogenesis were observed in four
of seven patients: "normospermia" in two patients and impaired
spermatogenesis in the other two. One additional patient fathered a
child 24 months after discontinuation of CVB.

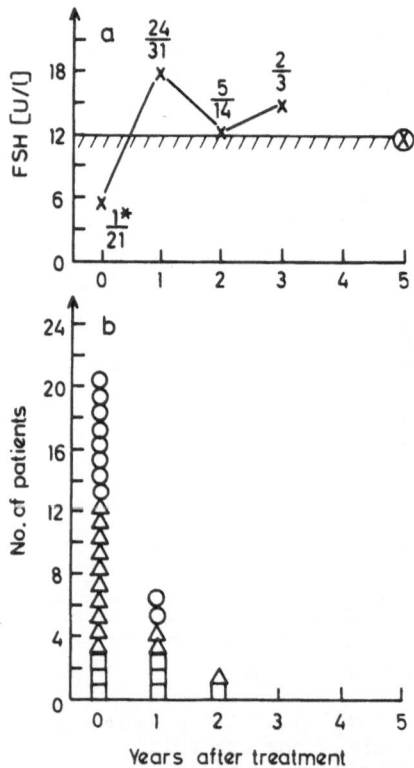

⊗ Upper limit of the normal range
□ Azoospermic patients
△ Patients with impaired sperm analysis
○ "Normospermic" patients
* No. of patients with elevated/evaluated FSH levels

Fig. 4.   Mean serum FSH levels (a) and sperm analysis (b) in
          orchiectomized patients with testicular cancer before and
          after CVB chemotherapy + retroperitoneal surgery.

After CVB and/or irradiation and/or cytostatic treatment (37 patients) (Figure 5)

There were marked serum FSH elevations and azoospermia in the majority of patients during the first 3 years after this treatment. There was no tendency to reduction of the mean serum FSH levels. However, occasionally one could observe signs of spermatogenesis in two individuals. In addition, 29 months after discontinuation of 6 cycles of CVB, two of which were combined with Adriamycin and Cyclophosphamide, one patient fathered a healthy child.

DISCUSSION AND CONCLUSIONS

As recorded by other authors[7] a significant number of patients with malignant germ cell tumors have reduced spermatogenesis after orchiectomy and before further treatment. In accordance with these observations, increased serum FSH values may be observed in some of

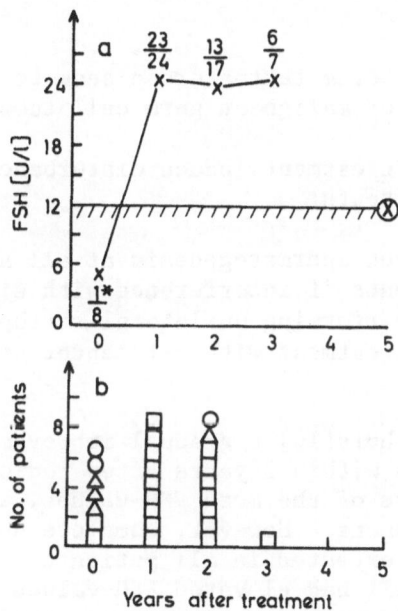

⊗ Upper limit of the normal range
□ Azoospermic patients
△ Patients with impaired sperm analysis
O "Normospermic" patients
* No. of patients with elevated/evaluated FSH levels

Fig. 5. Mean serum FSH levels (a) and sperm analysis (b) in orchiectomized patients with testicular cancer before and after CVB chemotherapy and irradiation/other cytostatic treatment.

these males. A low testosterone level also indicates a testicular hypofunction existing before further treatment. These observations, however, should not always be interpreted as signs of permanent hypo- or infertility. In our experience only 1 of 5 patients who sub- sequently achieved fatherhood were normospermic after orchiectomy and before further treatment. It is questionable how far a single sperm analysis, often taken under conditions of extreme stress, really mirrors the fertility status. The serum FSH level is probably less influenced by temporary environmental factors and may therefore be a better indicator of existing disturbances of spermatogenesis. This is supported by the observation that only one (slightly) elevated pre-treatment serum FSH level was found among 26 samples from patients who achieved fatherhood, whereas 26 of 36 serum FSH levels were raised from patients who were azoospermic. It is concluded from these results that an increased FSH value very probably indicates disturbed spermatogenesis. One of the reasons for the serum FSH increase during highly disturbed or absent spermatogenesis is the reduction of Inhibin, a hormone produced in the testicle during spermatogenesis[8]. Inhibin normally regulates the FSH-production via a feed-back mechanism.

Neither serum LH nor serum testosterone seem to be significantly affected by the treatment of malignant germ cell tumors.

The various types of treatment induce disturbances of sperm- atogenesis to different extents:

Surgery does not affect spermatogenesis at all and fertility can be preserved in many patients if interference with ejaculation can be avoided by, for example, performing unilateral retroperitoneal lymph node dissection[3] or by treatment with α-receptor stimulating drugs[9].

As shown by other authors[10] a gradual improvement of sperm- atogenesis is usually seen within 3 years after radiotherapy. This is combined with a decrease of the mean FSH-values, which reach their peak 12 months after treatment. However, complete normalization of spermatogenesis cannot be expected in all patients. As many as 10 of 26 irradiated patients still had elevated FSH-values 5 years after radiotherapy, indicating disturbed spermatogenesis in most of them.

In contrast to previously available chemotherapy for testicular cancer which usually contained alkylating agents[11], the CVB regimen seems to disturb spermatogenesis less and often only temporarily. The effect of CVB chemotherapy on spermatogenesis seems to be similar to that observed after irradiation; a decrease of spermatogenesis for 1 year after treatment is followed by a gradual recovery. Further observations in a larger series are desirable to confirm these pre- liminary results.

There is sufficient indication today that spermatogenesis will recover within the first three years after treatment in a significant number of patients treated with 3-4 courses of CVB chemotherapy[12]. However, if CVB chemotherapy is combined with radiotherapy and/or other types of cytostatic drugs, the risk of more permanent and more severe disturbance of spermatogenesis increases markedly. In our series this is indicated by long-lasting elevations of serum FSH in most such patients, and by only occasional signs of spermatogenesis. If possible, combinations of CVB chemotherapy and radiotherapy/other types of chemotherapy should therefore be avoided, at least in good-risk patients and in patients in whom preservation of fertility is important as a secondary objective of the treatment.

Our study also indicates that the question of post-treatment fatherhood and fertility should be considered when the primary treatment is discussed with the patient. The fact that 6 of 28 men became fathers within 3 years after surgery is evidence that a significant number of these young patients want to have children. Modern treatment of testicular cancer therefore should include this aspect of therapy and avoid "dry ejaculation" and azoospermia whenever possible.

REFERENCES

1. M. J. Peckham, A. Barrett, T. J. McElwain, and W. F. Hendry, Combined management of malignant teratoma of the testis, The Lancet, August II, (1979).
2. O. Klepp, R. Klepp, H. Host, G. Asbjørnsen, K. Talle, and A. E. Stenwig, Combination chemotherapy of germ cell tumors of the testis with vincristine, adriamycin, cyclophosphamide, actino-mycin D and medroxyprogesterone acetate, Cancer, 40:638 (1977).
3. S. D. Fosså, O. Klepp, S. Ous, H. H. Lien, A. E. Stenwig, T. Abyholm, and O. Kaalhus, Unilateral retroperitoneal lymph node dissection in patients with non-seminomatous testicular tumor in clinical stage I, Europ.Urol., submitted (1983).
4. O. Klepp, S. D. Fosså, S. Ous, H. Lien, J. T. Stenwig, V. Abeler, G. Eliassen, and H. Høst, Multi-modality treatment of advanced malignant germ cell tumors in males, I: Experience with cis-platinum-based combination chemotherapy, Scand.J. Urol.Nephrol., in press (1983).
5. S. D. Fosså, O. Klepp, S. Ous, H. Lien, J. T. Stenwig, V. Abeler, G. Eliassen, and H. Høst, Multi-modality treatment in males with advanced malignant germ cell tumors, II: Experience with surgery and radiotherapy following cis-platinum-based chemotherapy, Scand.J.Urol.Nephrol., submitted (1983).

6. S. D. Fosså, O. Klepp, K. Molne, and A. Aakvaag: Testicular
    function after unilateral orchiectomy for cancer and before
    further treatment, Int.J.Andrology, 5:179-184 (1982).
7. J. G. Berthelsen and N. E. Skakkebaek, Gonadal function in men
    with testis cancer, Fertility and sterility, 39:68-75 (1983).
8. B. P. Setchell, S. J. Main, and R. V. Davies, Effect of ligation
    of the efferent fucts of the testis on gonadotrophins and
    testosterone in rats, J.Endocrin., 72:13-14 (1977).
9. J. H. Lynch and W. C. Maxted, Use of ephidrine in
    post-lymphadectomy ejaculatory failure: A case report,
    J.Urol., 129,139 (1983).
10. E. W. Hahn, S. M. Feingold, L. Simpson, and M. Batata, Recovery
    from aspermia induced by low-dose radiation in seminoma
    patients, Cancer 50:337-340 (1982).
11. S. D. Fosså O. Klepp, A. Aakvaag, and K. Molne, Testicular
    function after combined chemotherapy for metastatic testicular
    cancer, Int.J.Andrology, 3:59-65 (1980).
12. J. Donohue, R. Drasga, S. Williams, L. Einhorn, D. Patel, and E.
    Stevens: Gonadal function after chemotherapy in testicular
    cancer patients, The XIX Int.Congress of the Société Inter-
    nationale d'Urologie, San Francisco, September 173:5-10
    (1982).

FERTILITY IN PATIENTS WITH CANCER OF THE TESTIS

BEFORE AND AFTER TREATMENT: FERTILITY AFTER SURGERY

A. Jardin

Hopital La Pitié
Paris
France

The effects of treatments for cancer of the testis on fertility throw a shadow over the remarkable results which can be obtained with surgery, radiotherapy and above all chemotherapy. This is made more distressing by the fact that cancer of the testis occurs in young men, most of whom have not yet had children. This delicate problem has become an important issue, especially over the last few years. The study of effects of surgery on fertility in men with cancer of the testis is biased from the start, since this cancer occurs in the gland responsible for fertility and develops at the expense of the germinal cells. Few studies have examined the fertility of men with cancer of the testis just before their tumor was discovered. We simply know that the rate of "paternity" in relation to age is essentially the same as for a normal population. It is reasonable to think that the cancer alters spermatogenesis on the affected side and perhaps on the opposite side by an unexplained mechanism. Examination of the seminiferous tubules of orchidectomy specimens for small cancers very often reveals a marked alteration of spermatogenesis that may be due simply to a mechanical action.

THE EFFECT OF ORCHIDECTOMY

In view of these reservations, it is difficult to evaluate the exact role of orchidectomy. However, after orchidectomy and prior to any other form of treatment, only 20% of patients are considered to have a normal fertility (more than 30 million spermatozoa, more than 50% mobile forms). These figures were observed in our series as in others[1,2,3].

The possibility of the storage of sperm before any form of treatment liable to aggravate the loss of fertility-lymphadenectomy, radiotherapy and chemotherapy should not be ignored. Sperm can be stored provided that, after one hour of freezing, there remains more than 100,000 mobile spermatozoa in the sample stored. Fifty four per cent of the samples studied by Czyglik,[2] including the patients from our series, fulfilled these criteria.

EFFECT OF LYMPHADENECTOMY

Lymphadenectomy, when it is extensive and bilateral, eliminates the sympathetic chains which carry the fibers which innervate not only the neck of the bladder, but also the smooth muscle of all of the spermatic excretory duct, i.e. this often results in anejaculation[4], or retrograde ejaculation[3,5]. It is not always easy to distinguish between anejaculation and retrograde ejaculation. The demonstration of spermatozoa in the urine after masturbation or sexual intercourse is a sign of retrograde ejaculation and the persistance of spermatogenesis. The presence of fructose or lactoferrin reflects secretion of the seminal vesicals.

It is difficult to quantify the importance of the sympathetic effects for several reasons: the extent of the lymphadenectomy varies from one surgeon to another and there are also major anatomical differences from one patient to another[3]. The reversibility of the disorder is assessed differently. Spontaneous return to normal ejaculation can be seen in 5%[3] to 45%[6] of cases and can occur between 3 months and 5 years after the operation. It is therefore difficult to give a prognosis for any particular patient.

Various "treatments" have been proposed including ejaculation with a full bladder, alpha-sympathomimetics such as ephedrine or derivatives and low doses of imipramine (large doses can cause anejaculation in some cases, in normal subjects)[6,7].

In the few cases which we have treated, we have not obtained the results reported in these publications and it seems that extensive lymphadenectomy is not without its disadvantages.

Let us hope that the increasing effectiveness of chemotherapy will minimize the need for operation and that where surgery is required, storage of sperm prior to treatment may become routine practice.

REFERENCES

1. R. B. Bracken, and K. O. Smith, Is semen cryopreservation helpful in testicular cancer? Urology, 15:581 (1980).

2. F. Czyglik, J. Auger, M. Albert, and G. David, L'autoconserv-
    ation du sperme avant thérapeutique stérilisante, <u>Nouv.Presse
    Med.</u>, 11:2749 (1982).

3. O. G Skinner, A. Melamud, and G. Lieskovsky, Complications of
    thoraco-abdominal retroperitoneal lymph node dissection,
    <u>J.Urol.</u>, 127:1107 (1982).

4. K. R. Kedia, C. Markland, and E. E. Fraley, Sexual function
    after high retroperitoneal lymphadenectomy, <u>Urol.Clin.North
    Am.</u>, 4:523 (1977).

5. P. Narayan, P. H. Lange, and E. E. Fraley, Ejaculation and
    fertility after extended retroperitoneal lymph node dis-
    section for testicular cancer, <u>J.Urol.</u>, 127:685 (1982).

6. J. M. Nijman, S. Jager, W. Boer, J. Kremer, J. Oldhoff, and H.
    Schraffordt Koops, The treatment of ejaculation disorders
    after retroperitoneal lymph node dissection, <u>Cancer</u>, 50:2967
    (1982).

7. D. Jonas, P. Linzback, and W. Weber, The use of midodrin in the
    treatment of ejaculation disorders following retroperitoneal
    lymphadenectomy, <u>Eur.Urol.</u>, 5:184 (1979).

TREATMENT OF TERATOMA AND EMBRYONAL CELL CARCINOMA

E. W. Vahlensieck

Bonn

A register and multicenter study for testicular tumors was founded in Bonn in September 1976. By August 1981 data from a total of 1508 patients with testicular tumors from 46 institutions partici- pating in Austria and the Federal Republic of Germany have been collected and evaluated.

In 970 cases the pathohistological classification was carried out by both the local pathologist and by our central reference path- ology. With regard to the validity of the histological classifi- cation, it should be noted that when, in addition to optimal workup, every tumor is appraised both by the local institute of pathology and by a reference pathology department with special experience improved results are achieved. Such cooperation chiefly helps to establish the diagnosis especially in the 10-20% of cases which are difficult to classify. In the interest of international comparability, all tumors in our register were classified histologically according to Dixon and Moore[1], Pugh[2] and Mostofi and Sobin[3]. This involved a great deal of hard work for our pathologists and it appears desir- able that the WHO classification is used throughout the world in the future.

As table 1 shows pure teratomas (mature and immature) are rare (17 cases) (1.9% of all germinal cell tumors, 1.7% of all testicular tumors). Pure mature teratomas (group III according to Dixon and Moore, TD according to Pugh) are rare tumors[4] and tend to occur without metastases. This suggests a policy of 'orchiectomy-only' with close follow up and further treatment in case of relapse.

More common was the combination of embryonal cell carcinoma with teratoma (group IV according to Dixon and Moore, MTI according to

257

Table 1. Histological distribution in 970 testicular tumors and
          proportion in children.

Histology	Total No.	Children-(up to 16 years)	
1. Germinal cell tumors	887	24	(2.7)
seminoma	362	-	
yolk sac tumor	6	3	
teratoma	17	7	
embryonic carcinoma	99	1	
chorionic carcinoma	2	-	
embryonic carcinoma + teratoma	79	2	
chorionic carcinoma + other	75	8	
other combinations	247	3	
2. Nongerminal tumors	83	11	(13.3)
stroma tumors	24	2	
gonadoblastoma	2	-	
malignant lymphoma	22	1	
secondary tumor	2	-	
adenomatoid tumor	13	0	
rhabdomyosarcoma	3	3	
other sarcomas	6	3	
unclassifiable	2	0	
repidermoid cyst	9	0	
3. Total	970	35	(3.6)

Pugh). We found this combination in 79 cases (8.9% of all germinal
cell tumors, 8.1% of all testicular tumors). The treatment options
in the management of these tumors are the same as for the other
non-seminomatous tumors. That is also the case for pure embryonal
carcinomas which we found in 11.2% of all germinal cell tumors and in
10.2% of all testicular tumors.

Difficulties in the exact evaluation of treatment results in
spite of the same treatment are to be expected in that we have regis-
tered 247 patients, with variable histological combinations, (27.8%
of all germinal cell tumors, 25.5% of all testicular tumors).

A very important additional point is the treatment of testicular
tumors in childhood. As table 1 shows among a total of 970 patients
with testicular tumor, there were 35 children (3.6%). One may thus
assume that the incidence of testicular tumors in childhood is also
correspondingly low in the entire population. Of interest is the
distribution - 24 germinal cell tumors (2.7% of all germinal cell
tumors, 2.5% of all testicular tumors) and 11 nongerminal tumors
(13.3% of all nongerminal tumors, 1.1% of all testicular tumors).

Table 2. Follow up (median observation time 19 months) in
55 stage I testicular tumor patients after
modified RLND. In the 4 patients with
progression complete remission was achieved with
chemotherapy (2 cases) or surgery and chemotherapy (2 cases).

No	NED	Progression
55	51	4
		(after 3, 7, 11, 14 months)

The high incidence of pure teratomas, which we found in 29% of
all germinal cell tumors in childhood, is well known. In these cases
we prefer 'orchiectomy-only' with close follow up and further treatment in cases of relapse. In all other cases of non-seminomatous
testicular cancer in children we have so far carried out lymphadenectomy and adjuvant chemotherapy if retroperitoneal metastases were
found. For the future we think that it is reasonable in clinical
stage I disease to do 'orchiectomy-only' with close follow up and
further treatment in cases of relapse. If we are to carry out 'orchiectomy-only' in stage I patients we should try to find out criteria to identify the risk group of 15-20% who relapse under these
conditions[4]. If the proximal spermatic cord is not free of tumor
or elevated serum alphafetoprotein and/or human chorionic gonadotropin levels do not revert rapidly to normal after excision of the
primary tumor we think it is necessary to do lymphadenectomy or
deferred chemotherapy. For the other cases it has to be emphasized
that a high degree of accuracy in clinical and pathohistological
staging, and the feasibility of meticulous follow-up are absolutely
indispensable prerequisites for the wait and watch policy.

In recent years in clinical stage I patients we have carried out
a limited, i.e. ipsilateral form of retroperitoneal lymph node dissection (RLND) which reduces the morbidity and preserves the inferior
hypogastric smypathetic plexus with prevention of the loss of ejaculation in most cases. If immediate histological examination of
resected lymph nodes during surgery of the regions where metastases
first occur is negative the lymphadenectomy is limited to these
regions, i.e. the paraaortic region, the left testicular vein and the
area of the left iliac vessels, if the primary tumor was in the left
testicle. For right-sided tumors para- and precaval and intercavoaortic, lymphadenectomy is undertaken together with extirpation of
the right testicular vein and dissection in the region of the right
iliac vessels.

Table   3.   Follow up (median observation time 24 months) in 24
             patients with bulky disease testicular tumors after
             chemotherapy and salvage-lymphadenectomy. 2* patients are
             currently NED after a second-look-lymphadenectomy.

Primary	No.	Fibrosis	Teratoma	Malignant	NED	AWD[1]	DOD[2]	DWD[3]
T1–4 N3 M0	7	3	1	3	5	–	1	1
T1–4 N3 M1	17	9	5	3	10	2*	5	–

1. alive with disease
2. dead of disease
3. dead with disease

        Between 1977 and 1982 we have had 55 patients without evidence
of metastases at the time of the modified RLND. These patients did
not receive any further therapy and were closely followed up. Median
observation time was 19 months. 51 patients remained with no evidence of disease (NED). Four patients relapsed 3, 7, 11 and 14
months respectively after surgery. Two cases with pulmonary progression in January 1983 are still under therapy. The third patient
showed inguinal and supraclavicular metastases 11 months after modified RLND. Treatment consisted of induction therapy with 4 courses
of vinblastine, bleomycin and cisplatinum (VBP), surgery, and reinduction therapy. Complete remission (CR) was achieved, and has so
far lasted for 19 month. The fourth patient had pulmonary and mediastinal metastases 14 months after RLND. CR was achieved with 4
courses VBP plus Ifosfamid and the patient has now been NED for 22
months.

        For stage II patients we recently organized a prospective multi-
center study. If on RLND a solitary metastasis with a diameter below
2 cm (Stage IIA) is found radical RLND will be done after which
patients are randomized to follow up without further therapy in one
group and to treatment with 2 courses of VBP in the other group. If
progression occurs induction or reinduction (further) chemotherapy
will be given and eventually a second-look-lymphadenectomy.

        If on RLND one or more metastases with a diameter between 2 and
5 cm (Stage II B) are found radical RLND will be done with subsequent
randomization to treatment with 2 or 4 courses of VBP.

        In patients with bulky disease we always start with chemotherapy
followed by salvage-lymphadenectomy and if necessary with reinduction
chemotherapy. As table 3 shows the results of chemotherapy in these
cases are really encouraging and may be improved by second-look-
lymphadenectomy as well as by developing highly effective reinductive
chemotherapy.

REFERENCES

1. F. J. Dixon, R. A. Moore, Tumors of the male sex organs, Atlas
    of tumor pathology, Washington, Armed Forces Inst.Pathol.,
    (1952).
2. R. C. B. Pugh, Pathology of the testis, Oxford, Blackwell Sci.
    Publ. (1976).
3. F. K. Mostofi, L. H. Sobin, Histological typing of testis tumors
    in: "Internat. Histol. Classif. of Tumors, No. 16," Geneva
    World Health Organization (1977).
4. M. J. Peckham, Non-seminomatous germ cell tumors of the testis:
    treatment options in: " Germ Cell Tumors," p.69 C. K.
    Anderson, W. G. Jones, A. M. Ward, eds, Taylor A Francis,
    London (1981).

EXPERIENCE WITH TREATMENT OF NONSEMINOMATOUS

GERM CELL TUMORS OF TESTIS IN TURKEY

Atif Akdas, Dogam Remzi, Ali Ergen,
Ziya Kirkali, and Yalcin Ilker

Department of Urology, Hacettepe University
Ankara, Turkey

INTRODUCTION

Nonseminomatous testicular tumors formerly had a very poor prognosis[1] but in the last 3 decades great progress has been made in the diagnosis, surgery and chemotherapy of these neoplasms. Although a large percentage can now be cured with proper treatment and follow-up, too many patients are still dying because of this disease.

In this article we present a review of our recent experience, which is unique in Turkey, in the treatment of nonseminomatous testis tumors.

MATERIALS AND METHODS

From September 1979 to September 1982, 26 patients with non-seminomatous germ cell tumors were evaluated histopathologically and treated at the University of Hacettepe Departments of Urology and Pathology in Ankara, Turkey.

The age of the patients ranged from 20 to 46 years; 17 (65%) had their orchidectomy elsewhere before referral to our hospital.

Pretreatment and follow-up evaluation consisted of complete physical examination, complete blood count, platelet count, urinalysis, chest x-ray, serum calcium, uric acid, total bilirubin, alkaline phosphatase, serum transaminases (SGOT, SGPT), lactate dehydrogenase (LDH), blood urea nitrogen (BUN), serum electrolytes, excretory urogram, abdominal ultrasound. In addition, most patients had a bipedal lymphangiogram and whole lung tomograms as part of their initial evaluation.

Serum tumor markers; alpha-fetoprotein (AFP) and beta-subunit human chorionic gonadotropin (β HCG) were followed in all patients.

The patients who had chemotherapy as primary or additional treatment also had pulmonary function tests, audiological tests and creatinine clerances. Each patient was followed and evaluated monthly in the first year, two monthly in the second year and then every three months.

All patients were staged according to the following scheme:

Stage   I - tumor confined to the testis
Stage II - tumor extends beyond the testis but not beyond the retroperitoneal draining lymph node region
Stage III - metastatic disease beyond the retroperitoneum, and/or palpable abdominal mass or metastases in abdominal organs.

Bilateral retroperitoneal lymph node dissection (RPLND) as described by Staubitz et al.,[2] and Mallis and Paton[3] were performed in Stage I and Stage II disease and in one patient who had stage III disease with a single lung metastasis.

The histopathological and clinical staging of these 26 patients are outlined in Table 1.

Six patients who had stage II disease (>5 lymph nodes) had chemotherapy after RPLND. Cis Platinum, vinblastine and bleomycin were used as the chemotherapy regimen for 4 cycles in stage II and 4-6 cycles in stage III disease (Table 2).

Response to treatment was recorded as complete remission (CR) when all measurable disease disappeared and partial remission when there was a measurable decrease in metastatic disease.

The patients who had partial remission (PR), stable disease or progression after chemotherapy regimen, had 4 cycles of Cis Platinum and VP-16213 (Vepeside) as described in Table 2.

RESULTS

The median follow-up of patients with stage I disease is 24 months and all the 11 patients (100%) remained disease free.

Five of the six patients with stage II disease were disease free for a median follow-up of 30 months; one patient had progression to

Table 1. Histological and Clinical Staging of the Tumors

Stage	Number of Patients	Embryonal Carcinoma	Terato-carcinoma	Chorio-carcinoma	Mixed Type
I	11 (42.3%)	2	4	–	5
II	6 (23.1%)	1	1	–	4
III	9*(34.6%)	5	1	1	2
Total	26 (100%)	8	6	1	11

*All patients but one had clinical staging.

Table 2.   Chemotherapy Regimens

A. Regimen 1
   Cis Platinum (P):   20 mg/m² I.V.   days 1-5   repeat
   Vinblastine   (V):   0.15 mg/kg.I.V. days 1 and 2   every 3 weeks
   Bleomycin   (B):   30 mg. I.V.   day 2   for 12 weeks

B. Regimen 2
   Cis Platinum :   20 mg/m² I.V.   days 1-5   repeat
   Vepeside   :   100 mg/m² I.V.   days 1,3 and 5   every 3 weeks

stage III disease after RPLND. He had a partial remission after chemotherapy and then progressed again despite further chemotherapy and eventually died of carcinomatosis 21 months after orchiectomy.

Two of the nine patients with stage III disease remained disease free. One of three patients with a complete remission clinical response, subsequently progressed. Also one patient had a partial remission. Five patients who had progression and the above mentioned patient who had progression after complete remission died.

The median follow-up in stage III is 16 months (Table 3). All 6 patients who died had bulky abdominal disease and their tumor markers were high during their treatment.

Clinical response in 15 patients, with stage II and stage III disease, after 4 cycles of chemotherapy were also outlined in Table 5. Complete remission response was 83.3% in stage II and 33.3% in stage III, and the over all complete remission rate was 53% (8/15 patients).

All 15 patients who had chemotherapy developed alopecia, and had nausea and vomiting during the treatment. Major complications were seen in only 9 (60%) patients (Table 4).

Table 3. Results of Treatment

Stage	No.of Patnt.	Disease free Survival	Survival with Disease	Pro- gression	Death and Mean Time to Death (months)	Median Follow-up (months)
I	11	11(100%)	-	-	-	8-44 (24 months)
II	6	5(83.3%)	-	1(16.7%)	1(16.7%) (21 months)	21-41 (30 months)
III	9	2(22.2%)	1(11.1%)	-	6(66.7%) (13 months)	3-33 (16 months)

Table 4. Results of Chemotherapy (Clinical Response after 4 cycles of Treatment)

Clinical Response	Stage II (6 patients)	Stage III (9 patients)	TOTAL	
CR	5 (83.3%)	3 (33.3%)[b]	8/15 (53.3%)	} 73.3%[c]
PR	1 (16.7%)[a]	2 (22.2%)	3/15 (20%)	
Progression	-	3 (33.3%)	3/15 (20%)	
Disease Free Survival	5/6 (83.3%)	2/9 (22.2%)	7/15 (46.6%)	

[a] Relapse occurred after PR

[b] One patient relapsed after 6 weeks of CR

[c] Eleven of fifteen patients (73.3%) had CR or PR clinical response.

DISCUSSION

Retroperitoneal lymph node dissection is the modern surgical treatment of nonseminomatous testicular tumors[4-8].

Most and Cuneo, and later Jamieson and Dobson, and Roviere defined the primary lymphatic drainage of the testis in the retro-peritoneal lymph nodes around the aorta and vena cava[8]. Since then increasing attention has been focussed on excision of these nodes[9]. Staubitz and coworkers[2] emphasized in their paper that RPLND is a staging and a therapeutic procedure.

Although some authors prefer unilateral dissection, the relapse rate in this group is higher than after bilateral dissection [4-8, 10-12]. The treatment failures in patients after surgical treatment are most commonly due to pulmonary metastases[8].

Table 5.   Complications of Chemotherapy (seen in 9
            patients)

Complications		No. of Patients
Severe Leucopenia	(1000/mm^3)	3
Trombolytopenia	(<100.000/mm^3)	5
Anaemia	(Hb $\leq$13g/dl)	6
Fever		7
Sepsis		2

There is a recent treatment policy of orchidectomy and follow-up
in stage I disease. Peckham and coworkers[13] reported that the
relapse rate in stage I embryonal carcinoma after orchidectomy and
follow-up was 42.8%. We think that it is justifiable to perform
RPLND in stage I disease particularly for the patients with embryonal
carcinoma. There is no indication for adjuvant chemotherapy in stage
I.

There is a definite trend towards better response rates and
survival using cis Platinum, vinblastine and bleomycin combination
therapy after RPLND in stage II disease and as Vugrin and coworkers
[14] and others have advocated, combined chemotherapy should be given
in stage II-B.

Great progress has been made within the last 10 years in the
chemotherapy of testicular tumors due to the availability of effective new chemotherapeutic agents. In order to reduce toxicity without compromising response rates, attempts have been made to modify
the dose schedule.

Although we have seen toxicity, it was transient, and we had no
deaths due to toxicity following chemotherapy.

In spite of the high percentage of complete response in stage
III disease, bulky visceral metastases continue to be responsible for
a large percentage of the patients not responding to therapy. In our
small series 6 of the 9 patients with stage III disease who died had
bulky disease. We are currently using bleomycin, cis Platinum and
VP-16213 (vepeside) as the first choice of treatment in this group of
patients.

SUMMARY

Twenty six patients with nonseminomatous testicular tumor were
treated between September 1979-September 1982. Bilateral retroperitoneal lymphadenectomy was performed for stage I and II disease.

Chemotherapy was also given as an adjuvant treatment in stage II and as primary treatment in stage III disease.

The patients who died had high volume abdominal disease and markedly elevated tumor markers.

REFERENCES

1. D. A. Culp, Testicular neoplasms: an analysis of 113 cases, J.Urol., 70:282, (1953).
2. W. J. Staubitz, I. V. Magoss, M. H. Lent, E. M. Sigman, and J. T. Grace, Surgical management of testicular tumors, New York J.Med., 59:3959 (1959).
3. N. Mallis and J. F. Patton, Transperitoneal bilateral lymphadenectomy in testis tumor, J.Urol., 80:501, (1958).
4. W. J. Staubitz, K. S. Early, I. V. Magoss, and G. P. Murphy: Surgical treatment of non seminomatous germinal testis tumor, Cancer, 32:1206, (1973).
5. W. J. Staubitz, K. S. Early, I. V. Magoss, and G. P. Murphy, Surgical management of testis tumor, J.Urol., 111:205, (1974).
6. D. G. Skinner, Non-seminomatous testis tumors: a plan of management based on 96 patients to improve survival in all stages by combined therapeutic modalities, J.Urol., 115:65, (1976).
7. J. P. Donohue, L. M. Einhorn, and J. M. Perez, Improved management of non-seminomatous testis tumors, Cancer, 42:2903, (1978).
8. W. F. Whitmore, Jr., Surgical treatment of adult germinal testis tumors, Semin.Oncol., 6:55, (1979).
9. J. E. Pontes, I. V. Magoss, W. J. Staubitz, and G. P. Murphy, The treatment of stages I and II nonseminomatous testicular tumors, The Buffalo Experience from 1970 to 1979, J.Urol., 128, 1201, (1982).
10. W. J. Staubitz, I. V. Magoss, J. T. Grace, and W. G. Schenk, III: Surgical management of testis tumors, J.Urol., 101: 350, (1969).
11. D. G. Skinner, and W. F. Leadbetter, The surgical management of testis tumors, J.Urol., 106:84, (1971).
12. R. J. Babaian, and D. E. Johnson, Management of stage I and II nonseminomatous germ cell tumors of the testis, Cancer, 45:1775, (1980).
13. M. J. Peckham, A. Barrett, J. E. Husband, and W. F. Hendry, Orchidectomy alone in testicular stage I nonseminomatous germ-cell tumors, Lancet, 11:678-680 (1982).
14. D. Vugrin, W. F. Whitmore, Jr., and E. Cvitkovic, Adjuvant chemotherapy with VAB-3 of stage II-B testicular cancer, Cancer, 48:233, 1981.

TREATMENT OF MALE CHORIOCARCINOMA

Sophie Dorothea Fosså

The Norwegian Radium Hospital
Oslo
Norway

INTRODUCTION

Choriocarcinoma is distinguished by the presence of two types of cells:

1. Cells with large nuclei and light cytoplasm resembling cytotrophoblasts, usually in closely packed masses.
2. Cells with darker, smaller nuclei and eosinophilic or amphophilic cytoplasm which resemble syncytiotrophoblasts.

There should be a demonstrable attempt at villus formation with the two cell types forming adjacent layers. Hemorrhages and necrosis are constant features. The demonstration of β-HCG production in choriocarcinoma is essential for diagnosis.

The incidence of choriocarcinoma among patients with malignant germ cell tumors is low. In a series of 6000 testicular tumors Mostofi[1] found only 0.3% were pure choriocarcinomas, whereas about 10% contained elements of choriocarcinoma combined with other tumors. The demonstration of choriocarcinoma, either pure or mixed, is usually regarded as a bad prognostic sign.

The present report deals with the experience of The Norwegian Radium Hospital (NRH) and the Swedish-Norwegian Testicular Cancer Project (SWENOTECA) in treating patients with malignant germ cell tumors, which contained choriocarcinomatous elements.

PATIENTS AND METHODS

Incidence

From 1974-1980 592 patients with malignant germ cell tumors were admitted to the NRH, representing about 95% of all Norwegian patients with this diagnosis within this period. About 50% of the patients had non-seminomatous and 50% had seminomatous tumors. Seventeen patients (mean age 31,5 years) were recorded (by the computerized patient registry) as having choriocarcinoma (pure - 4, mixed - 13). Extragonadal germ cell tumor was diagnosed 5 times (pure choriocarcinoma - 3, mixed - 2). In these cases histology was performed on biopsies from large retroperitoneal tumors[4] or from metastases in the left supraclavicular fossa[1,2].

The SWENOTECA Project was started in April 1st 1981. Nearly all patients with non-seminomatous testicular cancer from Norway and Sweden (except from the Stockholm area) are included in this study. Patients with extra-gonadal tumors are not included in the project. By April 1983 208 patients had entered the SWENOTECA Project. Seventeen of these had choriocarcinomatous elements in their primary tumor (pure choriocarcinoma - 1, mixed choriocarcinoma - 16). Table 1 a) and b) show the stage distribution in the 34 patients at the time of the initial presentation, using the Royal Marsden staging system[3]. More than half the patients presented with an advanced stage or developed advanced stages during follow-up.

Treatment

Patients with stage I disease received prophylactic irradiation to the subdiaphragmatic retroperitoneal lymph nodes or were not treated at all until reactivation of the disease (1974-1978). Later on patients with clinical stage I disease underwent retroperitoneal lymph node dissection with adjuvant CVB (see below) for patients with pathological stage II disease. These treatment principles were also adopted by the SWENOTECA Project.

From 1974 to 1978 the treatment of advanced malignant germ cell tumors at the Norwegian Radium Hospital consisted of combination chemotherapy with vincristine, adriamycin, cyclophosphamide, actinomycin D and medroxyprogesterone acetate (VACAM)[4]. Later on, chemotherapy was based on a slightly modified Einhorn regimen containing cis-platinum, vinblastine and bleomycin (CVB)[5,6]. Patients in the SWENOTECA Project were treated with a similar cis-platinum-based combination chemoptherapy. Second line surgery was increasingly used as consolidation treatment after chemotherapy (together with radiotherapy as third line treatment)[7].

Table 1. Patients with Malignant Germ Cell Tumors containing
choriocarcinoma (NRH 1974-1980).

Initials + Age	Stage	Extra-gonadal	Treatment	Response	Observation time (months)	Status
a) "Pure choriocarcinoma"						
M.J., 57	IV CL_3	No	VACAM$^\oplus$	CR	70	Alive NED
O.M., 22	IV CL_3	Yes	CVB$^\bullet$	CR	66	Alive NED
F.N., 23	IV CL_3	Yes	VACAM	CR	102	Alive NED
E.C., 41	III C	Yes	CVB	PR	36	Alive Disease activity
b) "Mixed choriocarcinoma"						
T.A., 30	CS×I	No	No	–	106	Alive NED
R.A., 30	CS I → IV OL_1	No	Irrad.→ VACAM+ CVB	CR	36	Dead
T.S., 23	CS I → IV CL_2	No	No → VACAM	PR	40	Dead
J.S., 37	CS I → IV BL_2	No	No → VACAM	CR	108	Alive NED
P.H., 34	II C	No	CVB	CR	36	Alive NED
T.J., 22	III C	No	CVB	CR	13	Dead
O.H., 17	IV CL_1	No	CVB	CR	30	Alive NED
J.A.E., 28	IV CL_3 H+	No	CVB	PR	3	Dead
K.T., 22	IV CL_3	No	CVB	PR	< 1	Dead
K.S.N., 36	IV CL_3	Yes	VACAM	PR	7	Dead
M.J.C., 43	IV BL_3	No	VACAM	No change	4	Dead
L.T., 29	IV CL_3	Yes	CVB	CR	54	Alive NED
T.P., 41	IV CL_2	No	VACAM + CVB	PR	75	Alive disease activity

CR: Complete response. PR: Partial response. $\oplus$ Vincristine,
Adriamycin, Cyclophosphamide, Actinomycin D, Medroxy progesterone
acetate. $\bullet$Cis-platinum, Vinblastine, Bleomycin.   Clinical stage.

RESULTS

From the NRH series (mean observation time: 46 months) 10 of 17
patients are alive, 2 of them with active disease, whereas 8 have no
evidence of disease (NED) (Table 2). Three of the 4 patients with
pure choriocarcinoma are alive without disease 66, 70 and 102 months
after the initial diagnosis (treatment: VACAM - 2, CVB - 1). Of the
13 patients with mixed choriocarcinoma 7 are dead and 6 are alive
with NED. The levels of β-HCG do not seem to be related to survival,
though the data are limited for patients treated before 1978, when
only the urinary gonadotrophins were determined and quantitation was
done irregularily.

Of the 17 patients from the SWENOTECA Project one patient with
pure choriocarcinoma is dead (Table 2). 12 patients are alive with
NED (mean observation time: 9 months) whereas 3 patients are still
undergoing primary treatment. One patient is alive with disease
progression.

DISCUSSION

Choriocarcinoma is demonstrated in only relatively few patients
with malignant germ cell tumors, most often combined with other
histological elements. Each cancer center will therefore gain only
limited experience with this type of cancer. The published results
therefore often lack statistical significance. Neither can the
present study of 34 patients provide statistically relevant data
regarding stage distribution and optimal treatment. There is a need
for future cooperative retrospective and prospective studies with a
larger number of patients resulting hopefully in statistically significant information.

With these reservations the present study does not indicate that
the demonstration of choriocarcinomatous elements per se warrants a
poor prognosis in patients with non-seminomatous germ cell tumors.
This suggestion is in accordance with the observations of other
authors who have used modern treatment principles[8,9]. Even if only
patients with metastatic disease are considered, no difference of
survival can be found whether or not choriocarcinoma is demonstrated
in mixed malignant germ cell tumors[6].

However, patients with non-seminomatous germ cell tumors which
contain choriocarcinoma often present with far advanced disease, for
example large retroperitoneal tumors. Such advanced tumor stages are
generally combined with a bad prognosis for the patient, regardless
of whether choriocarcinoma is demonstrated or not. To define the
prognostic significance of choriocarcinoma, one should compare
patients with and without choriocarcinoma stage by stage, preferably
using the various substages.

Table 2. Patients with Malignant Germ Cell Tumors containing
          choriocarcinoma (SWENOTECA April 1980 - April 1983).

Pat. No	Stage	Observation time (months)	Status
a) "Pure choriocarcinoma"			
1	IV OL$_3$	6	Dead
b) "Mixed choriocarcinoma"			
2	PS* I	19	Alive NED
3	PS I	7	Alive NED
4	PS I	7	Alive NED
5	PS I	5	Alive NED
6	PS IIA	1	Under primary treatment
7	PS IIB	5	Alive NED
8	PS IIB	13	Alive NED
9	CS●IIB	10	Alive NED
10	CS IIB	7	Alive NED
11	IIIB		Under primary treatment
12	IIIB		Under primary treatment
13	IIIC	6	Alive NED
14	IV OL$_2$	4	Alive NED
15	IV BL$_2$	7	Alive NED
16	IV CL$_3$	22	Alive NED
17	IV OL$_3$		Alive with progression after primary treatment

PS*: Pathological stage.
CS●: Clinical stage.

The extent of choriocarcinomatous elements in the total tumor
volume is probably of prognostic significance. Quantitation of this
is difficult, especially if only a limited biopsy is available from a
large tumor mass. Even if such a biopsy shows pure choriocarcinoma,
the remaining tumor may contain other histological subtypes. The
fact that we did not find any difference in prognosis between
patients with pure or mixed choriocarcinoma could be explained by
this sampling error. Probably the level of serum β-HCG reflects the
amount of choriocarcinoma to a certain degree and may be an indicator
of prognosis. Future studies should evaluate this aspect.

There is no doubt that modern combination chemotherapy, even
before the cis-Platinum era, has improved the survival in patients
with choriocarcinomatous elements[10]. Today about 60% of patients

with metastatic mixed or pure choriocarcinomatous testicular cancer
will be cured. However, the fast proliferation rate and the fact
that most of the patients present with very advanced disease warrants
even more intensive chemotherapy than the "conventional" CVB regimen.

CONCLUSIONS

1. When using modern treatment principles the presence of
    choriocarcinoma per se does not seem to worsen the prognosis in
    patients with non-seminomatous germ cell tumors.
2. Treatment should be as intensive as possible, near the limit of
    tolerability, in patients with rapidly progressing tumors and/or
    far advanced stages.
3. Future retrospective and prospective studies should evaluate the
    prognostic and therapeutic significance of the demonstration of
    choriocarcinomatous elements in larger series of patients with
    malignant germ cell tumors. These projects need to take into
    account the different stages and substages of malignant germ cell
    tumors and, if possible, the amount of choriocarcinomatous
    elements and the serum level of β-HCG.

REFERENCES

1. F. K. Mostofi, Testicular tumors. Epidemiologic, etiologic and
    pathologic features, Cancer, 32:1186 (1973).
2. F. K. Mostofi and L. H. Sobin, Histological typing of testis
    tumors, in: "International Histological Classification of
    Tumors No 16", World Health Organization, Geneva (1977).
3. M. J. Peckham, T. J. McElwain, A. Barrett, and W. F. Hendry,
    Combined management of the malignant teratoma of the testis,
    Lancet 2:267 (1977).
4. O. Klepp, R. Klepp, H. Høst, G. Asbjørnsen, K. Talle, and A. E.
    Stenwig, Combination chemotherapy of germ cell tumors of the
    testis with vincristine, adriamycin, cyclophosphamide,
    adtinomycin D and medroxyprogesterone acetate, Cancer, 40:638
    (1977).
5. L. H. Einhorn, Testicular cancer as a model for a curable
    neoplasm: The Richard and Henda Rosenthal Foundation Award
    Lecture, Cancer Res., 41:3275 (1981).
6. O. Klepp, S. D. Fosså, S. Ous, H. Ken, J. T. Stenwig, V. Abeler,
    G. Eliassen, and H. Høst, Multi-modality treatment of
    advanced malignant germ cell tumors in males. I Experience
    with cis-platinum-based combination chemotherapy,
    Scand.J.Urol.Nephrol., in press (1983).
7. S. D. Fosså, O. Klepp, S. Ous, H. Lien, J. T. Stenwig, V.
    Abeler, G. Eliassen, and H. Høst, Multi-modality treatment in
    males with advanced malignant germ cell tumors. II Experience
    with surgery and radiotherapy following cis-platinum-based
    chemotherapy, Scand.J.Urol.Nephrol., submitted (1983).

8. M. L. Samuels, D. E. Johnson, B. Brown, B. Bracken, M. E. Moran, and A. V. Eschenbach, Velban plus continuous infusion bleomycin (VB-3) in stage III testicular cancer: Results in 99 patients with a note on high-dose Velban and sequential cis-platinum, in: "Cancer of the Genito-urinary tract", E. D. Johnson and M. K. Samuels, eds., Raven Press, New York, 159-172 (1979).

9. W. Jellinghaus, F. Weinschrod, and H. Frohmüller, Überlebenswahrscheinlichkeit von Patienten mit germinalen Hodentumoren, Fortschr.Med., 49:1643 (1981).

10. M. C. Li, W. F. Whitmore, R. Golby, and H. Grabstald, Effects of combined drug therapy on metastatic cancer of the testis J.A.M.A., 174:1291 (1960).

THE TREATMENT OF CHORIOCARCINOMA: DISCUSSION

J. P. Blandy

DEFINITION

Dr Mostofi began the discussion at the request of the panel, by
defining for the participants what was to be understood by the term
"choriocarcinoma". Clearly there were two distinct entities - pure
choriocarcinoma and choriocarcinoma associated with other cell types,
especially embryonal cell carcinoma, yolk sac tumor, and seminoma.
To these he added a third - foci of choriocarcinoma in situ, i.e.
within seminiferous tubules:-

1.   Pure: cyto- and syncytio-trophoblast only.
2.   Choriocarcinoma associated with other cellular elements.
3.   Foci in situ in tubules.

In each of these categories it was necessary to find both cyto- and
syncytio-trophoblast for which, Dr Mostofi admitted, special immunoperoxidase staining might be necessary.

NATURAL HISTORY

Dr Fossa described the experience of the Norwegian Radium Hospital and of the Scandinavian combined study. Of a total of 800
patients there were 5 'pure' choriocarcinomas (0.6%), and 29 (3.6%)
in the mixed group: both were rare, the 'pure' choriocarcinomas
exceptionally so. Many seemed to arise in extragonadal sites (5/17 =
29%) and their pattern of spread was largely systemic - findings
reinforced by Dr Javadpour whose experience had been much the same.

TREATMENT

Contrary to the prevailing opinion, treatment in both groups had been by no means hopeless: 24/34 (70.6%) of all types were cured by VACAM or CVB and 2/4 of the pure group, in Fossa's experience. One particular sub-group - those with brain metastases - posed a particular problem in this essentially systemic disease. Even so there was room for encouragement: Oliver reported 4/8 such patients with brain metastases that had been cured. Opinions differed as to the best way of effecting such a cure: Torti and Bagshaw preferred to use radiotherapy as an adjunct in treatment: Fossa referred her patients to the neurosurgeon, but Stoter's experience led him to mistrust a surgical approach - in his case the tumor had been found to be inoperable, attempts to remove it left a large cavity, and a drain had been left in for the purpose of irrigating the cavity with a chemotherapeutic agent.

SUMMARY

Taken altogether this exchange of views gave the participants (and the moderator) grounds for renewed hope in a field hitherto governed by gloom and despair. One question was not put to the panel by the audience, but would have received no answer - Why does choriocarcinoma in the male behave so much less favorably than in the female, when treated with modern chemotherapy? Perhaps the answer will never be known, since this latest review of international experience suggests that in fact, it does not do too badly after all.

NATURAL HISTORY OF RENAL CELL CARCINOMA: ASPECTS
OF TUMOR MORPHOLOGY, LYMPHATIC AND HAEMATOGENOUS
METASTATIC SPREAD

S. Hellsten*, T. Berge** F. Linell** and L. Wehlin***

Departments of Urology*, Pathology**, and Diagnostic
Radiology***, Malmö General Hospital
Malmö, Sweden

In an autopsy series comprising 16,294 autopsies performed
during a 12 year period of time 350 cases of renal cell carcinoma
were found. In 115 cases the diagnosis of a renal cell carcinoma was
made ante mortem whereas 235 tumors were clinically unrecognized, the
latter ones being the subject of the present study. Metastases were
found in 24% of the patients with a clinically unrecognized renal
cell carcinoma and were the main cause of death in 20%[1-3].

In patients with a clinically unrecognized renal cell carcinoma
the symptomatology was generally unhelpful. Cardiovascular disease
was predominant and caused the death in 44% while one or more malignant tumors other than the renal cell carcinoma were recognized in
33% and were the main cause of death in 20%.

At a survey of the records of the 235 patients 28 of them were
found to have undergone IVP within one year prior to death. About
two thirds of the overlooked renal tumors could be identified or
highly suspected on a re-examination of the X-ray pictures, the
majority being located in the cortex of the kidney and not involving
the calyceal system. They had escaped detection mainly because of
poor technical quality or misinterpretation of the X-ray pictures
underlining the importance of adequate bowel preparation and careful
examination of the kidney surfaces.

The number of metastasizing tumors increased significantly with
the size of the primary tumor. Local aggressiveness of the primary
in terms of vascular ingrowth or pericapsular growth was more common
in large tumors but was much more closely correlated to metastatic
spread than to size. Tumor ingrowth into the renal vein was significantly more common in metastasizing tumors as compared with
non-metastasizing tumors.

279

Lymphatic spread was revealed in 66% of patients with metastases. In 97% of these patients additional, non-lymphatic metastatic spread was seen. The most common sites of lymph node metastases were the retroperitoneal space (45%), the mediastinum (44%) and the supraclavicular fossa (30%). With involvement of these lymph node stations a high incidence of concomitant metastases in the lungs was seen (86%).

The present study confirmed that an analysis of the local aggressiveness of the primary tumor in the kidney and of the extent of the lymphatic spread was prognostically valuable and might be useful in defining the group of patients that may benefit from adjuvant treatment such as radiation therapy, chemotherapy and immunotherapy.

REFERENCES

1.   S. hellsten, T. Berge, and L. Wehlin, Unrecognized renal cell
           carcinoma, clinical and diagnostic aspects, Scand.J.Urol.
           Nephrol., 8:269 (1981a).
2.   S. Hellsten, T. Berge, and L. Wehlin, Unrecognized renal cell
           carcinoma, clinical and pathological aspects, Scand.J.Urol.
           Nephrol., 8:273 (1981b).
3.   S. Hellsten, T. Berge, and F. Linell, Clinically unrecognized
           renal carcinoma: Aspects of tumor morphology, lymphatic and
           haematogenous metastatic spread, Br.J.Urol., 55:166 (1983).

ADVANCES IN RENAL CANCER: ADVANCES IN DIAGNOSIS

R. Musumeci and J. D. Tesoro Tess

Istituto Nazionale Tumori
Milan
Italy

Although high dose urography with nephrotomography remains the main initial diagnostic investigation for suspected renal carcinoma, the wider use of the new imaging techniques has recently allowed a change in the diagnostic approach[1] as echography and computed tomography have demonstrated their capability to recognize the presence and diagnose the nature of a renal mass suspected to be neoplastic. The findings obtained can be confirmed, in selected cases, by means of fine needle biopsy. One or several of these tools in combination will often give additional information on the presence of a mass, differential diagnosis, tumor extension and the presence of local and/or distant extension. The value of the different procedures in the diagnosis and staging of renal carcinoma will be assessed.

ECHOGRAPHY

On the basis of abnormal urography or as a first step when an abdominal mass is clinically evident, echography can easily be performed in non obese, non distended patients[2-4]. It seems adequate in differentiating a cystic from a solid mass. Renal carcinoma is echogenic in about 75% of cases, transonic but without posterior wall enhancement in less than 10% and shows a mixed pattern in the remainder (Table 1).

From a theoretical point of view it would be possible to check the following points with echography in suitable patients: perirenal structures, lymph node metastases, vascular invasion of renal vein or inferior vena cava, invasion of the neighboring structures and metastases. Detailed knowledge of these conditions could make an exact preoperative staging of the disease.

Table 1.   Echographic Pattern of Renal
           Carcinoma (approximate
           percentages)

Pattern	%
Echogenic	75
Transonic	8
Complex	14

The current diagnostic possibilities of ultrasound are summarized in Table 2. With the procedure it is possible in the more
favorable cases to get valuable information about para-aortic lymph
node enlargement, invasion of the inferior vena cava and liver metastases. Thus echography is inadequate for staging because of its
inability to appreciate extracapsular extension and infiltration of
the adjacent structures. In fact, it is almost impossible with
echography to differentiate stage 1 tumors confined to the renal
parenchyma from stage II where tumor extends into the perinephric
space but is still within Gerota's fascia. Further a number of
echographically assessed stages I and II are surely understaged
because of the intrinsic difficulty of evaluating diffusion. On the
other hand at least 30-40% of those cases, echographically classified
as having stage III or IV disease, appear overstaged at operation.

ANGIOGRAPHY

Angiography is highly effective in the diagnosis of malignancy
but is a most unreliable method for staging renal carcinoma[3,5,6].
The angiographic diagnosis of external tumor spread is based mainly
upon demonstration of abnormal extrarenal vascularity and the extrarenal presence of dilated arteries which represent accessory blood
supply to the tumor. However, it is most difficult to differentiate
between extrarenal infiltration on the one hand and additional blood
supply from the perirenal circulation without extension beyond the
renal capsule on the other.

The incidence of involvement of the renal vein and inferior vena
cava by renal carcinoma is reported to range from 10 to 33%. Using
selective procedures, e.g., inferior vena cavography and/or right or
left renal vein phlebography, it is possible to demonstrate the
neoplastic thrombosis with some accuracy.

Aortography often reveals the presence of liver metastases, bone
metastases and lymph node metastases due to their hypervascular
appearance.

The diagnostic possibilities of vascular investigations are summarized in Table 2. The possibility of a correct staging compared with the pathology report appears good for some of the parameters and very poor for the others. On the basis of personal experience, in a group of patients with pathologically staged renal carcinoma, the clinical stage determined by vascular procedures was correct in about 75% of the cases, underestimated in 15% and overestimated in 10%.

COMPUTED TOMOGRAPHY

Computed tomography is accepted worldwide as the best single procedure for the diagnosis of renal masses and for complete staging of renal carcinoma[1,3,6,7,8,9].

The diagnostic possibilities of CT in staging the disease are reported in Table 2. It is clearly evident that this procedure shows the best theoretical possibilities in comparison with the other methods. The major difficulty is encountered in discovering the expansion of the renal tumor through the fibrous renal capsule and in differentiating thickening of Gerota's fascia due to neoplastic invasion from that indicating sequelae of trauma and inflammation. With CT it is always possible to disclose lymph node swellings but it is also impossible to differentiate between metastases and benign hyperplasia. On the other hand, CT is a reliable method of searching for neoplastic venous thrombi, but it must be remembered that these findings cannot be accurately evaluated without injection of contrast medium, preferably directly into the femoral vein.

In spite of the good results of the procedure in assessing individual prognostic factors, the overall results of CT staging when compared with pathologic findings, are still less than perfect. In

Table 2. Diagnostic Possibilities of Echography; Vascular Investigation and Computerized Tomography (CT) in Renal Carcinoma.

Parameter	Diagnostic Possibility		
	Echography	Vascular Investigation	CT
Perirenal structures	+/-	+/-	++
Lymph nodes	++	++	++
Renal vein invasion	+/-	+++	++
Vena cava invasion	++	+++	+++
Neighboring structures	+	+	++
Metastases	++[a]	++[a]	+++

[a] = liver metastases

fact the percentage of correctly staged patients ranges from 70 to
75%, with about 10% of underestimated and 20-25% of overestimated
cases.

In the evaluation of regional lymph node metastases from renal
cancer lymphangiography is not routinely performed[10]. The incidence of metastases is generally not high and surgery usually includes
as radical a lymphadenectomy as possible. Furthermore, renal lymphatic drainage is to the paraortic lymph nodes lateral to L_1-L_2; these
are often not opacified by pedal lymphangiography. Preoperative
demonstration of nodal involvement is not mandatory because the
staging of the disease is usually based on surgical and pathological
reports. Lymphangiography could be performed after surgery in those
cases not radically treated to appreciate the number and site of
metastatic lymph nodes and to help radiotherapeutic planning.

On the basis of these considerations, the diagnostic algorhythm
reported in Table 3 can be hypothesized. High dose urography with
tomography is usually the first procedure, often combined with echography. The discovery of a typical cyst can suggest fine needle
aspiration. A solid or equivocal mass makes a CT mandatory with the
possibility of fine needle biopsy. If a complete staging is possible, no further examination is requested. In equivocal cases CT may
be complemented with cavography for more detailed evaluation of tumor
thrombosis, or with angiography, as a method which may provide
additional information, important for staging as well as for surgical
intervention.

In the near future nuclear magnetic resonance (NMR) will become
available. This allows the representation of the internal structure
of human body "reading" the different tissue concentrations of water.
Though it achieved diagnostic importance only in 1980 there is al-

Table 3. Algorhythm for Renal Carcinoma.

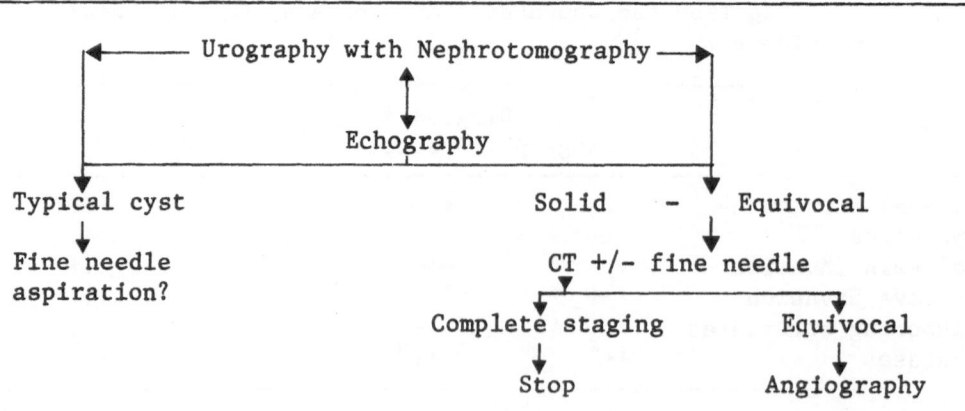

ready information on some 2,000 patients examined throughout the world and it offers very good prospects for the demonstration of small lesions in a completely innocuous manner[11].

REFERENCES

1. W. Karps, L. Ekelund, G. Olafsson, A. Olsson, Computed tomo-
    graphy, aniography and ultrasound in staging of renal car-
    cinoma, Acta.Radiol., 22:625 (1981).
2. E. Ceglia, N. Masciadri, Biopsia percutanea ed ecografia nella
    diagnosi dell'adenocarcinoma renale. Da "I tumori urologici,"
    Pag.31-37, C.E.A., Milano, (1982).
3. F. Garbagnati, M. Milella, La tomografia assiale computerizzata
    e l'angiografia nello staging dell'adenocarcinoma renale. Da
    "I tumori urologici" - pag. 27-30, C.E.A., Milano, (1982).
4. B. Green H. Goldstein, R. Weaver, Abdominal pansonography in the
    evaluation of renal cancer, Radiology, 132:421 (1979).
5. B. Bracken, K. Jonsson, How accurate is angiographic staging of
    renal carcinoma? Urology 14:96 (1979).
6. P. Weyman, B. McClennan, R. Stanley, R. Levitt, S. Sagel, Com-
    parison of computed tomography and angiography in the evalu-
    ation of renal cell carcinoma Radiology, 137:417 (1980).
7. E. Levine, K. Lee, J. Weigel, Preoperative determination of
    abdominal extent of renal cell carcinoma by computed tomo-
    graphy, Radiology 132:385 (1979).
8. E. Levine, N. Makland, S. Rosenthal, K. lee, J. Weigel, Com-
    parison of computed tomography and ultrasound in abdominal
    staging of renal cancer, Urology, 16:317 (1980).
9. L. Love, R. Churchill, C. Reynes, G. Shuster, R. Moncada, A.
    Berkow, Computed tomography staging of renal carcinoma, Urol.
    Radiol., 1:3 (1979).
10. R. Musumeci, M. Mauri, La linfografia in oncologia, Ilford, ed.,
    Origgio (Varese), (1980).
11. I. R. Young, D. R. Bailes, M. Burl, A. G. Collins, D. T. Smith,
    M. J. McDonnell, J. S. Orr, L. M. Banks, G. M. Bydder,
    Greenspan, R. E. Steiner, Initial clinical evaluation of a
    whole body nuclear magnetic resonance tomograph, J.of Computed
    Assisted Tomography, 6:1 (1982).

ADVANCED RENAL CELL CARCINOMA:

NATURAL HISTORY AND RESPONSE TO CHEMOTHERAPY

Frank M. Torti

Division of Medical Oncology
Stanford University Medical Center
California, USA

To place the chemotherapy of metastatic disease in perspective, certain features of the disease must be discussed.

Natural History of Metastatic Disease

Renal cell cancer is a disease with a quite variable natural history. Although most patients die within a few years of the diagnosis of metastatic disease, in one series, approximately 20% of patients lived 3 years or longer[1]. In another series, the median survival was 10 months, but 25% of patients lived 10 years or longer[2]. Neither therapeutic intervention nor spontaneous regression explain this prolonged survival of a substantial minority of patients with metastatic disease. Recently, the investigation of some of the host-tumor interactions which might explain this observation have been undertaken. Nonetheless, at this point the occasionally prolonged survival with metastatic renal cell carcinoma remains unexplained. Yet it is a critical factor to be incorporated into the design and interpretation of clinical trials in metastatic disease. For example, a 20% stabilization rate attributed to a hormonal or chemotherapeutic agent, even if sustained over a period of 2 or more years, would be suspect.

A number of authors have identified certain factors which influence the likelihood of prolonged survival after diagnosis of metastatic disease: 1) A prolonged interval from nephrectomy to first evidence of metastatic disease; 2) the absence of recurrent or persistent local tumor; 3) pulmonary metastases in the absence of other metastatic sites of disease. These factors should obviously be part of the stratification of randomized clinical trials, and should be incorporated into the reporting of phase II studies.

Table 1.   Chemotherapy with Vinca Alkaloids

DRUG AUTHOR (YEAR)	NO. OF PATIENTS	NO. OF CR & PR	(%)	REFERENCE NO.
Vinblastine				
Hahn (1977)	10	0		(15)
Horn (1967)	2	1		(16)
Hagan (1974)	35	11		(13)
Hahn (1978)	44	4		(14)
Frei (1961)	3	0		(17)
Smart (1964)	2	0		(18)
Wright (1963)	(?)	0		(19)
Hill (1961)	5	1		(20)
TOTAL	101	17	(17%)	
Vincristine				
Horn (1967)	1	0		(16)
Costa (1962)	3	0		(21)
Vindestine				
Wong (1977)	17	0		(22)
VM-26				
Hire (1979)	12	0		(23)
VP-16				
Hahn (1979)	31	1		(24)

Nor is the unusual natural history of metastatic renal cancer limited to those patients with clinically evident metastatic disease. McNichols et al.,[3] have shown that among 158 patients alive and disease-free at 10 years, 18 (11%) subsequently relapsed. Like breast cancer, there is a continued risk of relapse even after conventional estimations of 'cure' at 5 and 10 years.

Surgical Treatment of Metastases

If a complete surgical excision of a simple metastatic focus of renal cell cancer can be achieved, approximately 35% of these patients will be alive at 5 years[4]. This is perhaps the most successful treatment of metastatic disease, although this observation suffers from the lack of a control series, given the selected group in whom surgery is undertaken and the variable natural history of the disease discussed earlier. Nonetheless, the approach has been extended, with apparent success, to patients with more than one metastatic focus[5]. These sites are occasionally bone, but most frequently lung metastases. Whether these metastases occur simultaneously with or subsequent to the diagnosis of the primary disease has not always been carefully reported. Most series deal with the former situation[4], but the later group of patients may have a better prognosis[6].

Table 2.   Purine and Pyrimidine Antagonists

DRUG AUTHOR (YEAR)	NO. OF PATIENTS	NO. OF CR & PR	(%)	REFERENCE NO.
5-Fluorouracil				
Weiss (1961)	2	0		(25)
White (1962)	2	0		(26)
Khung (1966)	2	0		(27)
Moore (1968)	15	1		(28)
Rochlin (1962)	3	0		(29)
TOTAL	24	1	(4%)	
FUDR				
Ansfield (1962)	3	1		(30)
Wilson (1967)	21	1		(31)
Cytosine Arabinoside				
Frei (1969)	2	0		(32)
6-mercaptopurine				
Regelson (1967)	15	2		(33)
Lemon (1964)	1	1		(34)

Spontaneous and Surgically Induced Regressions of Metastatic Disease

Spontaneous regression of metastatic renal cell cancer occurs, but documented cases are rare. The most recent review of this subject by Fairlamb in 1981 found only 67 cases with adequate documentation in the literature[7]. Since many of these cases have had treatment of the primary tumor with nephrectomy, it has been argued that nephrectomy induces spontaneous regression[8]. However, with the possible exception of isolated bone metastases[9], the utility of routine nephrectomy has not been documented[1,10]. The rarity of spontaneous regression, and its occurrence without prior nephrectomy[11], argues against the routine use of nephrectomy on the large group of patients with asymptomatic primary tumors and metastatic disease. Whether the recent report of renal infarction followed by nephrectomy in patients with metastatic disease[12] will lead to the reutilization of nephrectomy in this setting remains unknown.

Chemotherapy

Overall, the chemotherapy of renal cell carcinoma is discouraging, with few responses in treated patients. This has led to the reporting of responses that are less than classical partial responses. This makes comparisons between reported studies difficult:

Table 3. Antibiotic Chemotherapeutic Agents

DRUG AUTHOR (YEAR)	NO. OF PATIENTS	NO. OF CR & PR	REFERENCE NO.
Adriamycin			
O'Bryan (1973)	15	0	(35)
Mitomycin C			
Watne (1967)	1	1	(36)
Kato (1979) (micro-encapsulated)	2	1	(37)
Actinomycin			
Watne (1960)	4	0	(38)
Hahn (1981)	61	1	(39)
Bleomycin			
Johnson (1975)	15	2*	(40)
Hahn (1977)	7	0	(15)
Mithramycin			
Kofman (1963)	2	0	(41)
Bisantrene			
Myers (1982)	37	2	(42)

*Mixed responses.

Some series report partial responses as greater than 25% reduction in tumor size, as opposed to the standard of greater than 50% reduction of the product of the perpendicular diameters. Most series do not document the site of response and some series allow mixed responses to be included as partial responses i.e., responses that occur at one site while the disease remains stable at other measurable sites of disease. Given this heterogeneity of reporting, combined with the large number of patients who have been reported out of broad drug-oriented phase II trials, it must be recognized that many of the agents listed in the following tables have had inadequate trials of chemotherapy by modern standards.

Among the chemotherapeutic agents studied, vinblastine has been the most carefully evaluated (Table 1). Of the 101 reported patients in the literature, 17 or 17% have had responses. These are usually of short duration although the duration of responses are not always reported. Further, it should be noted that the response rates are variably defined as well as quite variable from study to study, ranging from 31% in Hagan's series[13] in 1974 to 9% in Hahn's report[14] of 44 patients in 1978.

Table 4.   Alkylators and Related Compounds

DRUG AUTHOR (YEAR)	NO. OF PATIENTS	NO. OF CR & PR	(%)	REFERENCE NO.
CYCLOPHOSPHAMIDE:				
Solomon (1963)	1	0		(43)
Atkins (1962)	5	1		(44)
Anders (1961)	1	0		(45)
Fox (1965)	7	2		(46)
Bergsagel (1960)	2	1		(47)
Shnider (1960)	6	0		(48)
Dick (1961)	6	0		(49)
Wajsman (1980)	12	0		(50)
Kiruluta (1975)	10	0		(51)
Hahn (1979)	54	2		(24)
TOTAL	104	6	(5.7%)	
CHLORAMBUCIL:				
Moore (1968)	14	2		(52)
THIOTEPA:				
Hahn (1977)	7	1		(15)
AZETEPA:				
Choy (1967)	3	1		(53)
CCNU:				
Hahn (1977)	7	0		(15)
Mittleman (1973)	20	4		(54)
Merrin (1975)	23	4		(55)
TOTAL	50	8	(16%)	
DTIC:				
Luce (1970)	2	0		(56)
MeCCNU:				
Hahn (1978)	45	2		(14)
CHLORZOTOCIN:				
Gralla (1979)	21	0		(57)
CIS-PLATINUM:				
Rodriguez (1978)	23	0		(58)
PHOSPHAMIDE:				
Fossa (1980)	11	1		(59)
URACIL MUSTARD:				
Schumacher (1963)	1	1		(60)
TOTAL	103	4	(3.8%)	

Other vinca alkaloids and related compounds have had limited study.   There appears to be minimal to no activity for vindesine, VM-26, and VP-16.   Vincristine has never been adequately studied in this disease.

Among the purine and pyrimidine antagonists (Table 2), 5-FU has had evaluation primarily in drug-oriented phase II trials in which only one response in 24 patients has been observed.   FUdR appears to have no activity.   Cytosine Arabinoside has not been adequately

Table 5.   Folate Antagonists

DRUG AUTHOR (YEAR)	NO. OF PATIENTS	NO. OF CR & PR	REFERENCE NO.
Methotrexate			
Andrews (1967)	14	1	(61)
Baumgartner (1980)	8*	2	(62,63)
Methodichlorophen			
Hindmarsh (1979)	10	3	(64)
Triazinate (Baker's Antifol)			
Hahn (1981)	59	3	(39)
Bukowski (1980)	15	1	(65)

*Citrovorum factor rescue.

Table 6.   Miscellaneous Agents

DRUG AUTHOR (YEAR)	NO. OF PATIENTS	NO. OF CR & PR	(%)	REFERENCE NO.
HYDROXYUREA				
Falkson (1967)	1	0		(66)
Lerner (1965)	1	0		(67)
Nevinny (1968)	18	5		(68)
Stolbach (1981)	19	1		(69)
TOTAL	39	6	(15%)	
Methyl-Gag				
Knight (1979,1980)	23	3		(70,71)
Todd (1980,1981)	25	4		(72,73)
Zeffren (1981)	31	0		(74)
Killien (1980)	2	1		(75)
TOTAL	81	8	(9.8%)	
m-AMSA				
Schneider (1980)	20	1		(76)
Van Echo (1980)	16	0		(77)
Dianhydrogalactitol				
Hahn (1979)	35	0		(24)
Piperazinedione				
Pasmantier (1977)	20	0		(78)
Mitotane				
Hogan (1981)	12	0		(79)

Table 7. Multiagent Chemotherapy

AUTHOR (YEAR)	DRUGS	NO. OF PATIENTS	NO. OF CR & PR	REFERENCE NO.
I. Regimens which Include Vinblastine				
Davis (1978)	Vinblastine, CCNU	29	7	(80)
Merrin (1975)	Vinblastine, CCNU	6	1	(55)
Merrin (1975)	Vinblastine, MeCCNU	15	1	(55)
Levi (1980)	Vinblastine, bleomycin, HDMTX	14	5	(81)
Samson (1976)	Vinblastine, bleomycin, cis-platinum	4	0	(82)
Hahn (1981)	Vinblastine, CCNU	60	3	(39)
Knost (1981)	Vinblastine, bleomycin	15	2	(83)
II. Other Regimens				
Johnson (1975)	Vincristine + hydroxyurea	15	0	(84)
Baumgartner (1980)	Vincristine, bleomycin, HDMTX, + (cyclophosphamide or peptochemio)	12	2	(62,63)
Dana (1981)	Adriamycin, bleomycin, vincristine, cyclophosphamide, BCG	13	3	(85)
Werf-Messing (1974)	Cyclophosphamide, 5 Fluorouracil, methotrexate, vincristine	18	0	(86)

studied. Three responses in 16 evaluable patients have been reported
with 6-Mercaptopurine. This deserves further investigation.

Among the antibiotic chemotherapeutic agents (Table 3), Adriamycin has had limited evaluation with 0 of 15 response reported by
O'Bryan in a broad phase II study in 1973. Mitomycin has not had
adequate evaluation but some activity has been observed. There has
been minimal response rate to bleomycin and essentially no activity
for actinomycin D.

Among the alkylators and related compounds, only CCNU has significant activity with 8 of 50 patients, or 16%, responses reported
in the literature (Table 4). Although there is some evidence of a
response for cyclophosphamide, the overall response rate of 5.7%
suggests that this agent has no utility in metastatic disease.
Cisplatinum has had adequate evaluation and has no activity, nor does
chlorzotocin. DTIC has had an inadequate trial.

Table 8. Hormonal Therapy and Chemotherapy

AUTHOR (YEAR)	HORMONAL AGENT	CHEMOTHERAPEUTIC AGENTS	NO. OF PATIENTS	NO. OF CR & PR	REFERENCE NO.
Hahn (1978)	MPA	CCNU	38	4	(14)
Hahn (1978)	MPA	Vinblastine	38	8	(14)
Levi (1980)	Tamoxifen	HDMTX, vinblastine, bleomycin	14	5	(81)
Ishamael (1980)	Depo-provera	Adriamycin, vincristine, BCG	38	18	(87)
Talley (1979)	MPA	Cyclophosphamide, vinblastine, prednisone, hydroxyurea	42	8	(88)
Katakkar (1978)	MPA	Adriamycin vinblastine, hydroxyurea	8	2	(89)
Voiska (1978)	Delalutin	CCNU, vinblastine	17	0	(90)
Richards (1977)	Methylprednisolone	CCNU, bleomycin	16	1	(91)
Richards (1977)	Methylprednisolone	CCNU, bleomycin, Adriamycin	14	3	(91)
Patel (1978)	Testosterone + MPA	Vincristine, actinomycin-D, cyclophosphamide	4	0	(92)
Swanson (1980)	Estramustine	Estramustine	11	0	(93)
Dorn (1975)	MPA	Vinblastine	1	1	(94)

Among the folate antagonists, methotrexate has shown some activity with its utilization in the high-dose regimen with leucovorin rescue. There were 2 out of 8 responses reported by Baumgartner in 1980 (Table 5).

Other miscellaneous agents (Table 6) include hydroxyurea, which probably has minimal activity in this disease. Overall, Methyl-GAG also has limited activity. Although initial response rates of 13 and 16 percent were reported, more recent evaluation by Zeffren[74] in 1981 shows no responses in 31 patients. There is no activity for m-AMSA, dianhydrogalactitol, piperazinedione. Mitotane has only been studied in 12 patients; no responses were observed.

Lack of activity for single agents has prompted their utilization in combination to improve the response rate. Multi-agent chemotherapy, Table 7, shows limited activity with responses ranging from 5% to 35%. Those regimens which include vinblastine appear to have a somewhat higher response rate than those regimens which do not include vinblastine in the combinations. Nonetheless there is no convincing evidence that multi-agent chemotherapy is superior to single-agent chemotherapy in renal cell cancer.

Similar attempts to improve response rates have led to the simultaneous use of the hormone and chemotherapy in renal cell carcinoma. Table 8 lists the response rates for these combined modality treatments. As can be seen, there is no convincing evidence for additive or synergistic effects of hormonal therapy and chemotherapy in renal cell cancer, although some studies report response rates higher than might be anticipated from single-agent chemotherapy or hormonal therapy alone.

Summary

Vinblastine and CCNU remain the two agents with demonstrable activity in renal cell carcinoma. However, a number of agents have not been adequately studied in this relatively common disease. The variable natural history of the disease as well as the relatively gratifying responses to excision of limited pulmonary metastases needs to be considered in relation to the rather modest results of current chemotherapeutic approaches.

REFERENCES

1. J. B. DeKernion, K. P. Ramming, and R. B. Smith, The natural history of metastatic renal cell carcinoma: Computer analysis J.Urol., 120:148-152 (1978).
2. R. J. Papac, S. A. Ross, and A. Levy, Renal cell carcinoma: Analysis of 31 cases with assessment of endocrine therapy, Am.J.Med.Sci., 274:281-290 (1977).
3. D. W. McNichols, J. W. Segura, S. Wu, and G. E. Safire, Renal cell carcinoma: Long-term survival and late recurrence, J.Urol., 126:17-23 (1981).
4. B. M. Tolia and W. F. Whitmore, Jr., Solitary metastasis from renal cell carcinoma, J.Urol., 114:836-837 (1975).
5. D. G. Skinner, J. F. DeKernion, eds., in: "Genitourinary Cancer", W. B. Saunders Company, Philadelphia, London, Toronto, pp.128-129 (1978).
6. M. J. O'Dea, H. Zincke, D. C. Utz, and P. E. Bernatz, The treatment of renal cell carcinoma with solitary metastasis, J.Urol., 120:540-542 (1978).
7. D. J. Fairlamb, Spontaneous regression of metastases of renal cancer: A report of two cases including the first recorded regression following irradiation of a dominant metastasis and review of the world literature, Cancer, 47:2102-2106 (1981).
8. D. H. Garfield and B. J. Kennedy, Regression of metastatic renal cell carcinoma following nephrectomy, Cancer, 30:190-196 (1972).
9. J. E. Montie, B. H. Stewart, R. A. Straffon, L. H. W. Banowsky, C. B. Hewitt, and D. K. Montague, The role of adjunctive nephrectomy in patients with metastatic renal cell carcinoma, J.Urol., 117:272-275 (1977).

10. D. E. Johnson, K. E. Kaesler, and M. L. Samuels, Is nephrectomy justified in patients with metastatic renal cell carcinoma, J.Urol., 114:27 (1975).

11. S. Z. Freed, J. P. Halperin, and M. Gordon, Idiopathic regression of metastases from renal cell carcinoma, J.Urol., 118:538-541 (1977).

12. D. A. Swanson, S. Wallace, and D. E. Johnson, The role of embolization and nephrectomy in the treatment of metastatic renal carcinoma, Urol.Clin.N.Amer., 7:719-730 (1980).

13. K. Hagan, J. D. Trapp, R. K. Rhany, and V. H. Reynolds, Treatment of metastatic renal cell carcinoma, Southern Medical Journal, 67:1175-1178 (1974).

14. R. G. Hahn, N. R. Temkin, E. D. Savlov, C. Perlia, G. L. Wampler, J. Horton, J. Marsh, and P. P. Carbone, Phase II Study of Vinblastine, Methyl-CCNU, and Medroxyprogesterone in advanced renal cell cancer, Cancer Treat Rep., 62:1093-1095 (1978).

15. D. M. Hahn and S. C. Schimpff, Single-agent therapy for renal cell carcinoma: CCNU, Vinblastine, ThioTEPA, or Bleomycin, Cancer Treat Rep., 61:1585-1587 (1977).

16. Y. Horn and A. Hochman, The alkaloids of vinca rosea linn in malignant tumors, Oncology, 21:214-220 (1967).

17. E. Frei, III, A. Franzino, B. I. Shnider, G. Costa, J. Colsky, C. O. Brindley, H. Hosley, J. F. Holland, G. L. Gold, and U. Jonsson, Clinical studies of vinblastine, Cancer Chemothr. Rep., 12:125-129 (1961).

18. C. R. Smart, D. B. Rochlin, A. M. Nahum, A. Silva, and D. Wagner, Clinical experience with vinblastine sulfate (NSC-49842) in squamous cell carcinoma and other malignancies, Cancer Chemothr.Rep., 34:31-45 (1964).

19. T. L. Wright, J. Hurley, D. R. Korst, R. W. Monto, R. J. Rohn, J. J. Will, and J. Louis, Vinblastine in neoplastic disease, Cancer Res., 23:169-179 (1963).

20. J. M. Hill and E. Loeb, Treatment of leukemia, lymphoma, and other malignant neoplasms with vinblastine, Cancer Chemothr. Rep., 15:41-61 (1961).

21. G. Costa, M. M. Hreshchyshyn, and J. F. Holland, Initial clinical studies with vincristine, Cancer Chemothr.Rep., 24:39-44 (1962).

22. P. P. Wong, A. Yogada, V. E. Currie, and C. A. Young, Phase II study of vindesine sulfate in the therapy for advanced renal carcinoma, Cancer Treat Rep., 61:1727-1729 (1977).

23. E. A. Hire and M. K. Samson, Use of VM-26 as a single agent in the treatment of renal carcinoma, Cancer Clin.Trials, 2:293-295 (1979).

24. R. G. Hahn, M. Bauer, J. Wolter, R. Creech, J. M. Bennett, and G. L. Wampler, Phase II study of single-agent therapy with megestrol acetate, VP-16-213, cyclophosphamide, and dianhydrogalactitol in advanced renal cell cancer, Cancer Treat Rep., 63:513-515 (1979).

25. A. J. Weiss, L. G. Jackson, and R. Carabasi, An evaluation of
    5-fluorouracil in malignant disease, Ann.Intern.Med., 55:
    731-741 (1961).
26. J. E. White, W. N. Ricketta, and W. J. Strudwick, A clinical
    study of 5-fluorouracil in a variety of far advanced human
    malignancies, J.Natl.Med.Assoc., 54:315 (1962).
27. C. L. Khung, T. C. Hall, A. J. Piro, and M. M. Dederick, A
    clinical trial of oral 5-fluorouracil, Clin.Pharmacol.Ther.,
    7:527-523 (1966).
28. G. E. Moore, D. J. Brass, R. Audman, S. Nadler, R. Jones, N.
    Slack, and A. A. Rimm, Effects of 5-fluorouracil (NSC-19893)
    in 389 patients with cancer, Cancer Chemothr.Rep., 52:641-653
    (1968).
29. D. B. Rochlin, J. Shiner, E. Langdon, and R. Ottoman, Use of
    5-fluorouracil in disseminated solid neoplasms, Ann.Surg.,
    156:105-113 (1962).
30. F. J. Ansfield, J. M. Schroeder, and A. R. Curreri, A prelimin-
    ary comparison of 5-fluoro-2'-deoxyuridine administered by
    rapid daily intravenous injections and by slow continuous
    infusion, Cancer Chemothr.Rep., 16:289-390 (1962).
31. W. L. Wilson, H. F. Bisel, E. T. Krementa, R. C. Lein, and J. V.
    Prohaska, Further clinical evaluation of 2'-deoxy-5-fluoro-
    uridine (NSC-27640), Cancer Chemothr.Rep., 51:85-90 (1967).
32. E. Frei, III, J. N. Bickers, J. S. Hewlett, M. Lane, W. V.
    Leary, and R. W. Talley, Dose schedule and antitumor studies
    of arabinosyl cytosine (NSC 63878), Cancer Res., 29:1325-1332
    (1969).
33. W. Regelson, J. F. Holland, G. L. Gold, J. Lynch, K. B. Olson,
    J. Horton, T. C. Hall, M. Krant, J. Colsky, S. P. Miller, and
    A. Owens, 6-Mercaptopurine (NSC-755) given intravenously at
    weekly intervals to patients with advanced cancer, Cancer
    Chemother.Rep., 51:277-282 (1967).
34. H. M. Lemon, D. M. Miller, J. Smith, and E. E. Walker, Remission
    of metastases of erythropoietin-secreting renal cell adeno-
    carcinoma after 6-mercaptopurine (NSC-755) therapy, Cancer
    Chemother.Rep., 36:49-140 (1964).
35. R. M. O'Bryan, J. K. Luce, R. W. Talley, J. A. Gottlieb, L. H.
    Baker, and G. Bonadonna, Phase II evaluation of adriamycin in
    human neoplasia, Cancer, 32:1-8 (1973).
36. A. L. Watne, D. Moore, and B. Gorgun, Solid tumor chemotherapy
    with mitomycin C, Arch.Surg., 95:175-178 (1967).
37. T. Kato, R. Nemoto, H. Mori, and I. Kumagai, Correspondence:
    Microencapsulated mitomycin-C therapy in renal cell carcin-
    oma, Lancet, 1:479-480 (1979).
38. A. L. Watne, J. Badillo, A. Koike, T. Kondo, and G. E. Moore,
    Clinical studies of actinomycin D, Ann.NY Acad.Sci., 89:
    445-453 (1960).
39. R. G. Hahn, C. B. Begg, and T. Davis, Phase II study of
    vinblastine-CCNU, triazinate, and dactinomycin in advanced
    renal cell cancer, Cancer Treat Rep., 65:711-713 (1981).

40. D. E. Johnson, R. A. Chalbaud, P. Y. Holoye, and M. L. Samuels, Clinical trial of bleomycin (NSC-125066) in the treatment of metastatic renal carcinoma, Cancer Chemother.Rep., 59:433-435 (1975).

41. S. Kofman and R. Eisenstein, Mithramycin in the treatment of disseminated cancer, Cancer Chemother.Rep., 32:77-96 (1963).

42. J. W. Myers, D. D. Von Hoff, C. A. Coltman, J. G. Kuhn, D. Van Echo, S. Rivkin, and R. Pocelinko, Phase II evaluation of bisantrene in patients with renal cell carcinoma, Cancer Treat Rep., 66:1869-1871 (1982).

43. J. Solomon, M. J. Alexander, and J. Steinfeld, Cyclophosphamide: A clinical study, JAMA, 183:165-170 (1963).

44. H. L. Atkins, H. G. Gregg, and G. A. Hyman, Clinical appraisal of cyclophosphamide in malignant neoplasms, Cancer, 15: 1076-1080 (1962).

45. C. J. Anders and N. H. Kemp, Cyclophosphamide in treatment of disseminated malignant disease, Brit.Med.J., 2:1516-1523 (1961).

46. M. F. Fox, The effect of cyclophosphamide on some urinary tract tumors, Br.J.Urol., 37:399-409 (1965).

47. D. E. Bergsagel and W. C. Levin, A prelusive clinical trial of cyclophosphamide, Cancer Chemother.Rep., 8:120-134 (1960).

48. B. I. Shnider, G. L. Gold, T. Hall, M. Dederick, H. B. Nevinny, K. G. Patee, L. Lasagna, A. H. Owens, M. Hreschyshyn, O. Selawry, J. F. Holland, A. Franzino, C. G. Zubrod, E. Frei, and C. Brindley, Preliminary studies with cyclophosphamide, Cancer Chemother.Rep., 8:106-111 (1960).

49. D. A. L. Dick and A. F. Phillips, Clinical experience with cyclophosphamide in malignant disease, Canad.med.ass.J., 85: 974-986 (1961).

50. Z. Wajsman, S. Beckley, S. Madajewicz, and N. Dragone, Abstract: High dose cyclophosphamide (CPM) in metastatic renal cell cancer, Proc.AACR and ASCO., 21:423 (1980).

51. G. Kiruluta, A. Morales, and S. Lott, Response of renal adeno-carcinoma to cyclophosphamide, Urology, 6:557-558 (1975).

52. G. E. Moore, D. J. Brass, R. Ausman, S. Nadler, R. Jones, N. Slack, and A. A. Rimm, Effects of chlorambucil (NSC-3088) in 374 patients with advanced cancer, Cancer Chemother.Rep., 52:661-666 (1968).

53. D. S. J. Choy, J. Arandia, and I. Rosenbaum, Clinical evaluation of a new alkylating agent, azetepa, in one hundred and twenty-five cases of malignant tumors, Int.J.Cancer, 2:189-193 (1967).

54. A. Mittelman, D. J. Albert, and G. P. Murphy, Lomustine treatment of metastatic renal cell carcinoma, JAMA, 225:32-35 (1973).

55. C. Merrin, A. Mittleman, N. Fanous, Z. Wajsman, and G. P. Murphy, Chemotherapy of advanced renal cell carcinoma with vinblastine and CCNU, J.Urol., 113:21-23 (1975).

56. J. K. Luce, W. G. Thurman, B. L. Issacs, and R. W. Talley,
     Clinical trials with the antitumor agent 5-(3,3-dimethyl-1-
     triazeno)imidazole-4-carboxamide (NSC-45388), Cancer
     Chemother.Rep., 54:119-124 (1970).
57. R. J. Gralla and A. Yagoda, Phase II evaluation of chlorozotocin
     in patients with renal cell carcinoma, Cancer Treat Rep.,
     63:1007-1008 (1979).
58. L. H. Rodriguez and D. E. Johnson, Clinical trial of cis-
     platinum (NSC-119875) in metastatic renal cell carcinoma,
     Urology, 11:344-346 (1978).
59. S. D. Fossa and K. Talle, Treatment of metastatic renal cancer
     with ifosfamide and mesnum with or without irradiation,
     Cancer Treat Rep., 64:1103-1108 (1980).
60. H. R. Schumacher and J. P. O'Connell, The intravenous use of
     uracil mustard (U-8344), Cancer, 16:345-349 (1963).
61. N. C. Andrews and W. L. Wilson, Phase II study of methotrexate
     (NSC-740) in solid tumors, Cancer Chemother.Rep., 51:471-474
     (1967).
62. G. Baumgartner, R. Heinz, H. Arbes, R. Lenzhoffer, N. Pridun,
     and J. Schuller, Methotrexate-citrovorum factor used alone
     and in combination chemotherapy for advanced hypernephromas,
     Cancer Treat Rep., 64:41-46 (1980).
63. G. Baumgartner, R. Heinz and G. Linemayr, Methotrexate citro-
     vorum factor therapy in advanced hypernephromas, Wien Klin
     Wochenschr., 92(15):526-530 (1980).
64. J. R. Hindmarsh, R. R. Hall, and A. E. Kulatilake, Renal cell
     carcinoma: A preliminary clinical trial of methodichlorophen
     (D.D.M.P.), Clinical Oncology, 5:11-15 (1979).
65. R. M. Bukowski, A. LoBuglio, J. McCracken, and R. Pugh, Phase II
     trial of Baker's antifol in metastatic renal cell carcinoma:
     A Southwest Oncology Group Study, Cancer Treat Rep., 64:
     1387-1388 (1980).
66. H. C. Falkson and G. Falkson, A clinical trial with hydroxyurea,
     Med.Proc., 13:436-438 (1967).
67. H. Lerner and G. L. Beckloff, Hydroxyurea administered inter-
     mittently, JAMA, 192:138-140 (1965).
68. H. B. Nevinny and T. C. Hall, Chemotherapy with hydroxyurea
     (NSC-32065) in renal cell carcinoma, J.Clin.Pharmacol., 8:
     352-359 (1968).
69. L. L. Stolbach, C. B. Begg, T. Hall, and J. Horton, Treatment of
     renal carcinoma: A phase III randomized trial of oral
     medroxyprogesterone (Provera), hydroxyurea, and nafoxidine,
     Cancer Treat Rep., 65:689-692 (1981).
70. W. A. Knight, R. B. Livingston, D. Fabian, and J. Costanzi,
     Phase I-II trial of methyl-GAG: A Southwest Oncology Group
     Pilot Study, Cancer Treat Rep., 63:1933-1937 (1979).
71. W. A. Knight, R. B. Livingston, C. Fabian, and J. Costanzi,
     Methylglyoxal-bis-guanylhydrazone (methyl-GAG, MGBG) in
     advanced renal carcinoma, Proc.AACR and ASCO., 21:367 (1980).

72. R. F. Todd, III, M. B. Garnick, and G. P. Canellos, Abstract: Chemotherapy of advanced renal adenocarcinoma with methyl-glyoxal-bis-guanylhydrazone (methyl-GAG), Proc.AACR and ASCO, 21:340 (1980).

73. R. F. Todd, M. B. Garnick, G. P. Canellos, J. P. Richie, R. F. Gittes, R. J. Mayer, and A. T. Skarin, Phase I-II trial of methyl-GAG in the treatment of patients with metastatic renal adenocarcinoma, Cancer Treat Rep., 65:17-20 (1981).

74. J. Zeffren, A. Yagoda, R. C. Watson, R. B. Natale, M. S. Blumenreich, R. Chapman, and J. Howard, Phase II trial of methyl-GAG in advanced renal cancer, Cancer Treat Rep., 65:525-527 (1981).

75. J. Killien, D. Hoth, F. Smith, P. Schein, P. Woolley, and V. T. Lombardi, Methyl-glyoxal-bis-guanylhydrazone (NSC 32946) (methyl-G): Phase II experience and clinical pharmacology, Proc.AACR and ASCO., 21:368 (1980).

76. R. J. Schneider, T. M. Woodcock, and A. Yagoda, Phase II trial of 4'-(9-acridinylamino)methanesulfon-m-anisidide (AMSA) in patients with metastatic hypernephroma, Cancer Treat Rep., 64:183-185 (1980).

77. D. A. Van Echo, S. Markus, J. Aisner, and P. H. Wiernik, Phase II trial of 4'-(9-acridinylamino)methanesulfon-m-anisidide (AMSA) in patients with metastatic renal cell carcinoma, Cancer Treat Rep., 64:1009-1010 (1980).

78. M. W. Pasmantier, M. Coleman, B. J. Kennedy, R. Eagan, R. Carolla, R. Weiss, L. Leone, R. T. Silver, Piperazinedione in metastatic renal papac carcinoma, Cancer Treat Rep., 61: 1731-1732 (1977).

79. T. F. Hogan, D. L. Citrin, and B. L. Freeberg, A preliminary report of mitotane therapy of advanced renal and prostate cancer, Cancer Treat Rep., 65:539-540 (1981).

80. T. E. Davis and F. B. Manalo, Combination chemotherapy of advanced renal cell cancer with CCNU and vinblatine, Proc.AACR and ASCO, 19:316 (1978).

81. J. A. Levi, D. Dalley, and R. Aroney, Abstract: A comparative trial of the combination vinblastine (V), methotrexate (A) and bleomycin (B) with and without tamoxifen (T) for metastatic renal cell carcinoma (RCC), Proc.AACR and ASCO, 21:426 (1980).

82. M. K. Samson, L. H. Baker, J. M. Devos, T. R. Burker, R. M. Izbicki, and V. K. Vaitkevicius, Phase I clinical trial of combined therapy with vinblastine (NSC-49842), bleomycin (NSC-125066), and cis-dichlorodiammineplatinum(II) (NSC-119875), Cancer Treat Rep., 60:91-97 (1976).

83. J. A. Knost, R. K. Oldham, K. R. Hande, R. K. Rhamy, and F. A. Greco, Combination of vinblastine and bleomycin in metastatic renal cell carcinoma, Cancer Treat Rep., 65:349-350 (1981).

84. D. E. Johnson, L. Rodriguez, P. Y. Holoye, and M. L. Samuels, Combination vincristine (NSC-67574) and hydroxyurea (NSC-32065) for metastatic renal carcinoma, Cancer Chemother.Rep., 59:1159-1160 (1975).

85. B. W. Dana and D. S. Alberts, Combination chemoimmunotherapy for
     advanced renal carcinoma with adriamycin, bleomycin,
     vincristine, cyclophosphamide, plus BCG, Cancer Clin.Trials,
     4:205-207 (1981).
86. B. Van der Werf-Messing and J. Mulder, Metastatic kidney cancer
     treated with multiple drug therapy at the Rotterdam Radio-
     therapy Institute, Br.J.Cancer, 29(6):491-492 (1974).
87. D. R. Ishmael, R. Bottomley, and J. Geyer, Effect of nephrectomy
     on eventual response to chemotherapy in renal cell carcinoma,
     Proc.AACR and ASCO., 21:429 (1980).
88. R. W. Talley, N. A. Oberhauser, R. W. Brownlee, and R. M.
     O'Bryan, Chemotherapy of metastatic renal adenocarcinoma with
     a five-drug regimen, Henry Ford Hosp.Med.J., 27:110-112
     (1979).
89. S. B. Katakkar and C. R. Franks, Chemo-hormonal therapy for
     metastatic renal cell carcinoma with adriamycin, hydroxyurea,
     vinvlastine, and medroxyprogesterone acetate, Cancer Treat
     Rep., 62:1379-1380 (1978).
90. G. J. Vosika, M. J. Ryan, I. A. Fortuny, C. Meyer, D. T. Kiang,
     A. Theologides, and B. J. Kennedy, CCNU, vinblastine and
     delalutin therapy in renal cell carcinoma, Med.Pediat.Oncol.,
     5:89-91 (1978).
91. F. Richards, II, H. B. Muss, D. R. White, M. R. Cooper, and
     C. L. Spurr, CCNU, bleomycin, and methylprednisolone with or
     without adriamycin in renal cell carcinoma: A randomized
     trial, Cancer Treat Rep., 61:1591-1593 (1977).
92. N. P. Patel and R. W. Lavengood, Renal cell carcinoma: Natural
     history and results of treatment, J.Urol., 119:722-726
     (1978).
93. D. A. Swanson and D. E. Johnson, A clinical trial of estra-
     mustine phosphate (NSC 89199) in the management of metastatic
     renal cell carcinoma (Meeting Abstract), Proc.AACR and ASCO,
     21:346 (1980).
94. W. Dorn, III, M. P. Gladden, and E. A. Rankin, Regression of a
     renal-cell metastatic osseous lesion following treatment,
     J.Bone and Joint Surg., 57A:869-870 (1975).

TREATMENT OF RENAL CANCER INVADING

THE VENA CAVA AND RIGHT ATRIUM

A. Jardin

Hôpital La Pitié
Paris
France

One of the characteristics of renal cancer is its ability to
spread via the veins. Invasion of the intra-renal veins occurs in
50% of cases [1], of the main renal vein in 15% [2], of the vena cava
in 3 to 8% [1,3] and spread of a neoplastic thrombus into the right
atrium is seen in about 1% of cases. The diagnosis of venous spread
should be made pre-operatively, as it alters the surgical tactics and
puts the patient at a high risk of intra-operative migration of the
thrombus or a fragment of it into the pulmonary arteries by simple
mobilization of the kidney. Fortunately, this event is exceptional
being seen in 3% of cases or less[1,4].

The diagnosis of venous extension depends on two investigations.
The C.T. scan defines the site of the thrombus and its extension and,
in cases of caval invasion, cavography with injection via the superior vena cava provides good visualization of the upper limit of the
thrombus.

PRE-OPERATIVE PRECAUTIONS

In such cases, the surgical operation tends to be difficult
because of the large collateral circulation made up of fragile veins
and because of the risk of migration of the thrombus during the
operation. It is therefore useful to reduce the collateral circulation and to avoid mobilization of the kidney. For this reason,
these cases are a good indication for pre-operative embolization.
Very large tumors and tumors with venous extension are our only
indications for this technique. In cases of caval extension, the
upper limit of the thrombus needs to be clearly established in order
to decide on the surgical approach. Fragmentation of the tumor by

the embolization has been reported by some authors[2], but was never observed in our series.

SURGERY FOR CANCER OF THE KIDNEY WITH SUB-DIAPHRAGMATIC CAVAL EXTENSION

Sub-diaphragmatic caval extension hardly alters the initial approach to surgery for cancer of the kidney. In most cases, the cancer involves the right kidney. A vertical, oblique or tranverse, anterior approach is usually performed, depending on the personal preference of the surgeon. When the thrombus has extended as high as the supra-hepatic veins, a thoracophrenolaparotomy enables better mobilization of the liver in order to expose the vena cava.

Removal of a thrombus from the renal vein which is entering the vena cava is simple when a lateral clamp can be put onto the vein beyond the thrombus. No specific controls or dissection are required. However, when such an operation is not possible, the caval thrombus needs to be removed by cavotomy. For us, this implies clamping the left renal vein (being careful not to damage the reno-vertebral trunk) and the vena cava below the thrombus. The thrombus can usually be removed by a vertical cavotomy which is rapidly closed. When the tumor completely obstructs the vena cava we find it preferable to excise the kidney and the vena cava en bloc, after having ligated the left renal vein at its termination as well as the vena cava above and below the thrombus. The ligation of the left renal vein does not have any dire consequences for the patient.

Clamping of the vena cava above the thrombus is usually well tolerated because of the collateral circulation which ensures the venous return. When the vena cava has to be clamped above a mini-mally obstructive thrombus, the venous return to the heart is reduced by 80%. Massive compensation by the venous system allows a good venous pressure to be obtained, but does not prevent the possibility of a fall in arterial pressure. Clamping of the aorta limits such a phenomenon[5]. Another technique is to insert, into the vena cava above the thrombus, a partially occlusive plastic clip which is of the type usually used in the surgical prevention of pulmonary emboli.

Extension into the vena cava from a cancer of the left kidney is more exceptional. For us, this is the only indication for an initial anterior approach for cancer of the left kidney as we prefer a lumbotomy for inferior pole tumors and thoracoabdominal approach for superior pole tumors. Removal of a thrombus coming from the left renal vein is performed with the same precautions as for a thrombus coming from the right renal vein.

However, when a cavectomy is necessary, the problem is of re-establishing the venous return of the right kidney. We believe the

best solution to be an anastomosis between the right renal vein and
the portal vein, either directly or via a free graft. However, we
have never yet been faced with this situation in a series of more
than 300 extended nephrectomies for cancer.

SURGERY FOR CANCER OF THE KIDNEY WITH EXTENSION TO THE SUPRA-DIAPHRAGMATIC VENA CAVA AND THE ATRIUM

The supra- diaphragmatic inferior vena cava is so short and its
approach is so delicate, that we can consider an extension to this
level to be comparable with extension to the atrium. Several surgical techniques have been described but because of the rarity of this
extension, none have been tested in large numbers of patients. Some
authors use a cardiopulmonary by-pass[3], but we find this technique
too complicated. In 4 cases, we were easily able to remove a very
large thrombus with the following technique - a cardiac surgeon
approaches the right atrium by median sternotomy and controls the
orifice of the inferior vena cava with his finger; during this time,
with the same precautions described above, the thrombus is removed by
cavotomy with without any risk of migration. In our experience, such
a simple operation has always produced very good results and is worth
consideration.

RESULTS

The immediate results of this surgery are excellent: mortality
is less than 5% and morbidity is low. The long-term results however
are much less encouraging since whatever the type of venous invasion
and whatever the stage of the cancer, venous invasion markedly
worsens the prognosis[6-9].

INDICATIONS

A limited extension into the vena cava is an indication for
extended nephrectomy and removal of the thrombus, as cures can be
obtained. However, surgery for thrombus which has reached the atrium
is more debatable, as survival of more than 2 years is exceptional.
In our eyes, it is justified by the fact that the patients are often
young, the post-operative course is simple and such surgical excision
at least gives a hope of cure to the patient.

In conclusion, although extension into the vena cava severely
impairs the prognosis of cancer of the kidney, the precise diagnosis
of the exact location of the thrombus enables the most appropriate
operative technique to be selected which, in our view, should always
be the simplest practicable.

REFERENCES

1. A. Jardin, F. Richard, and P. Frantz, Aspects chirurgicaux de
   l'obstruction des veines rénales. <u>Sem.Uronéphrologie La
   Pitié</u>, p. 1, Masson, Paris, (1979).
2. R. Gittes, Locally extensive renal cell carcinoma, <u>in</u>:"Renal
   Tumors Proceedings of the 1st International Symposium on
   Kidney Tumors," 497, Alan. R. Liss, New York, (1982).
3. J. G. Paul, D. R. Rhodes, and J. R. Skow, Renal cell carcinoma
   presenting as right atrial tumor with successful removal
   using cardio-pulmonary by-pass, <u>Ann.Surg.</u>, 181, 471, (1975).
4. D. G. Skinner, R. F. Pfister, and R. B. Colvin, Extension of
   renal carcinoma into the vena cava: the rationale for
   agressive surgical management, <u>J.Urol.</u>, 207, 711 (1972).
5. J. Cinqualbre, J. M. Py, and C. Bollack, Renal cell carcinoma
   extending into the inferior vena cava. Technical problems.
   <u>in</u>: "Renal Tumors Proceedings of the 1st International Symposium on Kidney Tumors," 529, Alan R. Liss, New York, (1982).
6. G. H., Myers, L. G. Fehrenbaker, and P. P. Kelalis, Prognostic
   significance of renal vein invasion by hypernephroma,
   <u>J.Urol.</u>, 100, 420, (1968).
7. G. P. Kearney, W. B. Waters, L. A. Klein, J. P. Richie, and R.F.
   Gittes, Results of inferior vena cava resection for renal
   cell carcinoma, <u>J.Urol.</u>, 125, 769, (1981).
8. A. D. Beck, Renal cell carcinoma involving the inferior vena
   cava. Radiologic evaluation and surgical management,
   <u>J.Urol.</u>, 118, 533, (1977).

VINDESINE IN RENAL CANCER: AN EORTC STUDY

S. D. Fosså[1], L. Denis[2], A. T. van Oosterom[3],
M. de Pauw[4] and G. Stoter[5]

EORTC Urological Group and Data Center, Brussels
[1]General Department, The Norwegian Radium Hospital
Oslo, Norway
[2]Department of Urology, University Hospital
Middelheim, Antwerpen, Belgium
[3]Department of Clinical Oncology, University Hospital
Leiden, The Netherlands
[4]EORTIC Data Center, rue Heger-Bordet, 1, 1000 Bruxelles,
Belgium
[5]Oncological Department, University Hospital, Amsterdam,
The Netherlands

INTRODUCTION

Vinblastine has shown some cytotoxic activity in advanced renal cancer[1], a malignancy which usually is resistant to chemotherapy. The EORTC Genito-Urinary Tract Cancer Cooperative Group performed a phase II study with another vinca alkaloid derivate: Vindesine. In previous clinical studies[2,3] the drug has resulted in tumor regression and stabilization of the disease.

PATIENTS AND METHODS

Twenty-seven patients with advanced measurable renal cell carcinoma were entered into the study. Ineligibility criteria were: severe hepatic dysfunction, reduced bone marrow function, previous treatment with vinca alkaloids or brain metastases. Three patients were not evaluable for response. In 13 of the remaining 24 patients the disease had progressed during the last 2 months prior to the trial entry. Table 1 summarizes the patient characteristics.

Table 1. Patient Characteristics.

	No. of patients
Patients registered	27
Non-evaluable patients	3
Evaluable patients	24
Prior treatment:	
No treatment	2
Surgery only	16
Surgery + chemotherapy	1
Surgery + radiotherapy	3
Surgery + chemotherapy + radiotherapy	1
Surgery + hormone therapy	1
Marker lesions:	
Lung	15
Metastatic lymph nodes	4
Subcutaneous nodules	2
Primary tumor*	1
Metastatic lymph nodes + lung	1
Liver metastases* + primary tumor*	1

* Measurable by computer tomography

Vindesine (3 mg/m^2 i.v) was given weekly for 6 weeks and there-after every 2 weeks. The evaluation of response was done prior to the 5th Vindesine injection, thereafter every month, using the WHO criteria[4]. Reductions of the weekly dose were performed in patients with leukopenia ($2.0-3.0 \times 10^9/1$), thrombocytopenia ($50-100 \times 10^9/1$), paresthesiae or hepatic dysfunction. More severe toxicity led to discontinuation of the drug.

RESULTS

No objective remissions were observed in the 24 evaluable patients. In 10 patients (5 with prior progressive disease) no change was observed whereas progressive disease was seen in 14 patients.

The dominant side effect was cumulative neurotoxicity, which started after 3-4 injections. The neurological side effects were the main reason for discontinuation of the drug. Hematological and gastro-intestinal toxicity was usually mild to moderate (WHO, grade 1-2[4]).

DISCUSSION

The previously reported favorable effect of Vindesine in renal cancer[2,3] was not confirmed in the present study. "No change", "stabilization" or "minimal regression" is not considered to indicate a drug's efficacy in advanced renal cancer. The "natural history" of the disease is unpredictable: Periods of fast progression may alternate with periods of stable disease and even spontaneous remissions.

The weekly dose of 3 mg/m^2 is considered to be adequate even though the observed hematological side effects in the majority of patients were only mild to moderate. Indeed, the frequency and severity of neurological side effects would not have permitted the use of higher doses. In conclusion, Vindesine is ineffective in advanced renal cancer.

REFERENCES

1. W. J. Hrushesky and G. P. Murphy, Current status of the therapy of advanced renal carcinoma, J.Surg.Oncol, 9:277 (1977).
2. P. W. Wong, A. Yagoda, V. E. Currie, and C. W. Young, Phase II study of vindesine sulfate in the therapy for advanced renal carcinoma, Cancer Treat Rep., 61:1727 (1977).
3. M. Valdivieso, S. Richman, A. M. Burgess, G. P. Bodey, and E. J. Freireich, Initial clinical studies of vindesine, Cancer Treat Rep., 65:873 (1981).
4. A. B. Miller, B. Hoogstraten, M. Staquet, and A. Winkler, Reporting results of cancer treatment, Cancer, 47:207 (1981).

Other participants included: J. Mulder, RRTI, Rotterdam, The Netherlands, J. P. Bergerat, Hop. Hautepierre, Strasbourg, France, F. de Bruyne, Radboud, Nigmegan, The Netherlands, K. Roozendaal, O. L. V. Gasthuis, Amsterdam, The Netherlands, A. Bono, Osp. di Circolo e Fund., E. S. Macchi, Varese, Italy and W. Ten Bokkel-Huinink, A. V.L., Amsterdam, The Netherlands.

ADVANCES IN RENAL CANCER:

ARGUMENTS AGAINST HORMONE THERAPY

G. Pizzocaro, L. Piva, R. Salvioni, E. Ronchi
V. Cappelletti, G. Di Fronzo, and Lombardy Group
Istituto Nazionale Tumori*
Milan
Italy

ABSTRACT

Since July 1979, 96 evaluable patients from 6 Centers in Lombardy underwent radical nephrectomy for renal cell carcinoma. Sex steroid receptors (estrogen: ER; progestin: PgR; androgen: AR) were studied in the cytosol of both the tumor mass and the surrounding healthy parenchyma by the dextran coated charcoal technique. None of the 24 category M1 patients showed an objective response to high dose medroxyprogesterone acetate (MPA) or testosterone propionate for progestin failure, but 33% of patients did show disease stabilization for 4 to 22 months.

The 72 category M0 patients were randomly allocated to a control group and a group treated with high dose MPA for one year. After a median follow-up of 30 months, relapses were 28% in the treatment group versus only 12.5% in the 40 untreated. Patients frequently complained of toxicity from MPA, which was sometimes severe. Receptor studies demonstrated that sex steroid receptors were only occasionally found in the tumor tissue and that no correlation could be found between receptors levels and the clinical behavior of the disease.

This study does not support the advocated hormone dependence of human renal cell cancer.

---

*Istituto Nazionale per lo Studio e la Cura dei Tumori, Milano, with the cooperation of: Clinica Urologica Università di Milano; Urologic Departments of Brescia, Lecco, Magenta, Saronno District Hospitals.

INTRODUCTION

The response of human renal cell carcinoma (RCC) to hormone treatment is controversial: Bloom[1] reported objective responses in 16% of 80 patients with advanced disease treated with medroxyprogesterone acetate (MPA) followed by testosterone propionate for progestin failures, while Alberto and Senn[2] reported no objective response in 40 patients treated with either progestins or androgens. As far as post operative treatment is concerned, Bracci and Di Silverio[3] reported a significant decrease in the recurrence rate of patients treated with adjunctive MPA versus historical controls, while Bono et al.,[4] reported a negative trend in treated patients.

Concolino et al.,[5] reported that 81% of human RCCs were positive for either estrogen or progesterone receptors and postulated that tumors positive for at least one receptor were hormone dependent. In addition, the same authors[6] found that over 50% of RCCs were also positive for androgen receptors and postulated that MPA could counteract the promoting action of estrogens on tumor growth with an androgenic action of progestins. Also Bojar et al.,[7] were able to detect cytoplasmic components which specifically bound the progestin R5020 in 42% of 88 human RCCs studied, but despite the enormous case material investigated, they were unable to confirm the presence of receptor concentrations as high as those reported by Li and Li[8] in the estrogen induced RCC of the golden Syrian hamster. Besides, Bojar et al.,[9] were unable to demonstrate a higher R5020 binder concentration in humans after estrogen administration, while it was strikingly demonstrated in the hamster kidney[8].

Things being so controversial, we started a prospective multicentric study in 1979 in order to correlate clinical responses to hormone treatment and receptor studies in both advanced and localized RCC in humans.

DESIGN OF THE STUDY

Good risk patients with RCC underwent transperitoneal radical nephrectomy; preoperative embolization was avoided. Cytoplasmic receptors to estrogens (ER), progestins (PgR) and androgens (AR) were studied in both the tumor mass and the surrounding healthy parenchyma not infiltrated by the tumor by the dextran coated charcoal technique as described elsewhere[10,11]. Category MO patients were randomly allocated to a control group and to a group receiving adjunctive MPA 500 mg three times per week for one year, after stratification according to sex and intrarenal (T1-2, NO, MO) or extrarenal tumors (T3-4 or N1-2 or V1-2). Category M1 patients were treated with MPA 500 mg daily i.m. for at least two months. At progression they were treated with testosterone propionate 100 mg three times per week i.m.

RESULTS

Twenty eight consecutive patients with metastatic disease (M1) from six centers in Lombardy entered the study from July 1979 to June 1982. There was no post operative mortality nor significant morbidity. MPA toxicity was documented in 5 cases: 4 patients had an increase of the body weight of 10 kg or more and one suffered recurrent thrombophlebitis. Four patients were not evaluable for clinical response either for protocol violations or because of inadequate follow-up. No true objective response was observed in the 24 evaluable patients, but disease stabilization did occur in 8 of them (33%) for 4 to 22 months. The disease progressed rapidly in the remaining 16 patients and subsequent testosterone therapy achieved disease stabilization in 3 of 9 evaluable patients for 4 to 12 months.

Sex steroid receptor studies were carried out in 23 of 28 patients with advanced disease. Only 6 of 23 tumor cytosols (26.1%) had detectable amounts of binding sites either for extradiol, progesterone or androgens versus 13 of normal tissues (56.5%) and rarely did values exceed 10 f.mol/mg.protein. Also, there was no correlation between clinical response and receptor studies[11].

From July 1979 to December 1981, 83 patients with category M0 RCC entered the study. There was no post operative mortality and 12 patients (17%) developed 15 surgical complications. Eleven patients are not evaluable: 3 had to interrupt adjuvant MPA because of severe toxicity and 8 because of an inadequate follow-up. The remaining 72 patients were followed for 16 to 44 months (median 30 months) and it is too early to evaluate survival. Relapses occurred in 14 patients: 9 in the 32 receiving adjunctive MPA (28%) and 5 of the 40 in the control group (12.5%). In addition, 21 side effects to MPA were recorded in 15 of the 32 treated patients. The negative trend in the treated patients was more striking in females: 6 relapses in 13 patients receiving adjunctive MPA (46%) versus only one in the 14 controls (7%). The study is now closed after an accrual of over 150 cases.

Receptor studies are available in 53 of the 72 clinically evaluable patients. Again cytosol binding sites were more often detected in the surrounding healthy parenchyma than in the tumor tissue and levels were usually very low. No correlation could be found between receptor studies, relapses and post operative treatment[10].

CONCLUSIONS

There is no convincing evidence that RCC is a hormone dependent or a hormone responsive tumor in humans. Objective responses to hormone therapy in advanced disease are only occasionally seen and it can be argued whether they are due to a true hormone mechanism or to some cytotoxic effect of high dose hormones. Adjunctive MPA in association with radical surgery in MO patients seems to be harmful since there is a negative trend in the recurrence rate and untoward side effects. The findings of very low levels of binding components for sex steroid hormones in the cytosol of human RCC is also of note, as Maillot et al.,[12] succeeded in demonstrating low levels of sex steroid receptors in many tumors of tissues generally considered to be hormone independent and because of the lack of correlation between receptor studies and clinical response.

Acknowledgement

        Supported in part by grant no. 82.01335.96, Progetto Finalizzato "Controllo della Crescita Neoplastica", CNR, Rome.

REFERENCES

1. H. J. C. Bloom, Hormone induced and spontaneous regression of metastatic renal cancer, Cancer, 32:1066 (1973).
2. P. Alberto and M. J. Senn, Hormonal therapy of renal cell carcinoma, Cancer, 33:1226 (1974).
3. U. Bracci and F. e Di Silverio, Attuali orientamenti nella diagnosi e terapia del carcinoma del rene. L'ormonodipendenza, Atti XLIX Congresso Soc.It.Urol., vol.I-1, p.167 (1976).
4. A. V. Bono, C. Benvenuti, E. Gianneo, G. C. Comeri, and A. Roggia, Progestens in renal cell carcinoma, Eur.Urol., 5:94 (1979).
5. G. Concolino, F. Di Silverio, A. Marocchi, and U. Bracci, Renal cancer steroid receptors. Biochemical basis for endocrine therapy, Eur.Urol., 5:319-322 (1979).
6. G. Concolino, A. Marocchi, V. Toscano, and F. Di Silverio, Nuclear androgen receptor as marker of responsiveness to medroxyprogesterone acetate in human renal cell carcinoma, J.Steroid Biochem., 15:397 (1981).
7. M. Bojar, K. Maar, and W. Staib, The endocrine background of human renal cell carcinoma, Urol.Int., 34:302 (1979).
8. S. A. Li and J. J. Li, Estrogen induced progesterone receptor in the Syrian hamster kidney, Endocrinol., 103:2119 (1978).
9. M. Bojar, K. Maar, and W. Staib, The endocrine background of human renal cell carcinoma II, Urol.Int., 34:313 (1979).
10. G. Pizzocaro, G. Di Fronzo, and L. Piva, Adjunctive medroxyprogesterone acetate to radical nephrectomy in category MO renal cell carcinoma, Eur.Urol., 9:202 (1983).
11. G. Pizzocaro, G. Di Fronzo, and V. Cappelletti, Hormone treatment and sex steroid receptors in metastatic renal cell carcinoma, Tumori, (in press) (1983).
12. K. V. Maillot, P. Lesmack, and H. H. Gentsch, Steroid receptors in tumor of tissues generally considered to be hormone independent. Studies of 120 cases, J.Cancer Res.Clin.Oncol., 93:77 (1979).

THE USE OF AN ANIMAL MODEL TO GAIN

INSIGHTS INTO BLADDER CANCER THERAPY

Mark S. Soloway
Department of Urology
Baptist Memorial Hospital
Memphis
Tennessee, USA

I have used a murine model of bladder cancer, which closely simulates the disease as observed in man, in an effort to answer some of the clinical dilemmas which urologists and oncologists face in the management of patients with bladder cancer. It seems appropriate to separate these clinical problems into those that deal with superficial bladder cancer and those that deal with advanced local or metastatic disease. The most distressing problem facing the patient with superficial bladder cancer is the high likelihood of a subsequent tumor. Although only approximately 10% of patients will ultimately develop a tumor which invades into the muscle or beyond, at least 50% will have a subsequent neoplasm after the first superficial tumor. Thus a major part of our laboratory effort is toward determining the reason for this high incidence of subsequent tumor and evaluating a variety of intravesical therapeutic agents hopefully to reduce or eliminate these new occurrences or true recurrences.

Unfortunately, the survival of patients who have tumors which have invaded the muscle or beyond has not dramatically changed over the past 20 years. Despite radiation therapy and/or radical extirpative surgery the chance of surviving five years averages only 50% for patients with category T2 - T4 bladder cancer. In patients who present with metastatic disease there is a 20-50% chance of obtaining a worthwhile objective tumor regression of disease with single or combination chemotherapy. However, the chance of a sustained complete response is low. To date these chemotherapeutic regimens have not produced an effective adjuvant chemotherapy program.

THE FANFT INDUCED ANIMAL MODEL

The tumor used in our laboratory is induced by the carcinogen N-[4-(5-nitro-2-furyl)2-thiazolyl] formamide (FANFT). FANFT is a nitrofuran which when added to the diet will induce urothelial tumors in virtually all mammals in which it has been tried and particularly in many strains of mice[1,2]. In a recent study we compared both a purified high nutrient diet and a standard diet both containing 0.1% FANFT[3]. The tumor incidence at 35 weeks was 40% in each group and at 52 weeks the incidence was approximately 100% with either diet. The tumors produced in this system closely resemble those in man both grossly and histologically[4]. The mice used in our laboratory are an inbred strain (C3H/He) and resultant tumors induced by the oral carcinogen can be transplanted to syngeneic mice without fear of rejection. To date we have established four transplantable tumor lines by taking the primary bladder tumor and transplanting into the hind limb of recipients. The tumor lines are MBT-2, 8, 683 and 409. The MBT-409 is a squamous cell carcinoma while the others are transitional cell (TCC).

These tumors metastasize although the incidence of dissemination is low when the tumor is placed into the flank[5]. Metastases are usually to the lung but may also appear in the liver.

SYSTEMIC CHEMOTHERAPY OF MURINE BLADDER CANCER

During the last few years there have been an increasing number of clinical trials on the use of systemic chemotherapy for advanced bladder cancer. Fortunately there has been some improvement in the response rates. Paralleling these clinical studies we have used the FANFT-induced murine model to screen single and combination systemic chemotherapy agents. Using the MBT-2 line, one of the early studies identified cis-platin, DDP, as an effective agent in murine bladder cancer[6]. The selection of this drug as the most active among a variety of chemotherapeutic agents helped spur clinical trials which confirmed the efficacy of this drug in human TCC[7,8]. It is currently the most effective single agent in bladder cancer. The close correlation between the results in the animal model and those in the clinic have led us to continue to use this animal model to screen both single and combination chemotherapy. Since there are now numerous platinum analogs we have used the transplantable TCC tumor line MBT-2 to compare the efficacy of DDP and several analogs, JM6, PHM, SHP and JM16. As yet we have not thoroughly evaluated some of the more recent analogs such as CHIP and CBCDA. In this study mice were randomized into 8 groups after receiving 1×10^4 MBT-2 tumor cells in the hind limb. Therapy was given intraperitoneally weekly for three weeks. Mice were monitored for tumor development and, if present, they were measured biweekly. The mean tumor diameter for each group was calculated and compared to untreated controls. DDP provided the

greatest reduction in diameter when measured on Day 25 and, import-
antly, provided the highest increase in life span (% ILS) compared to
the control group (Table 1). This study suggests that the analogs
may provide an advantage with respect to less toxicity, however,
superior efficacy is not anticipated. Hopefully, however, a difference in toxicity might be of clinical value.

Methotrexate (MTX) has some activity in patients with advanced
bladder cancer[9]. Methotrexate might be of added efficacy when used
in combination with DDP. Prior to pursuing a clinical trial this
hypothesis was evaluated using the murine model. Mice were randomized to receive either no chemotherapy, MTX, DDP, or the combination
of the two agents in different dosages. The study was done utilizing
both the MBT-409 (squamous) and the MBT-683 (TCC) tumors. Both
studies indicated that the greatest reduction in mean tumor diameter
was achieved by the combination of DDP and MTX (Table 2 and 3). This
study suggests that the combination be evaluated clinically. Both
the National Bladder Clinical Collaborative Group and the EORTC have
initiated clinical trials with this combination regimen.

One of the most interesting results from recent studies has been
the additive effect between DDP and radiation[10,11]. This combination might be considered when using preoperative radiation therapy
or full dose radiation therapy in patients with bladder cancer.
Preoperative radiation therapy may improve survival in patients
undergoing cystectomy but the failure rate, primarily from distant
metastases, remains high. Full dose radiation therapy also does not
treat the micrometastases that must frequently exist in patients
harboring T2-T4 tumors. An optimal chemotherapy program designed for
patients receiving either preoperative or full dose radiation therapy

Table 1. Effect of Platinum Analogs on Transplanted Murine
          Transitional-Cell Carcinoma. 1×10^4 MBT-2 Cells D-0.
          Therapy Day 7, 14, 21. Results Day 25.

Therapy (mg/kg)	No. of Mice	Mean Tumor Diameter (mm)	P	T-C^B (Days)	%ILS
Control	20	11.2	–	–	
JM-6 32	15	10.5	N.S.	1.2	98
PHM 45	14	9.8	0.025	2.5	98
SHP 3	14	8.6	0.001	4.9	106
JM-6 40	11	8.0	0.001	6.0	102
JM-16B 20	15	7.5	0.001	6.7	106
JM-16B 32	12	6.6	0.001	6.5	94
DDP 6	12	6.6	0.001	17.2	120

T-C^B = Days for mean tumor diameter of treated mice to reach 11.5 mm
         minus time for controls to reach this size.

Table 2.   Effect of Chemotherapy on murine Squamous Cell Carcinoma.
           5×10^4 MBT-409 Cells, Day 0. Therapy Days 7, 14, 23.
           Results Day 27

Therapy (mg/kg)	No. of Mice	% Tumors	Mean Tumor Diameter (mm)	Std. Dev.	P	% ILS[1]
Controls	14	93	13.5	2.03	---	---
MTX 32	13	100	13.2	1.83	N.S.	0
DDP 6	13	23	4.8	0.15	.001	39
DDP 5 + MTX 50	13	8	4.3	0	.001	54
DDP 6 + MTX 32	13	0*				54

*7 Tumors developed later in the study for a 54% ILS
[1]ILS = Increase in life span.

Table 3.   Effect of Chemotherapy on Murine Transitional Cell
           Carcinoma. 1×10^5 MBT-8 Cells, Day 0. Therapy Days, 7,
           14, 21. Results Day 25

Therapy (mg/kg)	No. of Mice	% Tumors	Mean Tumor Diameter (mm)	Std. Dev.	P	% ILS
Control	22	100	15.7	0.94	----	---
MTX 32	13	100	14.4	1.40	.001	0
DDP 5 + MTX 50	13	100	8.4	1.31	.001	62
DDP 6	13	92	7.5	1.43	.001	46
DDP 6 + MTX 32	13	100	7.3	1.40	.001	67

might not only incorporate an agent with demonstrated effectiveness
in TCC but also use one that is a radiation sensitizer. Other investigators have demonstrated a therapeutic synergism between DDP and
radiation therapy in murine L1210 and P388 leukemia[12,13] and thus
we sought to evaluate this combination in murine bladder cancer.

We utilized the MBT-2 tumor line. Mice received either chemotherapy alone (utilizing DDP or cyclophosphamide) or received one or
two doses of radiation therapy, either 250R or 600 R two times a week
for three weeks. The tumors were placed into the hind limb. Neither
cyclophosphamide nor DDP had any effect in reducing tumor growth
(Table 4). Animals receiving either 1500 or 3600 R had a 20% increase in their lifespan compared to the control group. The group
receiving the higher radiation therapy dose had a significant reduction in the tumor incidence as only 4 of 10 mice developed tumors.
The addition of cytoxan to radiation did not increase the median
survival time compared to the group receiving radiation therapy
alone. The most effective regimen consisted of the combination of

Table 4. Effect of Chemotherapy and/or Radiation Therapy on Transplanted Murine Bladder Cancer. Day 0.5 x 10^4 MBT-2 Cells. Days 7, 14, 21. Therapy

Therapy	No. of Mice	No. of Tumors	%	Median Survival Time (Days)	% ILS
Control	19	17	(89)	56	---
DDP	10	9	(90)	60	107
Cyclo	10	10	(100)	60	107
1500 RAD	9	8	(89)	67	120
3600 RAD	10	4	(40)*	>180	>100
1500 RAD & Cyclo	9	9	(100)	67	120
1500 RAD & DDP	9	7	(78)	77	138
3600 RAD & Cyclo	8	7	(88)	71	127
3600 RAD & DDP	9	2	(22)*	>180	>100

*$P < 0.01$.

the higher dose of radiation therapy plus DDP. Only 22% of mice in this group developed tumors and there was a significant reduction in the tumor growth compared to controls in those that were not cured.

This concept of utilizing one of the most effective drugs against advanced bladder cancer, DDP, with simultaneous radiation therapy, has already been initiated in limited clinical trials with encouraging results[14,15]. A preliminary phase 1-11 study has also been initiated by the National Bladder Cancer Collaborative Group.

PRIMARY FANFT-INDUCED BLADDER CANCER

Although transplantable tumors offer some insight into clinical problems referable to metastatic bladder cancer the use of primary tumors seems to better simulate patients harboring tumors in the bladder. Thus we have performed several studies in which groups of C3H/He mice are placed on the carcinogen, FANFT, and are monitored for tumor development. At the appropriate time therapy can be introduced and its effect on tumor development determined. Since these studies are expensive and require one year for tumor initiation and growth, we utilize a variety of methods to monitor tumor development. Excretory urography (IVP) can be performed in the mice by injecting 1.5 ml of 20% diatrizoate subcutaneously into the upper back and obtaining films at approximately 20-30 minutes. This will outline

the upper urinary tract to detect any obstruction and also will
demonstrate filling defects within the bladder. We have also performed retrograde cystograms to outline the bladder. This compares
favorably with our standard screening method – urinary cytology[16].
In fact, when comparing these two techniques to the gross bladder
examination both cystography and urinary cytology had a 79% accuracy.
When these two techniques were compared with histologic interpretation cytology provided accuracy of 79% vs. 67% for the x-ray examination.

We have used the primary tumors to evaluate the effect of inter-
mittent intravesical chemotherapy. Our first studies utilized the
most commonly employed agent in the United States, thio-tepa[17].
Two groups of C3H/He mice were place on FANFT and were randomly
subdivided into two groups. Some received thio-tepa and others
intravesical saline. In the first group the animals received either
thio-tepa, 5 mg/kg, or saline weekly for five weeks. This was initiated before the expected onset of tumor – that is, the urinary cytology was not yet positive. The second group, however, did not
begin therapy until the cytology was noted to be positive in a high
percentage of the mice. They also received thio-tepa or saline
weekly for five weeks beginning at week 47. The animals in the first
groups were sacrificed at week 45 and those in the cytology positive
groups at week 54.

When therapy we initiated prior to evident tumors those receiving thio-tepa had a tumor incidence of 67% while 73% of the
controls had tumor (Table 5). This was not a significant difference.
However, when the incidence of low grade (1, 11) and low stage (0-A)
carcinomas were evaluated they occurred in 62% of the thio-tepa
treated group vs. only 22% in the saline treated animals and this
difference was highly significant (P=0.01). This suggests that
although the overall incidence of tumor was not reduced with the use
of thio-tepa, it did prevent the rapidity of growth and reduced the
number of invasive tumors.

When treatment was delayed until tumors were present as evi-
denced by positive cytology, the tumor incidence was approximately
90% in each group and there was no difference between the groups in
those with low grade, low stage tumors.

A subsequent evaluation of intravesical chemotherapy included a
group receiving systemic cyclophosphamide. This was included because
of an interesting report by England, et al.,[18] on the beneficial
effect of cyclophosphamide in patients with carcinoma in situ. We
compared the efficacy of systemic cyclophosphamide to intravesical
thio-tepa and mitomycin C (MMC).

The strain utilized in these studies was B6C3F1. Therapy was
initiated after 47 weeks on FANFT. The animals were randomly distri-

Table 7. Comparison of Intravesical Chemotherapeutic Agents in the Treatment of Primary FANFT-Induced Bladder Cancer

Therapy mg/kg	No. of Mice	No. Tumors (%)	Median Bladder Wt. (Range)	Mean Bladder Wt.	Mean Ln.[1]	P
Control Saline	27	18 (67%)	43.0 (21.51-67.86)	41.92	3.73	---
ADR-5	34	17 (50%)	48.18 (24.70-99.14)	53.91	3.98	0.05
TTP-5	33	13 (39%)	53.43 (26.62-79.76)	41.79	3.73	N.S.
MMC-10	36	11 (31%)	54.48 (34.58-606.06)	135.72	4.91	0.025
BCG-10^6 Cells	35	9 (26%)	55.63 (17.28-518.28)	102.40	4.6	N.S.

[1]LN = Mean log bladder weights to base e.

Table 5.   Effect of Thio-Tepa on Murine Bladder Cancer

Group A	Tumors	Gr. 1-11, Stage 0-A
Thio-Tepa	67%	62%
Saline	73%	22% P=0.01
Group B		
Thio-Tepa	90%	49%
Saline	89%	45%

buted in four groups of 33 mice each and received either saline;
cytoxan 50 mg/kg interperitoneally; MMC, 10 mg/kg intravesically; or
thio-tepa, 5 mg/kg intravesically. Therapy was administered weekly
for four weeks and then every two weeks for further doses. The
animals were sacrified at week 72.

The incidence of tumor in the control group was 76% (Table 6).
The only modality which significantly reduced tumor incidence was
MMC. Animals receiving intravesical MMC had a tumor incidence of 32%
and only two animals had invasion into the lamina propria. Although
systemic cyclophosphamide reduced the size of the resultant tumors as
indicated by the mean bladder weight, the tumor incidence was 94%.
Given the lack of benefit from cyclophosphamide and since it is a
urothelial carcinogen[19], I would be reluctant to utilize this for a
long period of time in patients already demonstrating an abnormal
urothelium.

A very recent study using the primary tumor model placed 172
mice which had ingested FANFT for 44 weeks on either adriamycin 5
mg/kg, thio-tepa 5 mg/kg, MMC 10 mg/kg, or BCG, 10^6 cells. The
control group received intravesical saline. All of the agents were
given by the intravesical route for 6 weeks. The animals were
sacrified at week 54.

The control group had a tumor incidence of 67% (Table 7). Each
drug was effective in reducing the incidence of subsequent tumor. An
analysis of the mean bladder weight indicates that this was significant only with adriamycin and MMC. The high mean bladder weight in
the BCG group may reflect the inflammatory reaction with infiltration
of mononuclear cells.

OTHER ASPECTS OF INTRAVESICAL THERAPY

One aspect of the use of intravesical chemotherapy that has not
been explored is the safety of these compounds with particular refer-

Table 6. Effect of Systemic and Intravesical Chemotherapy on Murine
          Bladder Cancer

Therapy	No.	No. of Tumors (%)	Mean Bladder Weight	P
Control	29	22 (76)	79.33	---
Cyclophosphamide	33	31 (94)	60.63	0.05
Thio-Tepa	26	29 (73)	112.07	N.S.
Mitomycin	25	8 (32)	50.91	0.05

ence to their potential carcinogenicity. Mitomycin C has been shown
to be carcinogenic in certain animal strains and in certain
instances[20-22]. These studies have all been done by repeated
intraperitoneal, subcutaneous, or intravenous drug administration.
We thought it would be worthwhile to determine whether the inter-
mittent instillation of intravesical chemotherapeutic agents would
produce tumors in the urinary bladder and we selected MMC as the
first drug to evaluate.

120 C3H/He female mice were randomized into a control group and
given water and three other groups who received MMC 0.2 mg/mouse; MMC
0.4 mg/mouse; and MMC 0.6 mg/mouse. The instillations were continued
every other week for a total of 13 instillations. At week 24 four
mice from each group were sacrificed and the remaining mice were
sacrificed and cystectomy performed during week 53. All bladders
were subjected to histologic review. I emphasize that these mice
were normal and did not receive any other compound.

All the control bladders were normal although many did show some
degree of denudation. A high percentage of the treated animals had
marked denudation, and two invasive tumors were found in mice which
received the lowest MMC dose. One animal in the high dose group had
focal carcinoma in situ. Since murine bladder tumors are extremely
rare in normal mice, one must conclude that these tumors were induced
by the repeated instillation of MMC. This study must be repeated
before any conclusions can be drawn as to the relative carcinogenicity of MMC. Other intravesical drugs such as adriamycin and
thio-tepa must now be evaluated.

MMC is currently the second most frequently utilized drug for
intravesical chemotherapy in the United States. It has been shown to
be effective in trials in the United States[23,24], Europe[25] and
Japan[26]. Although this agent has primarily been utilized for
patients who have failed thio-tepa or have had myelosuppression
related to thio-tepa, recent studies have used this as a first line
agent both for definitive treatment and for prophylaxis following
resection of all evident tumor.

One of the drawbacks with MMC treatment is the expense of this agent. We thus conducted some studies to determine whether MMC was stable in human urine at body temperature and to determine the amount of MMC which could be recovered in the urine following a two hour intravesical administration[27]. In addition we were able to utilize our animal model to determine whether the voided urine-MMC mixture retained the drug's antineoplastic activity.

The method utilized to quantitate the amount of MMC present in the urine following the two hour instillation in 15 consecutive patients was high pressure liquid chromatography. We observed a single peak corresponding to MMC. The amount of MMC in the urine samples ranged from 34-210 µg/ml. In the 15 patients the average total MMC voided was 20.2±4.5 mg or approximately 50% of the 40 mg dose instilled (Table 8).

There are several explanations which may account for our recovery of only 50% of the instilled drug. The most likely explanations are the attachment of the drug to the urothelial surface or its entry into cells. An alternative explanation is that the compound is absorbed and rapidly enters cells and thus serum levels are not detectable.

Does the recovered MMC retain its anti-tumor activity? If it does one could theoretically reutilize it. We thus compared the anti-tumor activity of the MMC-urine solution to MMC obtained directly from a new vial utilizing one of our transitional cell tumor lines, MBT-683. Animals received 7.5×10^4 viable cells in the hind limb and were randomized into a control and three treatment groups. Chemotherapy was given intraperitoneally on day 7 and repeated on days 14 and 21. Group 1 received intraperitoneal MMC obtained from a stock solution with a concentration of 3mg/kg. Animals in group 2 received the voided urine from MMC treated patients. This was titrated so that each animal was given 3mg/kg of this recycled MMC. Animals in group 3 were given urine obtained from healthy volunteers in equal volumes to that administered to animals in group 2.

Table 8. Amount of MMC in Urine 2 hours after Intravesical Therapy (15 samples)

	Average Concentration mcg/ml	Average Volume ml	Total Amt. Recovered mg
Average	135.23	169	20.19
S.D.	45.92		4.51
Range	34.5-209.69	76-345	11.9-29.15

Table 9. Effect of Human Urine Stock MMC and Recycled MMC on
         Murine TCC. 7.5×10^4 Viable MBT-683 Cells Injected Day 1.
         Therapy Days 7, 14, and 21

Therapy	Average Tumor Diameter (Mean ± SD)		
	11 Days	19 Days	29 Days
Control	4.99 ± 0.43	11.20 ± 1.18	17.59 ± 1.77
Human Urine	4.81 ± 0.35	10.13 ± 2.12	16.76 ± 2.14
Recycled MMC (3 mg/kg)	4.90 ± 0.21	7.18 ± 1.61+*	12.41 ± 3.34+*
Stock MMC (3 mg/kg)	5.03 ± 0.48	6.89 ± 1.71+	10.19 ± 3.52+

+ $P < 0.05$ vs Control
* $P < 0.05$ vs Human Urine.

The stock MMC and the recycled MMC each yielded statistically significant growth inhibition of the tumor at all times at which the tumors were measured (Table 9). The mean tumor diameter was not significantly altered in animals receiving normal urine. Thus one can conclude that in an appropriate situation the MMC urine mixture could be frozen and subsequently re-used. Although we do not advocate this procedure, it would theoretically be possible when cost was a major factor.

Acknowledgements

Supported in part by PHS Grant CA18643 by the National Cancer Institute, National Institute of Health and by Research Funds from the Veterans Administration.

REFERENCES

1. E. Erturk, S. M. Cohen, and G. T. Bryan, Urinary bladder carcinogenecity of N-[4-(5-nitro-2-furyl)-2-thiazolyl] formamide in female swiss mice, Cancer Res., 30:1309-1311 (1970).
2. E. Erturk, S. M. Cohen, J. M. Price, and G. T. Bryan, Pathogenesis, histology and transplantability of urinary bladder carcinomas induced in albino rats by oral administration of N-[4-(5-nitro-2-furyl)-2-thiazolyl] formamide, Cancer Res., 29:2219 (1969).
3. M. S. Soloway, I. Nissenkorn, and L. McCallum, Comparison between a purified and standard diet on FANFT-induced murine bladder cancer, Urology, 18:482-484 (1981).

4. A. J. Tiltman and G. H. Friedell, The histogenesis of experi-
    mental bladder cancer, Invest.Urol., 9:218-226 (1971).

5. W. N. Crabtree, M. S. Soloway, R. B. Matheny, and W. M. Murphy,
    Metastatic characteristics of FANFT-induced murine bladder
    tumors, Urol., (in press).

6. M. S. Soloway and W. M. Murphy, Experimental chemotherapy of
    bladder cancer - systemic and intravesical, Sem.Oncol., 6:
    166-183 (1979).

7. A. Yagoda, R. C. Watson, J. C. Gonzalez-Vitale, H. Grabstald,
    and W. F. Whitemore, Cis-dichlorodiammine platinum (11) in
    advanced bladder cancer, Cancer Treat.Rep., 60:917-923
    (1976).

8. M. S. Soloway, Cis-diamminedichloroplatinum (11) (DDP) in
    advanced urothelial cancer, J.Urol., 120:716-719 (1978).

9. R. T. D. Oliver, Methotrexate as salvage or adjunctive therapy
    for primary or invasive carcinoma of the bladder, Cancer
    Treat.Rep., 65:179-181 (1981).

10. M. S. Soloway, C. R. Morris, and B. Sudderth, Radiation therapy
    in cis-diamminedichloroplatinum in transplantable and primary
    murine bladder cancer, Int.J.Rad.Oncol.Biol.Phys., 5:1355
    (1979).

11. M. S. Soloway, M. Ikard, M. Scheinberg, and J. Evans, Concurrent
    radiation and cis-platinum in the treatment of advanced
    bladder cancer: A preliminary report, J.Urol., 128:1031
    (1982).

12. E. B. Douple, W. L. Eaton, Jr., and M. E. Tulloh, Skin radio-
    sensitization studies using combined cis-dichlorodiammine-
    platinum (11) in radiation, Int.J.Rad.Oncol.Biol and Phys.,
    5:1383 (1979).

13. P. C. Merker, I. Wodinsky, J. Mabel, A. Branfman, and J. M. L.
    Venditti, A comparison of combination chemotherapy and
    combined modality studies with cis-platinum-diamminedichlor-
    ide using in vivo animal tumor models, J.Clin.Hematol.Oncol.,
    7:301-320 (1977).

14. H. M. Pinedo, A. D. Karim, W. H. Van Vliet, G. B. Snow, and
    J. B. Vermorken, Daily cis-dichloro-diammine platinum as a
    radio enhancer: A preliminary toxicity report, J.Cancer Res.,
    Clin.Oncol., 105:79-82 (1983).

15. G. Jakse, H. Frommhold, and H. Marberger, Combined cis-platinum
    radiation therapy in bladder cancer stage pT3 and pT4. A
    pilot study, J.Urol., (in press).

16. I. Nissenkorn, L. McCallum, W. N. Crabtree, W. M. Murphy, and
    M. S. Soloway, Evaluation of cystography for detection of
    bladder carcinoma in mice - comparison with urinary cytology,
    Urology, 18:68 (1981).

17. W. M. Murphy and M. S. Soloway, The effect of thio-tepa on
    developing and established mammalian bladder tumors, Cancer,
    45:870-875 (1980).

18. H. R. England, E. A. Molland, R. T. D. Oliver, and J. P. Blandy,
    Systemic cyclophosphamide in flat carcinoma in situ of the

bladder, in: "Bladder Cancer: Principles of Combination Therapy", R. T. D. Oliver, W. F. Hendry, and H. J. G. Bloom, eds., Butterworths, London, pp.97-105 (1981).

19. R. M. Pearson and M. S. Soloway, Does cyclophosphamide induce bladder cancer? Report of a case and review of the literature Urology, 11:437-447 (1978).

20. J. H. Weisburger, D. P. Griswold, J. D. Prejean, A. E. Casey, H. B. Wood, and E. K. Weisburger, The carcinogenic properties of some of the principle drugs used in clinical cancer therapy, Recent Results in Cancer Research, 52:1-17 (1975).

21. R. Ikegami, Y. Akamatsu, and M. Harota, Subcutaneous sarcomas induced by mitomycin C in mice, Acta.Path.Jap., 17:495-501 (1967).

22. D. Schmahl and H. Oswald, Experimental studies on carcinogenic effects of anticancer chemotherapeutics and immunosuppressives, Arzniem-Forsch, 20:1461-67 (1970).

23. R. B. Bracken, D. E. Johnson, A. C. Von Eschenbach, D. A. Swanson, D. DeFuria, and S. Crooke, Role of intravesical mitomycin C in management of superficial bladder tumors, Urol., 16:11-15 (1980).

24. M. S. Soloway and K. S. Ford, Subsequent tumor analysis of 36 patients who have received intravesical MMC for superficial bladder cancer, J.Urol., (in press).

25. H. Huland and U. Otto, Mitomycin installation to prevent recurrence of superficial bladder carcinoma, Eur.Urol., 9:84-86 (1983).

26. T. Mishina, K. Oda, S. Murata, H. Ooe, Y. Mori, and T. Takahashi, Mitomycin bladder instillation therapy for bladder tumors, J.Urol., 114:217 (1975).

27. S. C. Hopkins, R. G. Buice, R. Metheny, and M. S. Soloway, Stability and anti-tumor activity of recycled intravesical mitomycin C, Cancer, (in press).

FLOW AND STATIC CYTOMETRIES

M. Grattarola[1], G. Giannetti[1], G. Bonanno[2]
P. Carlo[2] and A. Raveane[1]

Institutes of Electrical Engineering[1] and Pharmacology[2]
University of Genoa[2], Italy

INTRODUCTION

Automated cell flow microfluorometry and automated cell image analysis are two recent complementary approaches to cell biology, with far-reaching implications concerning both basic research and clinical applications.

The two main features of flow microfluorometry are rapidity of analysis, with consequent extremely good statistics, and the capability of describing in detail the whole distribution of cell populations, instead of simply giving average values. These two features make other techniques, like autoradiography and many bulk biochemical and immunological assays almost obsolete. On the other hand, the two main features of cell image analysis are the capability of quantitatively measuring morphological and densitometric parameters of single cells or tissues, and the permanent storage not only of numerical data but also of the complete images of already analyzed cells and tissues. These two features represent, at least theoretically, a tremendous improvement over any kind of subjective (i.e., visual) inspection of biological samples.

Obviously, in order to utilize these two powerful techniques correctly a rigorous standardization of the processes of preparing and staining the biological material is needed, besides some knowledge of the physico-chemical phenomena involved.

This crucial point will be better clarified below, where the two techniques will be described in some detail and their performances will be exemplified by referring to applications carried out with a cell sorter (FACS III) and the ACTA image analyzer, at the Biophysi-

329

cal and Electronic Engineering Division of the Electrical Engineering
Institute, University of Genoa.

AUTOMATED FLOW MICROFLUOROMETRY

The first step in the use of the flow microfluorometry technique
is the addition of a fluorescent dye to a cell suspension. The
suspension of stained cells is then forced to flow in a capillary
tube, where one cell at a time interacts with a beam of light in
front of an optical window. The light excites the dye trapped inside
the cell or on the cell surface. As a result of this interaction, a
forward diffuse light (forward scattering) and a 90-degree emitted
light of longer wavelength (fluorescence) are detected as two independent signals which finally result in two separate histograms.
Such histograms can be displayed and subsequently stored in a
computer for further analyses. The scattering signal is roughly
proportional to the cell volume, while the fluorescence signal is a
measure of the number of fluorescent molecules per cell, which,
according to the chosen dye, can be proportional to cellular parameters, like DNA content (and chromatin structure), RNA content and
membrane receptors. Since an average of 1000 cells per second are
usually analyzed by the above systems, histograms describing the
distribution of the chosen parameter inside the cell suspension can
be generated in a few minutes, with extremely good statistics. If
the sorting option is present, subpopulations of the whole distribution can be selected, physically sorted, and utilized for further
analyses.

Besides the problems related to the preparation of homogeneous
suspensions, the critical step in this method is represented by a
good control of the staining procedure. In fact, the measured data
retain a physical meaning which is related to a biomolecule spatial
conformation and/or amount only if the dye molecule is specific
(binds only) to the biomolecule under investigation. The staining
must be quantitative, i.e., the binding must be characterized by
known stoichiometry. If possible, the stain should be vital, so that
the cells will be available for further studies after optical
measurements.

Without the fulfilment of these conditions, results can be
confusing, and the great potentiality of the system useless. As a
typical case, we recall the confusion still existing about the dye
acridine orange (AO), whose properties as a probe for nucleic acids
have been described in the physico-chemical literature for about
twenty years[1]; nevertheless, it is still regarded as an unpredictable stain in the biomedical world.

Typical applications of flow microfluorometry are in the field
of immunological identification of cell subpopulations by means of

monoclonal antibodies[2], and in the field of cell cycle research
(cancer research, hematology, ageing), where DNA distributions are
used as a tool for following the cycling state and the ploidy of cell
populations.

Figures 1 and 2 show two examples of the last two applications.

The dimmer histogram refers to the stimulated cells, showing a
remarkable amount of cells in the S–G$_2$ compartment. The cells were
stained with AO, according to a method already described[3,4].

Note that the quiescent cells (enhanced histogram) show two
peaks which can be easily identified as G$_0$ and G$_1$ compartments, and
which are characterized by the same amount of DNA but by a different
accessibility to the intercalating probe[5,6].

Figure 2 shows scatter (left) and DNA (right) distributions of
rat liver cells of young animals (top) and old animals (bottom),
respectively. The dye used was again AO. This example shows the
detection of an increase in the ploidy of hepatocytes as a function
of age.

AUTOMATED CELL IMAGE ANALYSIS

In principle, automated cell image analyzers are simply optical
microscopes equipped with a TV camera (or photomultiplier) and connected with electronic computers via memory banks. The basic structure (i.e., the microscope) is exactly the same as routinely utilized

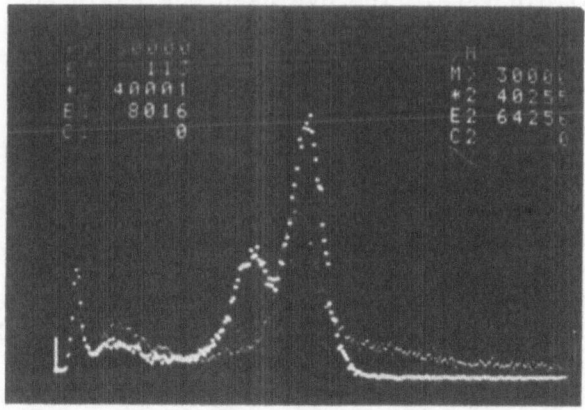

Fig. 1.   Two superimposed DNA histograms show the effect of the
          phytohemagglutinin (PHA)-induced stimulus on human lympho-
          cytes cultured in vitro for 72 hours and analyzed by means
          of our FACS III system.

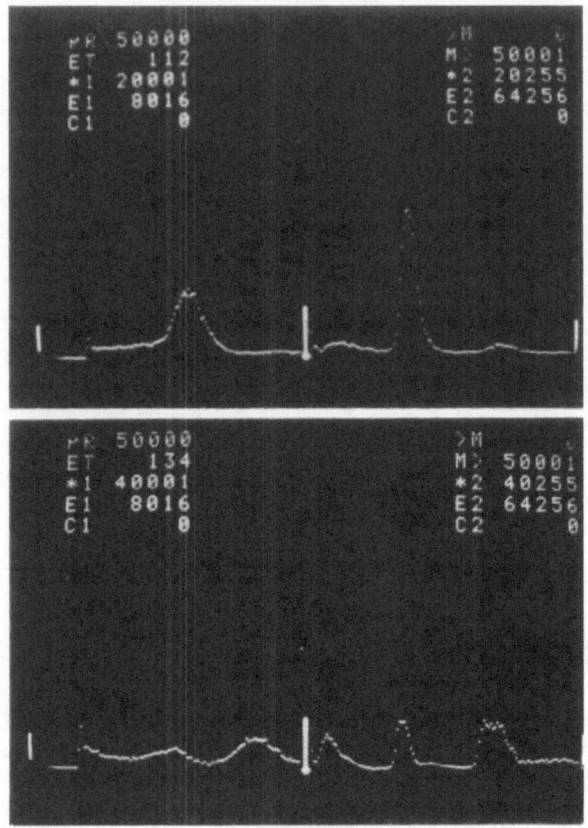

Fig. 2.   DNA distributions of rat liver cells of (a) young animals
          and (b) old rats. The dye used was again AO. This example
          shows the detection of an increase in the ploidity of
          hepatocytes as a function of age.

by cytologists and pathologists, and the optical measurements are
made according to the classic methods of phase-contrast, absorption
and fluorescence microscopy. The fundamental improvements introduced
by TV cameras and computers are twofold. First, it is possible to
quantitatively discriminate between tissues, cells and subcellular
structures in terms of areas, perimeters, optical density values, and
other geometrical and densitometric parameters. Second, once the
images of the samples have been digitalized and stored in the computer memory, they can be analyzed, measured, compared, and elabor-
ated over and over again, with virtually no deterioration of their
quality.

     This technique is very accurate and very slow, allowing the
analysis of only a few hundred images per hour. From this point of

view, its features can be optimized if it is used for the analysis of a few cells previously selected and quickly identified and sorted by means of a flow microfluorometer.

Typical applications of this technique are the study of DNA-chromatin structure and amount in Feulgen-stained nuclei, under absorption conditions, and the quantification of membrane receptors by means of fluorescence microscopy.

Also in the case of the Feulgen method, the staining procedure represents one of the most critical steps. In fact, Feulgen staining does not yield a simple measure of the DNA amount but it is influenced by the compactness of the chromatin structure, as already reported in the literature[7,8], so that physico-chemical parameters, like HCl normality and the hydrolysis duration and temperature, must be carefully monitored in order to get reliable results. As an example, an image of a Feulgen-stained nucleus of a human lymphocyte is shown in Figure 3. This image was obtained with the ACTA system[9], equipped with a Leitz absorption microscope connected, through a Plumbicon TV camera, with a memory (Tesak) and with a minicomputer (HP 21MX). The image has previously been filtered by means of software routines in order to eliminate dirtiness in the background and light imbalances due to uneven illumination.

The histogram superimposed on the nuclear image represents the light intensity distribution ("gray levels") of the image, starting

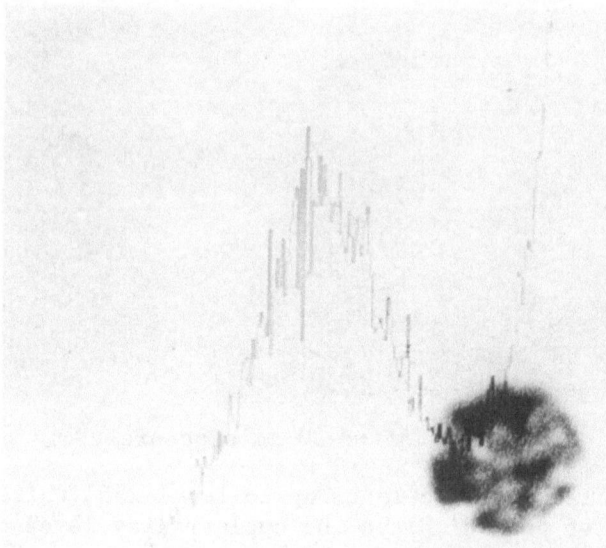

Fig. 3. Image of a Feulgen stained nucleus of a human lymphocyte with superimposed histogram to show the light intensity distribution.

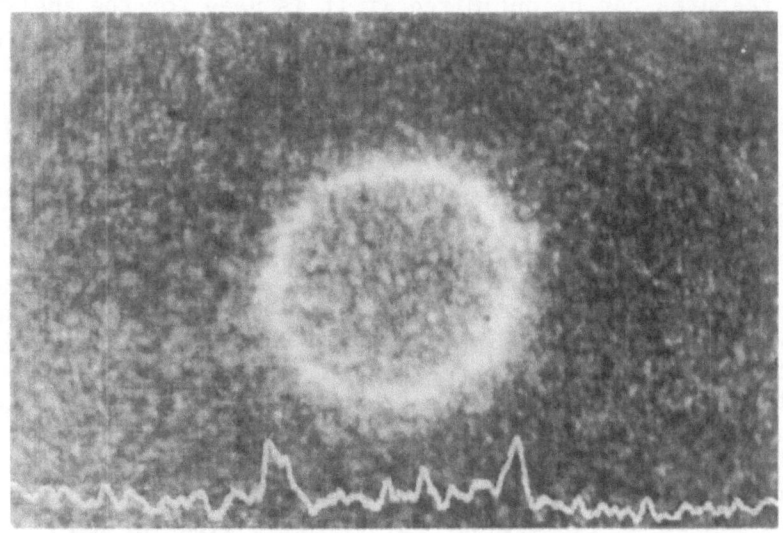

Fig. 4. Membrane receptors of human lymphocytes at point of
         exposure to fluorescein tagged Concanavolin A.

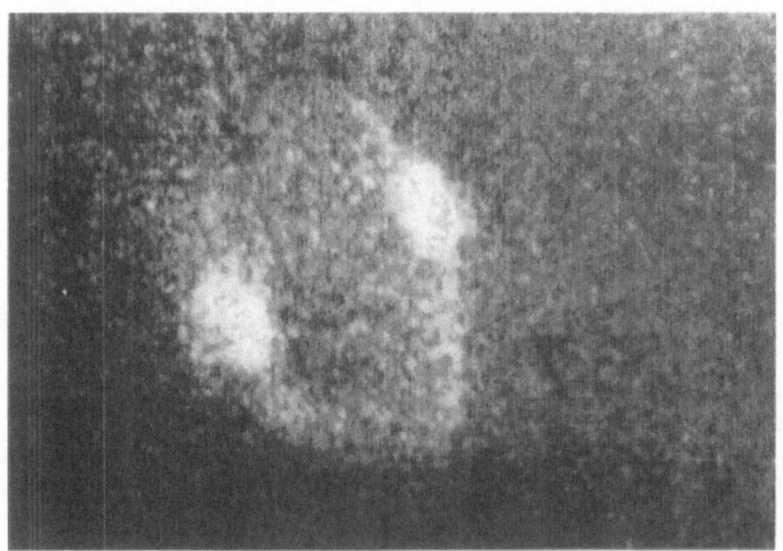

Fig. 5. After 30 mm exposure

from level 0 (black) on the left, up to level 255 (white) on the
right. The first peak displays the nuclear gray level distribution,
while the rising curve on the right represents the background.

     Finally, Figures 4 and 5 illustrate an application in the field
of membrane receptors.

Both images have been obtained with the ACTA system, equipped with a fluorescence Leitz microscope and an intensified ISIT TV camera. The use of this highly sensitive TV camera allows the detection of very low levels of light, so that dim and/or rapidly fading fluorescent objects can be acquired and analyzed[10]. The images show the membrane receptors of human lymphocytes at the beginning of the exposure to fluorescein tagged Concanavolin A (Figure 4), and after 30 minutes of exposure (Figure 5), when the formation of typical localized patches is evident. The line below the image in Figure 4 represents the profile of the gray levels (fluorescence intensity) along a line intersecting the image of the fluorescent cell.

Acknowledgement

This work has been supported by the National Research Council (CNE) of Italy.

REFERENCES

1. A. Blake and A. R. Peacocke, "The interaction of Aminoacridines with Nucleic Acids", Biopolymers, 6:1225 (1968).
2. J. Ortaldo, S. Sharrow, T. Timonen, and R. Herberman: Determination of surface antigens on highly purified human NK cells by flow cytometry with monoclonal antibodies, J.Immunol., 127:2401 (1981).
3. C. Nocolini, A. Belmont, S. Parodi, S. Lessin, and S. Abraham, Mass action and acridine orange staining: static flow cytofluorometry, J.Histochem. and Cytochem., 27:102 (1979).
4. F. Beltrame, A. Chiabrera, M. Grattarola, P. Guerrini, G. Parodi, D. Ponta, G. Vernazza, and R. Viviani, Reevaluation of acridine orange stain for flow cytometry, Bioelectrochemistry and Bioenergetics, 8:387 (1981).
5. L. Smets, Activation of nuclear chromatin and the release from contact-inhibition of 3T3 cells, Experimental Cell Research., 79:239 (1973).
6. S. Abraham, E. Vonderheid, S. Zietz, F. M. Kendall, and C. Nicolini, Reversible (G_0) and nonreadily reversible (Q) non-cycling cells in human peripheral blood, Cell Biophysics, 2:353 (1980).
7. R. Rasch and E. Rasch, Kinetics of hydrolysis during the Feulgen reaction for deoxyribonucleic acid. A reevaluation, J.Histo chem. and Cytochem., 21:1053 (1973).
8. F. Beltrame, A. Chiabrera, G. Giannetti, M. Grattarola, G. Parodi, D. Ponta, G. Vernazza, and R. Viviani, Re-evaluation of Feulgen stain for image cytometry, Bioelectrochemistry and Bioenergetics, 9:645 (1982).
9. F. Beltrame, A. Chiabrera, M. Grattarola, P. Guerrini, G. Parodi, D. Ponta, G. Vernazza, and R. Viviani ACTA automated

image analysis system for absorption, fluorescence and phase-contrast studies of cell images, Proceedings 2nd Annual Conference of the Engineering in Medicine and Biology Society, 58 (1980).

10. D. Taylor and Y. Wang, Fluorescently labelled molecules as probes of the structure and function of living cells, Nature, 284:405 (1980).

CELL MARKERS IN BLADDER CANCER

N. Javadpour

The Surgery Branch, National Cancer Institute
National Institute of Health
Bethesda, Maryland, USA

Perhaps the most encouraging progress in cancer immunology has
been in the field of immunodiagnosis; mainly, development of specific
and sensitive immunocytochemical techniques to measure and localize
tumor markers in the sera and cancer cells of cancer patients. In
this chapter, we will discuss the developments and utilization of
tumor markers in bladder cancer with emphasis on more clinically
established and useful markers[1-19]. Nonspecific serum markers such
as lactic dehydragenase, polyamines and a number of oncofetoproteins
are under study[14].

CELL MARKERS IN BLADDER CANCER

The major problem in diagnosis and management of a superficial
noninvasive bladder cancer has been the lack of criteria to assess
its potential invasiveness. The conventional histopathologic examination of the primary does not completely predict the potentiality of
this cancer to invade. During the last few years it has become
apparent that cellular differentiation in certain cancers is reflected in the presence or absence of certain cell surface antigens,
chromosomal features and potentially T-antigen and tumor associated
antigens.

There are a number of cell antigens that are normally expressed
on the cells of most urothelium that have been utilized as tumor
markers. Among these cell antigens are: 1) ABO(H) antigens, 2)
T-antigen, 3) Tumor-associated antigen(s), and 4) DNA and RNA.

ABO(H) ANTIGENS IN BLADDER CANCER

Surface antigens identical to those that designate the ABO(H) blood groups are present on the surface of normal urothelial cells as well as cells from a variety of tissues (Figure 1). Davidsohn[1] and co-workers developed the specific red cell adherance test (SRCA) to determine the presence or absence of these cell surface antigens. In applying the test to cancer of the cervix, they found that as cells from the cervix undergo malignant dedifferentiation from atypia to anaplasia, there appears to occur a progressive loss of the ABO(H) cell surface antigens. Subsequent reports by a number of laboratories revealed that lesions greater than Stage A consistently showed absence of the ABO(H) cell surface antigens and that loss of these antigens could be correlated with advance in histologic grade. This potential ability of the SRCA test to predict which early bladder tumors might be destined to invade the bladder muscle or to metastasize prompted the investigators to examine the association between ABO(H) cell surface and bladder cancer.

All of the inflammatory lesions of the bladder studied so far have retained ABO(H) cell surface antigens. This is particularly

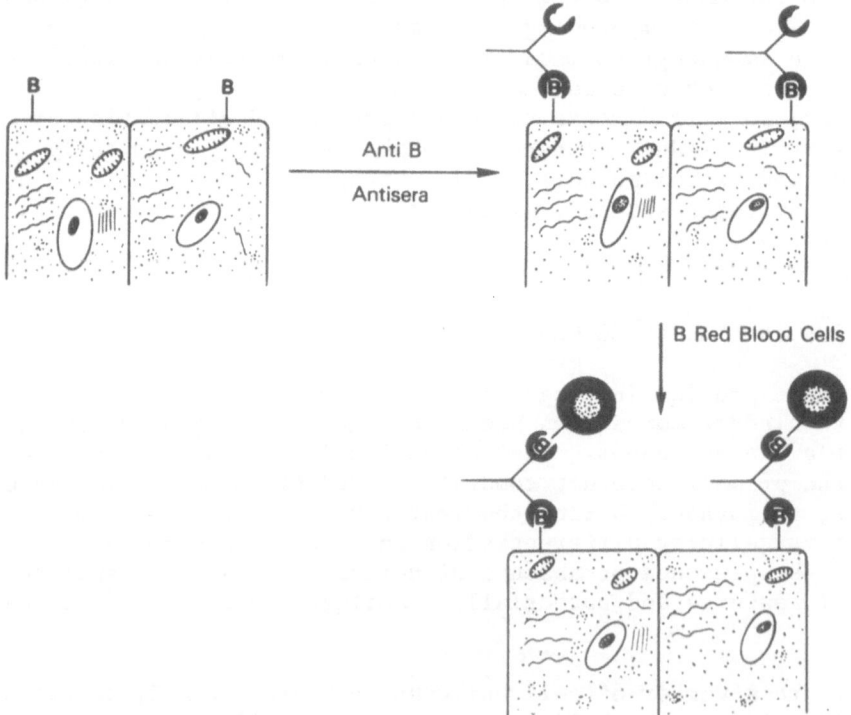

Fig. 1. Schematic drawing of the technique of red cell adherance
         test (SRCA).

important in view of the fact that inflammatory conditions such as cystitis cystica, cystitis glandularis, and squamous metaplasia are frequently seen in association with cancer and might, therefore, seriously influence interpretation of the SRCA test. On the other hand, squamous metaplasia, a nonmalignant lesion, demonstrated consistent loss of the surface ABO(H) antigens. Although this is not surprising in view of previous studies that have shown the ABO(H) antigens not to be present on squamous cells from other areas of the body, it emphasized the necessity of carefully distinguishing benign squamous metaplasia from areas of carcinoma when applying these antigens to bladder lesions.

SRCA in Detection of ABO(H) Antigens

The technique of SRCA has been described before briefly[4]. Slides are deparaffinized in two 5 minute xylene washes, washed in two 5 minute absolute alcohol baths and re-hydrated in three 0.05 M Tris buffer (pH 7.4) baths. Slides are then incubated with Ulex Europus extract for 30 minutes for O(H) and antisera to A, B and AB for the remaining blood group. The Ulex extract is prepared by homogenizing 20 g of Ulex seeds in 100cc buffered saline at 0°C. This is then centrifuged at 0°C, 7000 rpm, for 15 minutes. The supernatant is recentrifuged in like manner and kept frozen until needed. The excess Ulex extract is then rinsed in three 5 minute Tris buffer washes. The slides are then incubated at room temperature with O banked blood and subsequently inverted in Petri dishes so that the buffer touches the slide, washing off untreated RBC, and read after 5 minutes. Endothelium provides a positive internal control and connective tissue and muscle negative internal controls. Currently, a form of immunoperoxidase may be used because it appears to be more reproducible with low false negative rates.

Peroxidase Anti-Peroxidase (PAP) in Detection of ABO(H) Antigens

The technique of PAP has previously been discussed[15]. Briefly, two 5 micron serial sections of formaldehyde-fixed tumor are deparaffinized in xylene. These sections are exposed to 3% hydrogen peroxide in methanol for 25 minutes (Figure 2 and 3). After washing 3 times with Tris solution the slides are covered with 3% normal goat serum to block the non-specific staining. After serial washing with Tris solution the slides are incubated with rabbit antiulex or other antisera to A or B (1:100) in a humid chamber for 60 minutes. The control slides are incubated with normal rabbit serum. After several washings with Tris solution the slides are incubated with goat and anti-rabbit IgG (1:20) for 30 minutes in room temperature and then washed with Tris solution 3 times and exposed to freshly prepared peroxidase anti-peridoxidase (1:50) in 1% normal goat serum for 30 minutes at room temperature. After appropriate washing the slides

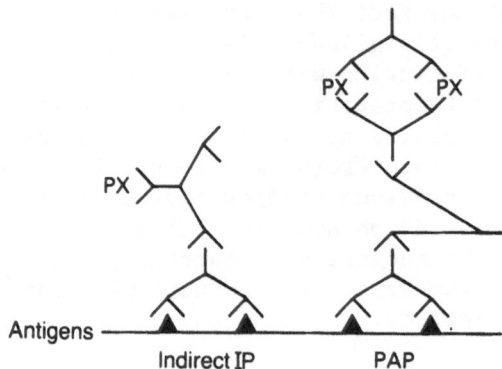

Fig. 2.   Various techniques of immunoperoxidase (IP) including direct
          IP, indirect IP and peroxidase-antiperoxidase (PA).

are covered with diaminobenzidine solution (40 mg/100 ml of Tris plus
1.75 ml of H_2O_2) for 7 minutes. The slides are then counter stained
with hematoxylin for 60 seconds, mounted with cover slides and interpreted in a double blind fashion. The controls consist of internal
controls and the slides that are treated with absorbed rabbit antiulex, anti-A and anti-B antisera.

    Any given test for predicting the prognosis of cancer including
cancer of the bladder should be most specific, sensitive, fast,
inexpensive, reproducible, practical, quantitative and easy to perform with a high accuracy rate.

    In a double blind study we reported that in terms of prognosis,
SRCA was positive in only 1 of 8 patients with a single tumor, whereas PAP was positive in 5 of 8. In patients with recurrent tumors,
all low stage, SRCA was positive in 1 of 15 whilst PAP was positive

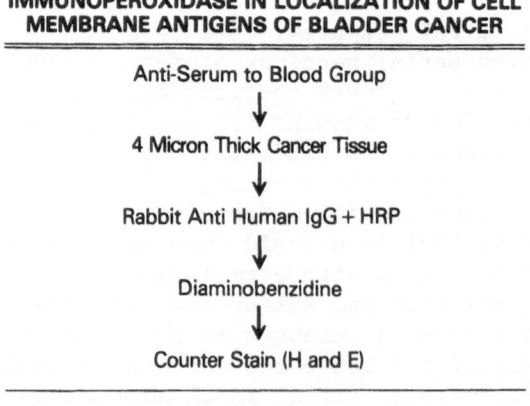

Fig. 3.   Schematic tabulation of PAP.

in 6/15. In invasive lesions both PAP and SRCA were negative in all 7 patients. Thus, prognostically PAP appears to be a better predictor of prognosis than SRCA in Group O patients.

The results of this study demonstrate that PAP decreases the rate of false negative results (Table 4) as compared to SRCA in patients with group O(H) isoantigen. We have previously reported that grade II bladder cancers have generally lost their cell surface antigen. These data also support this finding. Furthermore, it demonstrates the superiority of PAP over the SRCA technique in decreasing the false negative results.

Reports from autopsy material of normal ureters in group O patients utilizing SRCA have been found to be falsely negative in more than a third of studied specimens. Since about 45% of the population expresses the O(H) antigen, we need a more sensitive and reproducible technique than SRCA to detect this antigen.

In conclusion we have demonstrated that PAP correlates better with prognosis and invasiveness of bladder cancer in group O patients than SRCA in a double blind study.

Furthermore, the convenience, permanence and ready availability of the PAP technique should encourage its use in conventional histopathologic laboratories.

CORRELATION OF THESE ANTIGENS WITH STAGE AND GRADE

In seventy-six bladder tumors of various stage and grades examined with the SRCA technique for the presence of the A, B, O(H) cell surface antigen, 70% of the grade I lesions were positive for the cell surface antigen, and none of the 26 grade III tumors retained their antigens (Table 1). When correlated with clinical stage, the tumors showed no antigens present on those of Stages B to D, whereas 12 of 18 Stage A lesions were positive for the antigen (Table 2). When Stage A lesions were studied and the findings correlated with recurrence and metastasis/invasion rates, in only one lesion (which recurred at an invasive stage) was the cell surface antigen present on the initial tumor. Therefore, it is concluded that cell surface antigens are helpful in predicting malignant potential in low-grade, low-stage cancer of the bladder. These antigens may offer the capability of selecting low-grade, low-stage bladder tumors that are destined to invade or metastasize while they are at stages curable by cystectomy.

The findings of cell surface ABO(H) antigens in a limited number of various lesions examined for these antigens are shown in Table 3. Squamous metaplasia in either neoplastic or non-neoplastic lesions consistently demonstrated apparent absence of the ABO(H) cell surface

Table 1. Correlation of Grades and Cell Surface Antigens
          in 76 Patients with Bladder Cancer.

Pathologic Grade	SRCA	
	Positive	Negative
I	7	3
II	15	25
III	0	26

Table 2. Cell Surface Antigen in 76 Patients Correlated
          with the Stage of Tumor.

Clinical Stages	SRCA	
	Positive	Negative
A	12	16
B-D	0	48

antigens. Conversely, cystitis cystica and cystitis glandularis were
always positive for the presence of these cell surface antigens.
Areas of carcinoma within lymph nodes were uniformly negative for the
presence of ABO(H) antigens.

T-ANTIGEN

     The changes in cell surface antigens result in the interruption
of communication and interaction through the cell surface membrane.
These changes reflect an alteration in the genome which modifies the
biosynthetic pathways of the Golgi apparatus. The demonstration of
T-antigen expression in breast carcinoma and the loss of blood group
isoantigens (A, B, O, or H) in invasive transitional cell carcinoma
prompted inquiry into the correlation between the carbohydrate precursor of the human blood group system M.N. and the epithelial
surface blood group isoantigens, to determine if these antigens can
be used as complementary in bladder cancer.

     The T-antigen has been identified in breast carcinomas, colon
and gastric carcinomas and was felt to represent a precursor of the
human blood group MN glycoproteins. Subsequently, a plant lectin
derived from peanuts and termed peanut agglutinin was found to
possess high affinity for glycoproteins containing the terminal

Table 3.   SRCA Findings in Patients with Benign Lesions and Carcinoma
           In Situ[10].

Lesions	Patients (n)	SRCA	
		Positive	Negative
Cystitis cystica	8	8	–
Cystitis glandularis	8	8	–
Squamous metaplasia	4	–	4
Nodal metastases	4	–	4
Carcinoma in situ	8	1*	7

*Partial loss of cell surface antigen. This is considered as a
 negative.

sequence galactose (1-3)-N-Acetyl-D-galastomine. This concealed
antigen was so termed after the Thomsen-Friedenreich phenomenon which
was a result of exposure of the red blood cells to influenza viruses
which contained a receptor destroying enzyme known to be neuraminidase which released the terminal sialic acid exposing the T-antigen.
Expression of this antigen was noted in 17/22 breast carcinomas, all
of which were well differentiated. The five carcinomas that failed
to demonstrate the antigen were poorly differentiated, possibly
implying that the antigen may be further altered as the cell becomes
more anaplastic.

TECHNIQUE FOR DEMONSTRATION OF T-ANTIGEN

     A modified cell adherence assay has been utilized to demonstrate
the T-antigen on the cell surface membrane because of the ease and
simplicity of performing the test (Figure 4). It is primarily a
sandwich type reaction with the deparaffinized tissue sections as the
bottom layer, the indication erythrocytes as the top layer and the
specific anti-T lectin (Peanut lectin) as the middle layer. Demonstration of the indicator erythrocytes implied the cells in tissue
expressed the T-antigen. Peanut seeds (10 gms) are soaked in H_2O_2
for 24 hours and then homogenized in 25cc of buffered saline; this is
then centrifuged for five minutes at 3200 revolutions per minute.
The supernatant was decanted off and placed in a Beckman preparatory
ultracentrifuge at 48,000 revolutions per minute for one hour. The
supernatant is then frozen at minus 20cc until needed. T-antigen
activated erythrocytes are prepared by using outdated blood bank
cells washed in buffered saline three times. A solution of .5cc
mucopolysaccharide N-acetylneuraminylhydroinase (500 units/cc) plus
2.5cc of packed erythrocytes plus 2.5cc phosphate buffer at pH 7.4
was made and incubated at 37°C for four hours. The T-activated

erythrocytes were stored at 2-8° until used. Positive and negative
controls are made on each histologic section to be studied. The
positive control consists of placing a solution of neuraminidase and
phosphate buffer (50 units/cc) on a slide of the deparaffinized
section previously described for thirty minutes. The slides are
washed in three changes of phosphate buffer for five minutes in each
change. The T-activated erythrocytes are then incubated with the
slides for thirty minutes before inversion in Petri dishes containing
phosphate buffer. The negative control consists of incubating the
slide with peanut agglutinin for 30 minutes, then washing in phosphate buffer for five minutes three times. The section is then
incubated with non-T-activated erythrocytes for 30 minutes. The
actual test consists of a 30 minute incubation of the slide with
peanut agglutinin followed by washing in phosphate buffer for five
minutes three times. The slide is then incubated with T-activated
erythrocytes for 30 minutes.

CARCINOMA IN SITU

    Predicting the ultimate natural history of carcinoma _in situ_ is
of utmost importance in management of this lesion.

    The natural history of carcinoma _in situ_ remains controversial.
The high incidence of carcinoma _in situ_ with infiltrative tumors in
the bladder and the more ominous prognosis that this situation
implies have encouraged urologists toward more definitive early
treatments. However, in the absence of gross lesions, carcinoma
in situ may behave in a less aggressive fashion.

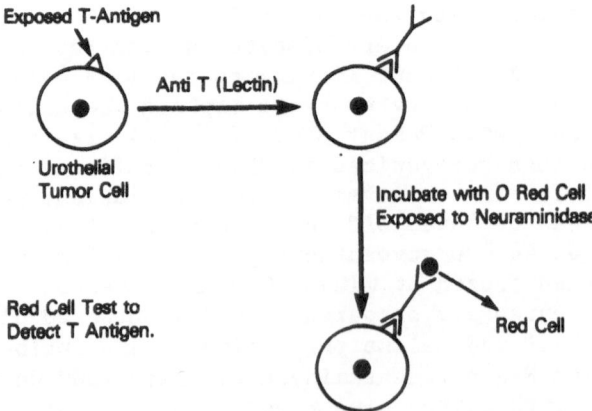

Fig. 4. Quality control and improvements for SRCA.

SIMULTANEOUS DETERMINATION OF T-ANTIGEN AND ABO CELL
SURFACE ANTIGENS IN TRANSITIONAL CELL CARCINOMA

Bladder tumors have been studied utilizing combined antigens
such as T-antigen and ABO antigens in an effort to assess the role of
combined antigens in bladder cancer. The red cell adherence assay to
detect T-antigen utilizing peanut lectin which has anti-T properties
as well as other nonspecific binding sites is layered over the
slides, washed, then covered with neuraminidase treated red cells.
In studying 17 patients, 14 demonstrated loss of their cell surface
ABO antigens. Of the 14, 28% demonstrated expression of T-antigen.
This would imply that T-antigen may be complementary to ABO(H) as in
predicting potential malignancies of bladder cancer. Combined
antigens in bladder cancer need more investigation. The data from
our laboratory indicates a lack of correlation of T-antigen to the
stage, grade and invasiveness of bladder tumor, whereas ABO(H)
antigens correlate well.

PERSPECTIVES AND LIMITATIONS

The development of various techniques including immunoperoxidase
to detect the presence or absence of the blood group ABO(H) or
T-antigen has been encouraging in predicting the natural history of
the pathologically low-grade, low-stage bladder cancer that may be
transformed into the high grade and high stage. Utilizing these
markers, one may predict the natural history of the bladder cancer
that would not be clear otherwise. The high-grade anaplastic tumor
of the bladder will lose the antigen, although the low-grades, low-
stages are immunologically well differentiated. These findings by us
and other investigators have important implications in management of
the bladder cancer.

Although attempts may be made to perform the SRCA with O blood
group, false negative results may occur. At the present time the
prolonged (30 minute) incubation and preparation of fresh reagents
with a pH of 7.4 diminished the false negatives (Table 4). Also in
patients receiving radiotherapy, a conversion from a negative to a
positive reaction has occasionally been demonstrated. In light of
the lack of these isoantigens on the surface of squamous metaplastic
cells of bladder, the value of this reaction appears to be limited in
squamous cell carcinoma of the bladder. This limitation of SRCA does
not affect the usefulness of the test, since the majority of bladder
cancers in the western hemisphere are of transitional cell origin.
Another limitation of SRCA is that it is a qualitative rather than a
quantitative test as are many other histopathologic tests. However,
one can make this test permanent and counterstain it with eosin in
order to recognize the cells that have lost their antigens in relation to the other cells in a section. It is important to realize
that any given test for predicting the prognosis of cancer, including

Table 4.   Quality Controls and Improvements.

SRCA[a]

1.   Conventional H&E stain

2.   Specificity controls
     a)   Positive control
     b)   Negative control
     c)   Inflammatory lesions

3.   Quality controls
     a)   Fresh antisera and indicator RC
     b)   Incubation - Ulex extract
     c)   pH of TBS (7.4)

---

[a]H and E, hematoxylin and eosin.

cancer of the bladder, should be most specific, sensitive, fast, inexpensive, reproducible, practical, quantitative and easy to perform with a high accuracy rate. At the present time, such a test is not available for any cancer. However, continuous effort to improve the test and to accumulate accurate knowledge is essential in evaluating any given test, including the specific red cell adherence (SRCA) test. Currently, the immunoperoxidase and peroxidase/antiperoxidase (PAP) tests are being explored in my laboratory in an attempt to improve and to standardize this test. The data indicate the superiority of PAP over SRCA in the detection of O(H) isoantigen.

Potentiality for Radioimmunotherapy in Bladder Cancer

There are a number of cell surface antigens that are normally expressed on the cells of most urothelium or urothelial cancer. The antigens that are expressed on bladder cancer cells are T-antigen and tumor associated antigens. These antigens are concealed in the normal transitional cell of the bladder and expressed during carcinogenesis. T-antigen has been detected on bladder cancer utilizing a specific red cell adherence test or immunoperoxidase in bladder cancer (Fig. 5). It appears attractive to attempt to label a gamma-emitting radioactive material to the T-antigen or other tumor associated antigen for the purpose of localization or perhaps chemo- or radioimmunotherapy of metastatic bladder cancer. These approaches are attractive because of lack of reliable techniques to localize the lymphatic involvement in terms of diagnosis and the radiosensitivity of this cancer. With the advent of monoclonal antibodies (Fig. 6) one may prepare large amounts of pure antibodies for diagnostic and therapeutic purposes.

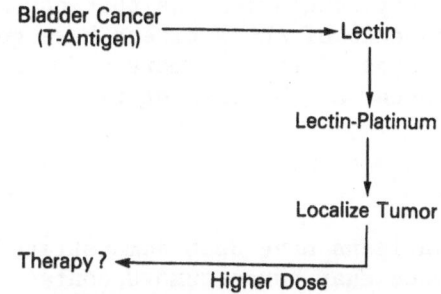

Fig. 5. Affinity therapy for bladder cancer.

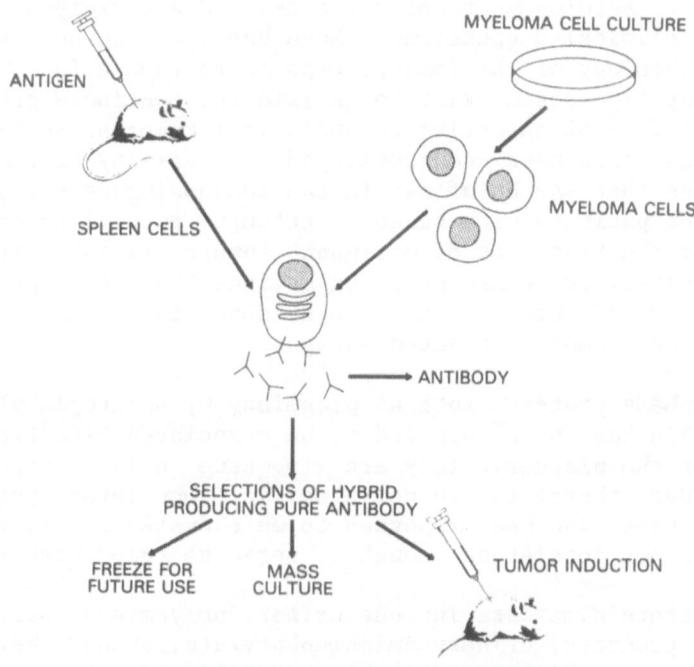

Fig. 6. General scheme of production of monoclonal antibodies.

COMBINED CELL SURFACE ANTIGENS AND CHROMOSOMAL ANALYSIS

Chromosomal abnormalities also appear to be a sensitive bladder tumor marker. Chromosomal changes appear to be of prognostic value in low-grade, low-stage bladder cancer in predicting the rate of recurrence and invasiveness of a clinically superficial tumor. Tumors with normal chromosomes are usually noninvasive. The combination of the SRCA, chromosome testing, and computed tomography has been of value in staging and determining the natural history of bladder cancer. In a prospective study of 20 patients with bladder cancer in which cell surface antigens and chromosomal analysis were employed, the preliminary conclusion was that chromosomal analysis and the presence or absence of ABO antigens have good correlations with the grades and stages of these tumors and are the most reliable markers for bladder cancer at the present time.

TUMOR-ASSOCIATED ANTIGEN

Tumor-specific antigens have been demonstrated on a variety of animal tumors. Evidence that human tumors contain tumor-specific antigens comes from many sources including the increased incidence of cancer in immune deficiency states, the in vitro demonstration of cellular and humoral immune responses to tumor antigens, reactivity of patients to autologous tumor extracts, and a variety of circumstantial clinical suggestions. Much has been learned about the basic immunobiology of the immune response to tumors in animals and humans. Many hypotheses exist to explain the continued growth of the tumor in the face of an active immunological response against it. Immunological tests have been developed to a variety of tumor-associated antigens that may be of use in the immunodiagnosis and immunoevaluation of patients with cancer. Attempts to utilize the immune response for the treatment of malignant tumors are in their earliest stages. However, there are many suggestions that this approach may be of value in the future. There is evidence that human bladder cancer may have tumor-associated antigen.

Acute phase proteins such as plasminogen, hepatoglobulin, and immunoglobulin have been reported to be associated with transitional carcinoma of the bladder. They are also seen in inflammatory lesions of the bladder, therefore are not specific. Oncofetoproteins such as CPE and HCG have also been reported to be elevated in bladder cancer, but they are not consistent enough to serve as tumor markers.

Other protein markers include urinary polyamines, urinary fibrin degradation products, urinary aminoisobutyrate, urinary beta-glucuronidase and urinary tryptophan metabolic products have been occasionally found in bladder cancer but they are not specific.

Flow Cytometry for DNA and RNA

Flow cytometry is based on passing the urinary sediments from patients with bladder cancer through a focused Argon laser beam in single file. Prior to this testing, the urinary sediments are stained with acridine orange which causes DNA to fluoresce green and RNA red. In a study of 48 patients with papilloma of the urinary bladder, Klein[11] and co-workers have concluded that for patients at risk of developing tumor of the bladder, a combination of DNA and RNA flow cytometry appears to offer greater diagnostic sensitivity than flow cytometry based on DNA content alone.

In following 97 patients with a history of low grade, low stage
bladder cancer utilizing flow cytometry these authors[12] have also
concluded that flow cytometry is a sensitive, quantitative method of
detecting abnormal cells in bladder irrigation specimens. Therefore,
it appears that flow cytometry of DNA and RNA will play a more important role in diagnosis and follow-up of patients with bladder
cancer in the future.

Future Perspectives

The role of cell surface antigens in low grade superficial
bladder cancer is encouraging. Utilization of immunoperoxidase in
detecting cell surface antigens especially in patients with blood
group O has yielded a low false negative making this test more
attractive and reproducible. Furthermore, application of chromosomal
studies with T-antigen detection has provided additional parameters
in the diagnosis and management of bladder cancer. Flow cytometry
also appears to be useful in the follow-up of patients with low
grade, superficial bladder cancer. Utilizing multiple markers
especially ABO(H) when detected with the immunoperoxidase or PAP
technique and chromosomal abnormality, will predict the clinical
course of a superficial bladder cancer in terms of prognosis of
future recurrence and invasiveness.

It must be emphasized that cell markers should be related to
other parameters such as grade, stage, multiplicity in time and
space, recurrence and available therapeutic options. It is our
experience that when superficial bladder tumors lose their ABO(H) and
have abnormal chromosomes, they are highly at risk for recurrence,
invasion and metastases.

REFERENCES

1. S. Bergman and N. Javadpour, The cell surface antigen ABO(H) as
    an indicator of malignant potential in Stage A bladder
    carcinoma, Preliminary report, J.Urol., 119:49 (1978).
2. J. S. Coon and R. Weinstein, Variability in the expression of
    the O(H) antigen in human transitional epithelium, J.Urol.,
    125:301 (1981).
3. J. S. Coon and R. S. Weinstein, Detection of the ABH tissue
    isoantigen by immunoperoxidase methods in normal and neo-
    plastic urethelium, Am.J.Clin.Path., 76:163 (1981).
4. I. Davidsohn, Early immunologic diagnosis and prognosis of
    carcinoma, Am.J.Clin.Path., 57:715 (1972).
5. B. A. Diamond, E. Dale, B. A. Yelton, and M. D. Scharff, Mono-
    clonal antibodies - a new technology for producing serologic
    reagents, N.Engl.J.Med., 304:1344 (1981).

6. R. C. Emmott, N. Javadpour, S. M. Bergman, and T. Soares, Corre-
    lation of the cell surface antigens with stage and grade in
    cancer of the bladder, J.Urol., 121:37 (1979).
7. C. Emmott, M. Droller, and N. Javadpour, Studies of ABO(H) cell
    surface antigen specificity: Carcinoma in situ and nonmalig-
    nant lesions of the bladder, J.Urol., 125:32-35 (1981).
8. D. R. Howard and J. G. Batsakis, Cytostructural localization of
    a tumor associated antigen, Science, 210:201 (1980).
9. N. Javadpour, The role of radioimmunodetection in urologic
    cancer (in press).
10. N. Javadpour, Cell surface antigen in bladder cancer, J.Urol.,
    March (1983).
11. F. A. Klein, M. R. Melamed, W. F. Whitmore, H. W. Herr, and P.
    C. Sogani, Characterization of bladder papilloma by 2 para-
    meters DNA/RNA flow cytometry: Proceeding of 1982 AMA
    Abstract 73.
12. F. A. Klein, H. W. Herr, W. F. Whitmore, and M. R. Melamed, Flow
    cytometry in follow-up of patients with low stage bladder
    tumors, Proceedings of 1981 AMA Abstract 548.
13. P. H. Lange, C. Limas, and E. E. Fraley, Tissue blood group
    antigens and prognosis in low stage transitional cell
    carcinoma of the bladder, J.Urol., 119:52 (1978).
14. M. Lippert and N. Javadpour, Cell surface antigen in renal
    pelvic tumors, Urology (in press).
15. R. McAlpine and N. Javadpour, Comparison of specific red cell
    adherence and immunoperoxidase in detection of ABO(H)
    antigens in normal urothelium - A double bline study, Urology
    (in press).
16. R. McAlpine and N. Javadpour, T-antigen in bladder cancer as
    detected by specific red cell adherence test and immuno-
    peroxidase, Urology (in press).
17. C. Milstein, K. Adetugbo, N. J. Cowan, et al., Somatic cell
    genetics of antibody-secreting cells: Cold Spring Harbor
    Symp. Quant.Biol., 41:793-803 (1977).
18. B. S. Stein and A. R. Kendall, Specific red cell adherence
    testing and radiotherapy, Cancer, 50:23-29 (1982).
19. J. Vafier, N. Javadpour, G. F. Worsham, and K. O'Connell, A
    double blind comparison of T-antigen and ABO(H) cell surface
    antigen in bladder cancer, Urology (in press).

LOCAL IMMUNE RESPONSE IN PRENEOPLASIA AND

IN SITU CARCINOMA OF THE URINARY BLADDER

G. Jakse, E. Rammal and F. Hofstaedter

Departments of Urology and Pathology
University of Innsbruck, Austria

INTRODUCTION

The infiltration of cancer tissue by host cells was termed stromal reaction by Russell in 1908[1]. This term should express the suggestion that the cell infiltrate indicates the resistance of the host of tumor invasion. The stromal reaction has only recently been mentioned as an important factor of cellular immunity to cancer. Moreover studies on cancer immunology have been almost exclusively concerned with the cell populations of the general circulation.

There are few studies concerning the host cell infiltration in patients with bladder cancer [2-6]; These studies indicate that round cell infiltration correlates with tumor differentiation and depth of infiltration[3-5]. Since these investigations deal with advanced or at least established tumors, the stromal reaction at this late stage of the tumor evolution may no longer reflect the capacity of the host to contain the tumor growth.

MATERIAL AND METHODS

Cold cup biopsies of patients with carcinoma in situ(Cis) of the urinary bladder were stained by Haematoxylim and Eosin (H.E.) and evaluated by light microscopy. The morphology of the local reaction was investigated in 63 sections. Lymphoid cells (Ly); polymorpho-nuclear cells (PMN), plasma cells (Pc) and blood vessels (V) were counted in 5 to 10 low microscopic fields. The values were expressed by numbers of cells per mm length of the urothelial lining. In addition T-cells were investigated by means of monoclonal antibodies (T101, Hybritech®) and the immunoperoxidase method.

RESULTS

The diagnosis of the urothelial changes and the amount of Ly, PMN, Pc and V per mm urothelial length are given on Table 1. There were significant differences (two-sided t-test) noted in the amount of stromal reaction in the three classes of urothelium except as stated in the table. There was an almost linear increase in the number of capillary vessels in the lamina propria from "normal" to Cis.

The differentiation of the lymphoid reaction by means of mono-clonal antibodies against surface T-antigen showed in preliminary data that (a) that 50-70% of the lymphoid cells are T-cells, (b) evidence of intraurothelial T-cells.

DISCUSSION

There are several mechanisms of tumor cell destruction including (a) non-specific - direct phagocystosis and proteolytic enzymes; (b) antibody mediated - IgG antibody promotes cell killing through killer cell activation of complement system; (c) cell mediated - T killer cells, armed macrophages, lymphotoxins. Therefore the cells of most interest in the morphological analysis of in situ expression of local immunity are lymphocytes, macrophages and plasma cells.

Since most of our study was done retrospectively we focused our interest on the lymphoid stromal reaction. It was interesting to note that the round cell infiltration was most prominent in dysplasia. This may indicate that dysplastic urothelia, considered as precursors of well or moderately differentiated tumors, express their surface antigenicity to a higher degree than do carcinoma in situ lesions. Although this is an oversimplification of a complex interaction between host and tumor it would at least be in accordance with the findings of Sarma and Mihatsch et al., demonstrating that

Table 1. Numbers of lymphoid cells (Ly), polymorphonuclear cells (PMN), plasma cells (Pc) and blood vessels (V) in the lamina propria given by absolute numbers per mm urothelium length. Statistical analysis was done using the two-sided t-test.

	Ly	PMN	Pc	V
"Normal"(n=11)	99±17.3	0.8±0.8	2±1.2	17±3.3
Dysplasia(n=34)	167±50.3	15±8.5)	10±6.7)	23±7.3
Cis(n=18)	118±65.9*	15±7.5)$^{n.s.}$	9±4.5)$^{n.s.}$	39±7.1

*Cis versus "Normal" is n.s.

well differentiated tumors are associated with a more intensive reaction than those which are poorly differentiated[3,4,5]. This finding would also lead to the suggestion that the pattern of the host reaction in established tumors closely resembles that in pre-neoplasia or in situ carcinoma.

Our data in regard to T-cells are only preliminary. However it is of great interest that we could demonstrate T-cells not only in the stroma but also in the urothelium. There is only one other report by Jacob et al., who showed the presence of lymphoid cells in normal urothelium by electron microscopy[7]. Catalona and coworkers showed on a very limited number of transitional cell carcinomas that the lymphoid cell population consists mainly of T-cells.

The accumulation of plasma cells in and around tumor tissues is considered as an indicator of a local humoral antitumor response. Plasma cell infiltrates were found especially in patients with highly differentiated squamous cell carcinomas of the lung and Hoachim et al.,[4] demonstrated by studies on lung cancer eluates, that specifically reactive antibodies were produced at the tumor site. Moreover this group of investigators were able to show that the amount of antibody production seems to be correlated with the amount of the plasma cell infiltration. In our own preliminary study it seems that the local humoral reaction as indicated by the presence of plasma cells is of minor importance as compared to that in squamous lung cancer.

The increase of capillary vessels from "normal" to ca in situ is another important finding of our study. The proliferation of the vessels may reflect the nutritional requirements of the rapidly growing urothelium or the presence of a biomedical tumoral factor promoting blood vessel formation as advocated by Folkman et al., or both factors may be active[9]. In this context it is of interest to remember the findings of Sarma, who noticed abnormal growth of capillaries in bladders bearing papillary tumors preceding actual tumor development[10].

SUMMARY

It is clear that the results of our study can only be considered as preliminary and cover only a small part of the local immune response, since we neglected such important cell populations as for example the macrophages. Taken together it seems that the pattern of stromal reaction in dysplasia and in situ carcinoma resembles that of the established tumors. A local humoral antitumor response indicated by the presence of plasma cells seems to be of limited importance. The increase in vascularity of the lamina propria from "normal" to in situ carcinoma is perhaps due to the presence of an angiogenic factor. The evaluation of the subsets of the lymphoid infiltration is part of an ongoing study.

REFERENCES

1. B. R. G. Russel, The nature of resistance to the inoculation of
    cancer, Third Scientific Rep.Imperial Cancer Res.Fund,
    3:341-358 (1908).
2. W. C. Catalona, R. Mann, F. Nime, C. Potvin, J. I. Harty, D.
    Gomolka, and J. C. Eggleston, Identification of complement
    receptor lymphocytes (B cells) in lymph nodes and tumor
    infiltrates, J.Urol., 114:915 (1976).
3. M. J. Mihatsch, M. Rist, T. Rompanen, and G. Rutishauser,
    Prognostic significance of peritumoral inflammation in
    invasive urothelial bladder cancer, Urol.Res., 7:97 (1979).
4. K. P. Sarma, The role of lymphoid reaction in bladder cancer,
    J.Urol., 104:843 (1970).
5. K. P. Sarma, Proliferative and lymphoid reactions in bladder
    cancer, Invest.Urol., 10:199 (1972).
6. T. Tanaka, E. H. Cooper, and C. K. Anderson, Lymphocyte
    infiltration in bladder carcinoma, Rev.Europ.Etudes Clin.
    Biol., 15:1084 (1970).
7. J. Jacob, C. M. Ludgate, J. Forde, and W. Selby-Tulloch, Recent
    observations on the ultrastructure of human urothelium,
    Cell Tiss.Res., 193:543 (1978).
8. H. L. Joachim, B. H. Dorsett, and E. Paluch, The immune response
    at the tumor site in lung carcinoma, Cancer, 38:2296 (1976).
9. J. Folkman and R. Cotran, Relation of vascular proliferation to
    tumor growth, Intern.Rev.Exptl.Pathol., 16:208 (1976).
10. K. P. Sarma, Genesis of Papillary tumors: histological and
    microangiographic study, Brit.J.Urol., 53:228 (1981).

EVALUATION OF BLOOD GROUP ANTIGENS A, B IN

TRANSITIONAL CELL CARCINOMA OF THE BLADDER

I. Tuncer, F. Simsek, and A. Akdas

Departments of Urology and Pathology
Hacettepe University Medical Faculty
Ankara, Turkey

Transitional cell carcinoma of the bladder is one of the important causes of death and in these patients a history of previous superficial transitional cell carcinomas is often found. It has been shown that 70% of superficial bladder carcinomas show recurrence within one year, 10-20% of these tumors eventually becoming invasive and metastasizing[1,2]. Many reports have been published about the biological potential of transitional cell carcinoma of bladder including studies of Urinary cytology[3], Chromosomal analyses,[4] chemicals in the urine such as aromatic amines, tryptophan metabolites polyamines[5,6] elevation of urinary enzymes like lactic dehydrogenase, alkalane phosphatase[7] and detection of increased levels of carcinoembryonic antigen[8]. All have been studied as indicators of tumor recurrence.

Tumoral cells can produce tumor-specific, new oncofetal antigens or lose their original antigenic characters[9]. Evaluation of these antigenic alterations can be a good guide in diagnosis, follow-up and detection of recurrences and metastases. Alterations in Blood Group antigens, A, B, and O (normal tissue antigens) are among those most investigated.

After Landsteiner's discovery of these antigens Coombs introduced more sophisticated techniques to show the presence of them in all body cells[10]. Decrease and loss of these antigens are also detected in tumors of different tissues[11,12,13]. The same antigenic variation can also be detected in transitional cell carcinoma of the bladder.

MATERIAL AND METHOD

Twenty eight patients with transitional cell carcinoma of bladder admitted to the Urology Department of Hacettepe University Medical Faculty, were chosen as subjects of this study. The diagnosis was confirmed by clinical, endoscopic and radiological means. Subjects were chosen among those of blood group A, B or A B. Patients with blood group O were excluded because of the difficulties in obtaining "Ulex Europeus" which was needed for detecting the H antigen. The subjects had undergone transurethral tumor resection within the years 1973-1981 and the tumoral tissues were available in blocks. The loss of the blood group antigens were then investigated in these tissues. Records of patients who had T_1 tumors originally were evaluated for recurrence.

The presence of the blood group antigens were detected with the specific red cell adherence technique (SRCA)[10] and the results were analyzed statistically.

RESULTS

Negative adherence of red cells in the neoplastic tissues were considered as a loss of blood group antigens (Figure 1) and this was observed in 19 out of 28 cases with transitional cell carsimoma of the bladder.

Distribution of the antigen loss according to the stage of the tumor is seen in Table 1. Antigen loss was found 64.7% of T1 and 72.7% of T3 patients. No statistical significance is present (P>0.05).

Loss of antigen was seen in one out of six (16.7%) patients with grade 1 lesions and in eleven out of thirteen (86.6%) grade 2 and seven out of nine (77.8%) grade 3 tumors statistical significance between the antigen loss and the histologic differentiation of tumor was found between the groups grade 1 and grade 2 and Grade 1 and Grade 3 patients (p<0.05).

Distribution of blood group antigens according to the grade of tumor in T_1 patients is shown in Table 3. Loss of antigen was seen in 16.7% of Grade 1, all grade 2 and 66.7% of grade 3 patients.

Patients with T_1 tumors who have been followed approximately 4,5 years were evaluated for the rate of recurrence according to the grade of tumors (Table 4). Tumor recurrences were seen in 16.7% of G 1, 87.5% of G 2 and 100% of G 3 tumors. There was a statistical significance between the recurrence and histologic grading of the tumor (p<0.05).

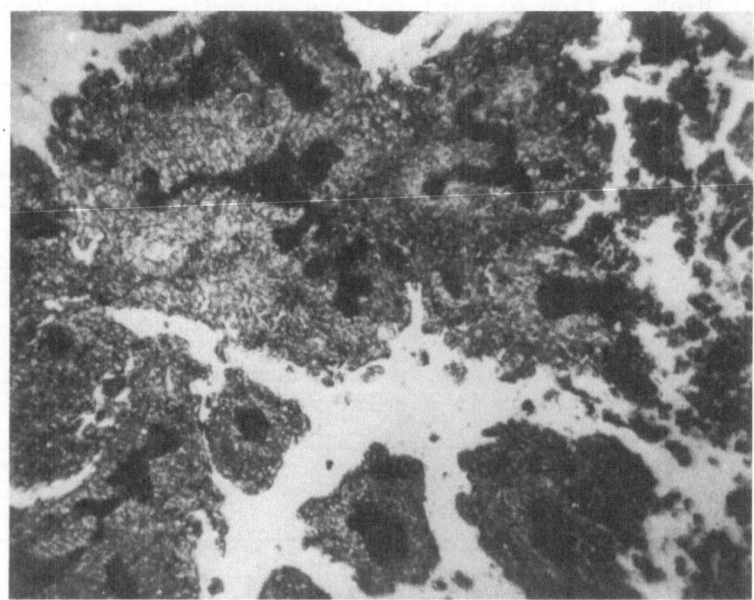

Fig. 1.   This micrograph shows the failure of red cells to adhere to
          tumoral cells, and indicates loss of antigen. Note that
          adherence of red cells to vascular endothelium in the
          bladder wall serves as a positive control.

        We also found statistical significance between the loss of
antigen and the recurrence of tumor, (P<0.05).   (Table 5).

DISCUSSION

        Prediction of the behavior of bladder cancers and application of
a definitive therapy at the time of diagnosis can be a difficult task
and detection of blood group antigens in tumoral tissues has played
its role within these studies.

        In this study, among the 17 T1 patients admitted to our clinic,
11 patients had antigen loss in their tumoral tissues (64.7%) as
compared to 8 out of 11 with T3 tumors (72.7%). No statistical
difference is present (P>0.05). In a study based only on clinical
staging, Emmot and Javadpour state that there is no antigen loss in
most of the patients, with superficial transitional cell tumors,
whereas antigen loss is seen in all the patients with invasive
tumor[14,15].

        Clinical staging alone is not sufficient to allow the use of
blood group antigen loss to predict the tumor behavior. The degree
of anaplasia must be considered as the most important criterion as

Table 1. Distribution of Antigen Loss According to the Stage of the
          Tumor.

Clinical Stage	No. of pts.	Antigen (+)	%	Antigen (-)	%
T_1	17	6	35,3	11	64,7
T_3	11	3	27,3	8	72,7

Table 2. Distribution of Blood Group Antigens According to the
          Grade of Tumor in all Patients.

Grade of Tumor	No. of pts.	Antigen (+)	%	Antigen (-)	%
G 1	6	5	83.3	1	16.7
G 2	13	2	13.4	11	86.6
G 3	9	2	22.2	7	77.8

Table 3. Distribution of Antigens According to the Grade of the
           Tumor in T_1 Patients.

Grade of Tumor	No. of pts	Antigen (+)	%	Antigen (-)	%
G 1	6	5	83.3	1	16.7
G 2	8	0	0	8	100.0
G3	3	1	33.3	2	66.7

Table 4. Presence of Recurrence According to the Grade of T_1
           Tumors.

Grade of Tumor	No. of pts	Recurrence (+)	%	Recurrence (-)	%
G 1	6	1	16.7	5	83.3
G 2	8	7	87.5	1	12.5
G 3	3	3	100.0	0	0

Table 5.   Recurrence in Patients with T_1 Tumors According to the Presence of Antigen

Blood Group Antigens	No. of pts.	Recurrence (+)	%	Recurrence (-)	%
Antigen (+)	6	2	33.3	4	66.7
Antigen (-)	11	9	81.8	2	18.2

this is more closely associated with loss of antigenic properties[16] and the loss of blood group antigens must be evaluated in parallel with the histological grading of the tumor.

We found a striking relationship between the loss of antigen and the grade of the tumor. 16.7% of grade 1, 86.6% of grade 2, and 77.8% of grade 3 tumors showed antigen loss with a significant statistical difference ($P<0.05$). This relationship has also been mentioned in other reports[14,15-17].

It is a well known fact that there is a parallelism between the stage and the grade in transitional cell carcinoma of the bladder [1,2]. As may be seen in Table 3, 6 of the tumors of T1 patients are grade 1, 8 of them are grade 2 and 3 of them are grade 3. If the antigen loss is evaluated within this distribution it will also be seen that 83.3% of the well diffentiated superficial tumors maintain their antigen. Grade 2 and grade 3 tumors showed loss of antigens in 100 and 66.7% respectively, similar to the results of other studies-[14,15,17].

The most important goal of detecting the blood group antigens in tumoral tissues is to predict the biological and clinical behavior of the tumor. In this respect the patients who have been followed for some time were selected and are evaluated. Tumor recurrence was detected in 16.7% of grade 1, 87.5% of grade 2 and 100% of grade 3 tumors over an average 4.5 years of follow-up. There is a statistical significance between the recurrence and histological grading. A similar proportion to the antigen loss in these groups. (Table 3). While 81.8% of tumors with antigen loss showed a recurrence, only 33.3% of the tumors with positive antigen did so. There also is a statistical significance within these groups ($P<0.05$). Therefore, it can be concluded that the possibility of recurrence of a tumor with antigen loss is quite high.

The relationship between the antigen loss, tumor recurrence and the type of recurrence is one other point that has been investigated. In study of this sort recurrence is found in 40% of tumors with antigen presence and in 73% of tumor with antigen loss. These figures are like ours[17].

The relationship between the invasive potential of recurrent tumor and antigen loss, shows contradictiory results. According to Kay and Wallace,[18] this relation is not clear whilst Decenzo et al.,[19] state that invasion is seen in 20% of tumors which are antigen positive and in 76% of antigen negative tumors. Similar results have also been obtained by Richie et al.,[20] and Johnson & Lamm[17].

In our survey among the T1 tumors 6 showed superficial and 5 showed invasive recurrence.

It seems that none of blood group antigens A and B in transitional carcinoma of the bladder can be a good indication of future invasive tumor recurrence. Javadpour et al., have shown a 5 year survival of 56% in patients with an invasive recurrence of a T1 tumor when treated by cystectomy or radiotherapy[21]. If the possibility of invasion can be predicted by detection of antigen loss, more radical methods of treatment can be applied from the beginning to increase life expectancy.

SUMMARY

The loss of blood group antigen A and B was detected in the tumoral tissues of 28 patients with transitional cell carcinoma of the bladder. 16.7% of grade 1, 86.6% of Grade 2, and 77.8% of Grade 3 tumors were found to have lost their tissue antigens. The correlation between the stage and the loss in blood group antigens was thought to be meaningful only when interpretated in association with the grade of the tumor.

The recurrence rate in the tumors that have lost their antigens was 81.8% and invasive recurrence was more likely in these cases.

The loss of blood group antigens in tissues can be easily detected in routine laboratory conditions by the specific red cell adherence test. Life expectancy of the patients with transitional cell carcinoma of the bladder may be increased by detecting the changes in blood group antigens in tissues and using these alterations in monitoring the treatment and follow-up.

REFERENCES

1. J. B. De Kernion and D. G. Skinner, Epidemiology, diagnosis and staging of bladder cancer, in: "Genituorinary Cancer," D. G. Skinner, J. B. Dekernion, eds., W. B. Saunders, Philadelphia (1978).
2. C. A. Olsson and R. W. White, Cancer of the bladder, in: "Principles and Management of Urologic Cancer," N. Javadpour, ed., Williams and Wilkins, Baltimore (1979).

3. J. Lessing, Bladder Cancer. Early diagnosis and evaluation of
    biologic potential. A review of newer methods, J.Urol.,
    120:1 (1978).
4. W. H. Falor and R. M. Ward, Prognosis in early carcinoma of
    bladder based on chromosomal analysis, J.Urol., 119:44 (1978).
5. O. Yoshida, R. R. Brown, and G. T. Bryan, Relationship between
    trytophan metabolism and heterotropic recurrences of human
    urinary bladder tumors, Cancer., 25:773 (1970).
6. E. J. Sanford, J. R. Drago, T. J. Rohner, G. F. Kessler, L.
    Sheehan, and A. Lipton, Preliminary evaluation of urinary
    polyamines in the diagnosis of genitourinary tract
    malignancy, J.Urol., 113:218 (1975).
7. J. D. Schmidt, Significance of total urinary lactic
    dehidrogenase activity in urinary tract disease, J.Urol.,
    96:950 (1966).
8. G. Reynoso, T. M. Chu, P. Guinan, and G. P. Murphy, Carcino-
    embryonic antigen in patients with tumor of the urogenital
    tract, Cancer., 30:1 (1972).
9. M. J. Weiss, and J. V. Klavins, The antigenic nature of
    mammalian cell surfaces mammalian cell membranes, in: "Mem-
    brane and Cellular Function," (Volume:4), G. A. Jamieson, and
    D.M. Robinson, eds, Butterworths, London, p.72 (1977).
10. R. R. A. Coombs, D. Bedford, and I. M. Roulland, A and B blood
    group antigens on human epidermal cells demonstrated by mixed
    agglutination, Lancet., 1:461 (1956).
11. H. E. M. Kay, A and B antigens of normal and malignant cells,
    Brit.J.Cancer., 11:409 (1957).
12. S. Kovarik, I. Davidhson, and R. Stejekal, ABO antigens in
    cancer, Arch.Pathol., 86:12 (1968).
13. I. Davidsohn, Early immunologic diagnosis and prognosis of
    carcinoma, Am.J.Clin.Pathol., 57:715 (1972).
14. S. Bergman and N. Javadpour, The cell surface antigen A. B. or H
    as an indicator of malignant potential in stage A bladder
    carcinoma: Preliminary report, J.Urol., 119:49 (1978).
15. R. C. Emmot, N. Javadpour, S. Bergman, and T. Soares,
    Correlation of cell surface antigens with stage and grade in
    cancer of the bladder, J.Urol., 121:37 (1979).
16. W. J. Catalona, Bladder carcinoma, guest editorial, J.Urol.,
    121:37 (1979).
17. D. J. Johnson and D. L. Lamm, Prediction of tumor invasion with
    the mixed cell agglutination test, J.Urol., 123:25 (1980).
18. H. E. M. Kay and D. M. Wallace: A and B antigens of tumors
    arising from urinary epithelium, J.Nat.Cancer Inst., 26:1350
    (1961).
19. M. J. Decenzo, P. Howard, and E. C. Irish, Antigenic deletion
    and prognosis of patients with stage A transitional bladder
    carcinoma, J.Urol., 114:874 (1975).
20. J. P. Richie, R. D. Blute, and J. Wajsman, Immunologic
    indicators of prognosis in bladder cancer, The importance of
    cell surface antigens, J.Urol., 123:22 (1980).

21. N. Javadpour, J. Roy, R. Bottomley, and K. Dagg, Combined cell
     surface antigens and chromosomal studies in bladder cancer,
     Urol., 19:29 (1982).

ECHOGRAPHY IN BLADDER CANCER STAGING

F. Micali, M. Porena, G. Vespasiani, and G. Virgili

Department of Urology
University of Perugia
Italy

There are three possible approaches when using ultrasonography
to scan the bladder: the abdominal, (Figure 1) the transrectal,
(Figure 2) and the transurethral (Figure 3). Although the method of
approach depends on the clinical picture and findings of other diagnostic examinations, various combinations of the three are often used
to give a more complete picture.

Bladder ultrasonography is indicated not only for tumors but
also for all those disorders leading to morphological alterations in
the bladder wall and its contents. Its use in cases of dysfunction
is being investigated. In tumor staging the choice of the best
ultrasonic method is suggested by the clinical picture and the findings of previous diagnostic examinations.

The abdominal approach. It is easy, rapid, non-invasive and
therefore repeatable, and bladder scans using different projections
are possible. This technique is becoming important as a preliminary
screening for other examinations which though requiring a shorter
preparation may have, of necessity, to be delayed for a few days or
for those which are potentially invasive such as radiographic, instrumental, ultrasonic or contrast techniques. It is therefore a
useful tool for the reliable diagnosis of tumors (Figure 4). It also
permits a direct approach to the differential diagnosis of hematuria,
the identification of an origin in the bladder or if negative a rapid
survey of the upper urinary tract.

However, the method has its limits: the definition is not always
optimal and there is a risk of false positives (clots, false images
resembling a third lobe) or false negatives (especially when the
lesions are small).

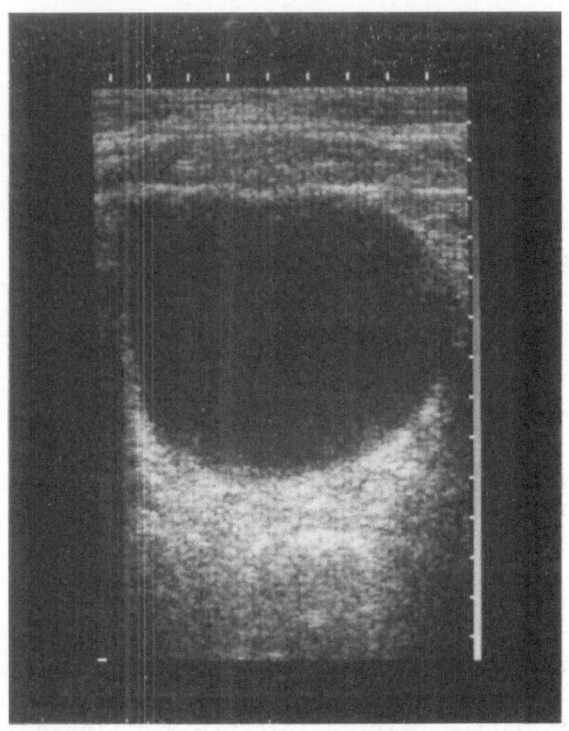

Fig. 1.   Transabdominal ultrasonography: transverse scan of normal bladder.

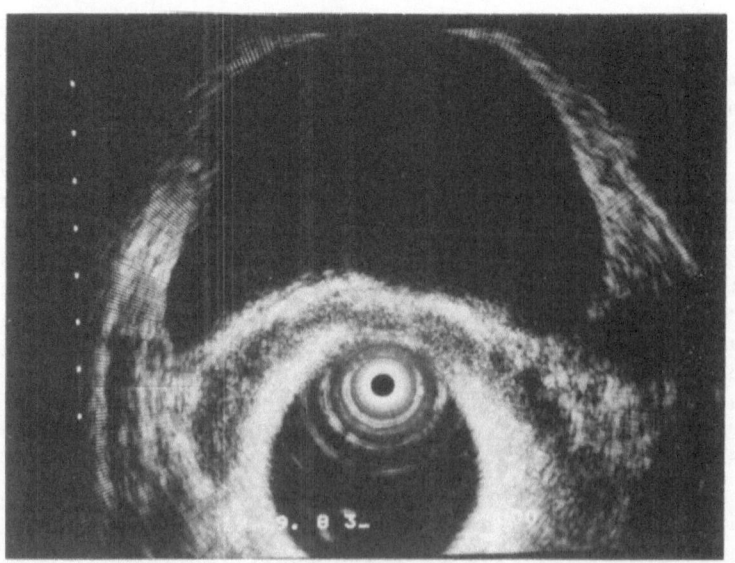

Fig. 2.   Transrectal scan of normal male bladder.

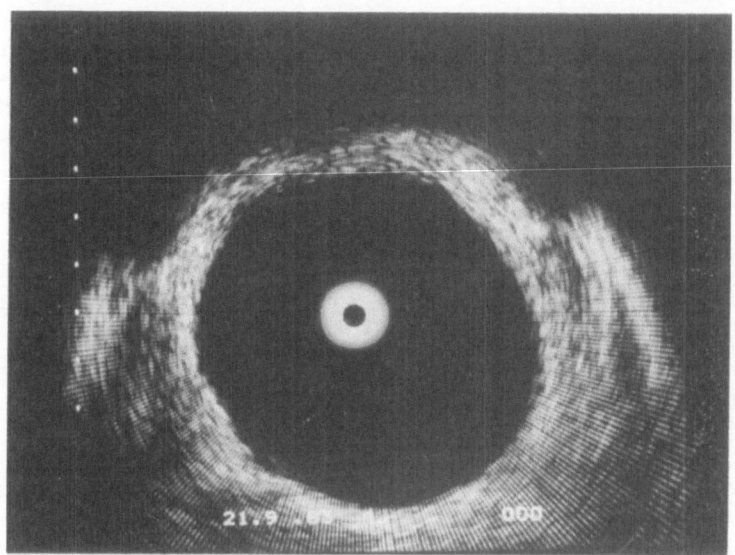

Fig. 3.   Transurethral scan of normal bladder.

Transrectal ultrasonography. Conceived and designed to study
prostatic pathology, it may be used to visualize the bladder neck and
ureteral ostia. it can give an excellent definition of the trigonal-
cervical region so permitting reliable differential diagnosis between
cervical tumors originating from the bladder (Figure 5) and prostatic
pathology (Figure 6).

On the other hand the difficulty in visualizing other regions of
the bladder make its use unsatisfactory at the present.

Transrectal ultrasonography is always carried out in association
with and subsequent to abdominal ultrasonography and it may precede
or indicate the need for the transurethral technique (Figure 7).

The transurethral technique. It is suitable for staging bladder
neoplasms and checking the bladder wall during endoscopic surgery.
Since it is essentially a staging and not a screening investigation,
it follows the other ultrasonic, radiological and morphological
investigations such as IVP and urinary cytology. It should especially be used as an examination to compare with the results of the
CT scan and staging by transurethral resection (TUR). For this
technique we use a probe 7 mm. in diam. introduced into the bladder
through a 22 or 24 French sheath (Figure 8). The tip of the probe
has a concave transducing disc with a resonant frequency varying from
5 to 7 MHz. The transducer, protruding a few mm from the sheath, can
rotate freely on its axis, the speed being regulated by the operator.
Probes with an angle of incidence which can vary from 0° to 90° are

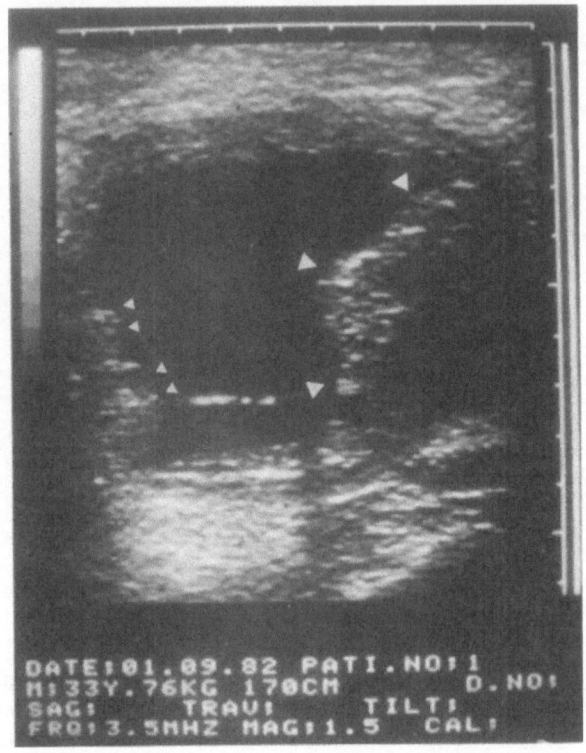

Fig. 4. Transabdominal scan: arrows indicate bladder tumors (trans-
        verse scan).

being developed and these will facilitate the examination of certain
bladder regions (neck and anterior wall), scanning them perpendicu-
larly at all points.

The latest models are equipped with a logarithmic amplifier
which gives a better definition by projecting the images on a cathode
ray tube and the probes can be introduced directly into the explor-
atory and cystoscopic sheaths so avoiding changes of instruments and
repeated introductions during examination or surgery.

We prefer to use general anesthetia for transurethral ultrasonic
exploration in male patients.

Once the probe has been positioned and the bladder filled with
water the transducer is rotated to indicate the echogenecity of a
circular section of the bladder wall. For complete exploration of
the bladder wall and the under-lying tissues the probe is retracted
slowly, scans are taken every 5 to 10 mm and the characteristics of
each section, up to the internal urethral meatus, are observed on the
monitor screen.

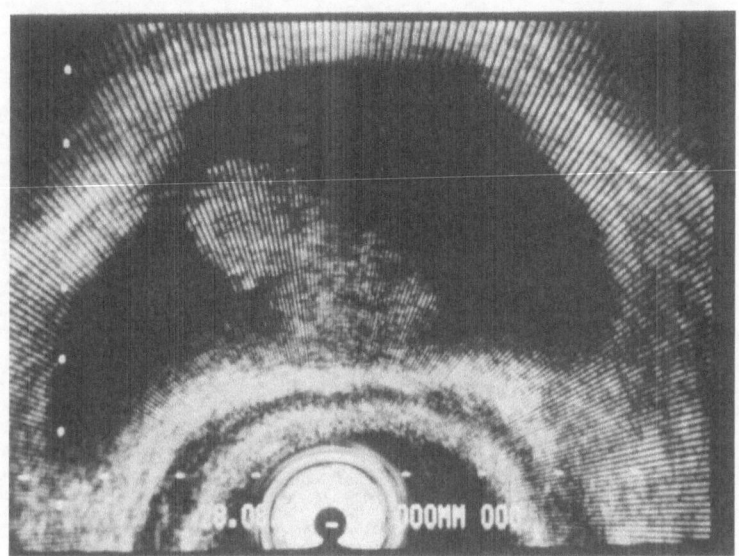

Fig. 5.   Transrectal scan: a large tumor from the bladder neck.

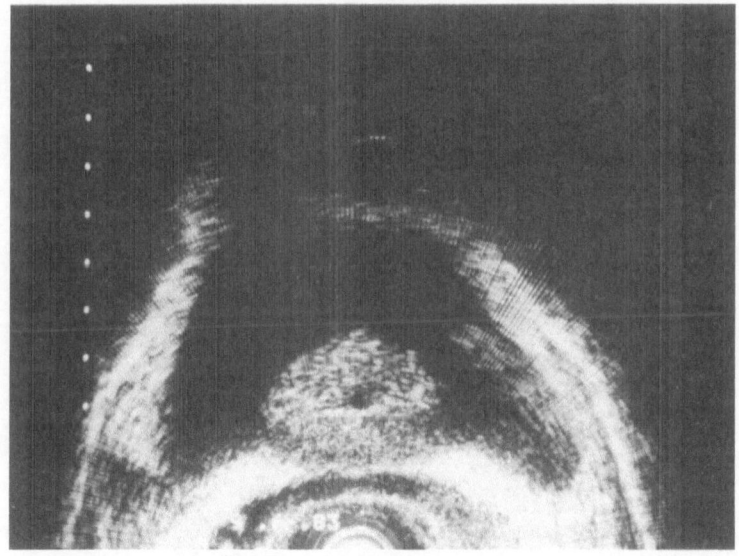

Fig. 6.   Transrectal scan shows benign hypertrophy of prostate.

Depending on the degree of bladder distension, the thickness of a normal bladder wall varies from 3 to 6 mm. The endoluminal profile appears smooth and regular whereas at the parietal level some irregularities in the distribution of the echoes are normally seen. For

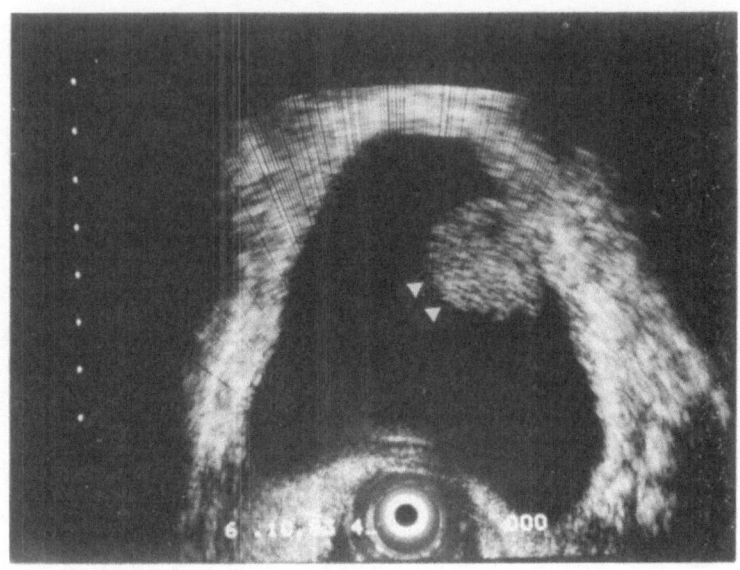

Fig. 7.   Transrectal scan: arrows indicate a large bladder tumor.

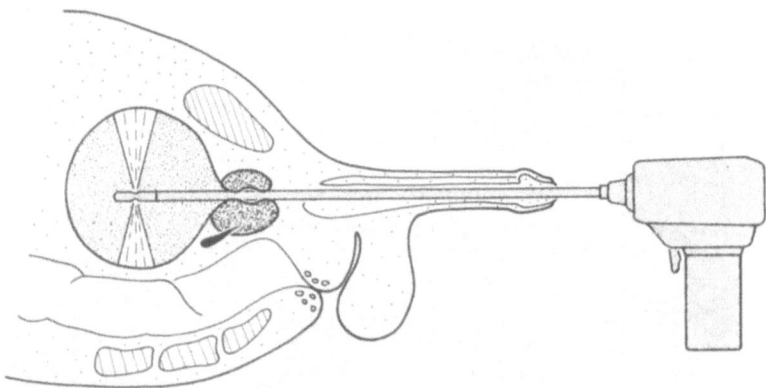

Fig. 8.   Transurethral scanning, procedural scheme.

the operator reflection of air bubbles situated near the dome, (Figure 9), and bilateral and symmetrical filling defects of the bladder floor corresponding to the ureteral orifices (Figure 10) are important reference points for localizing the region to be examined and for orienting the transducer.

If a suspicious area is seen during exploration the details of the image are brought into focus by adjusting and positioning the

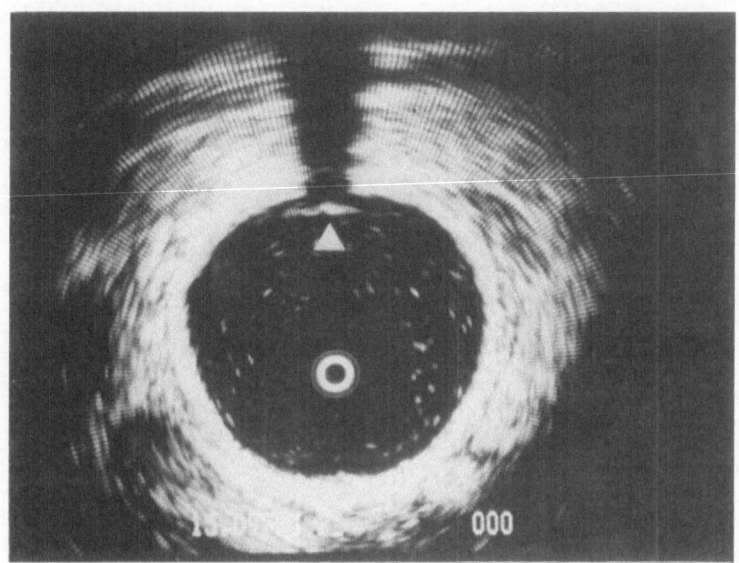

Fig. 9.   Transurethral scan in normal bladder: strong horizontal echo
reflection with ultrasonic shadowing behind at top of bladder originating from air bubble (arrow).

transducer and filling the bladder or mobilizing the wall (by insert-
ing a finger into the vagina or rectum or by pressing on the suprapubic region) to bring closer areas which are difficult to reach. In
a suspicious area it is always advisable to carry out a series of
tomograms not more than 3 to 4 mm apart so that any lesions will be
well seen.

     If and when exophytic lesions are found it is advisable to move
them and detach them from the wall in order to evaluate the character
of the implant base and pedicle.

DISCUSSION

     Traditional methods used to study bladder neoplasms have not
only given a certain precision in diagnosing their existence but also
provided important morphological data about the tumors. Radio-
diagnostic examinations such as IVP, cystography and double contrast
cystography have done much to define the site, volume, and intracavitary extent of the mass, the degree of rigidity or distensibility
of the bladder wall and the degree to which the upper urinary tract
is compromised. These findings can be confirmed and precisely defined by the diagnostic endoscopic findings and bimanual examination
under anesthetic. Exfoliative cytology or biopsy identify the histological grade of the tumor but the definition of the stage has always

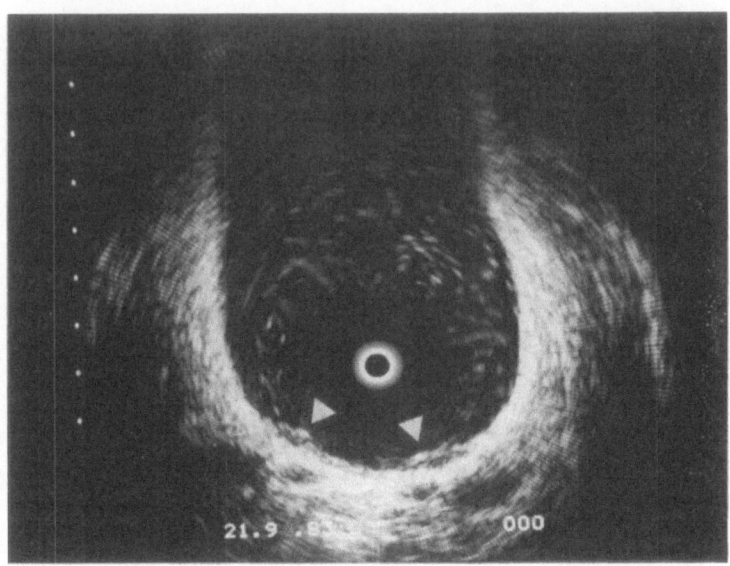

Fig. 10. Transurethral scan in normal bladder indicates the ureteral
          orifices (arrows).

been (and still is) less precise since no technique has been able to
analyze the innermost structures of the bladder wall or to recognise
and differentiate its anatomical layers.

        These diagnostic techniques are fraught with error, as much as
50% according to some investigators, when they are used for staging
bladder neoplasms.

        Although the IVP is indispensable for a morpho-functional class-
ification of the urinary system and is often the first test to indicate the type of bladder disorder present it cannot stage the tumor
since it gives sure indications only when the neoplasm is in a very
advanced stage and there is no doubt.

        Urinary cystography, and double contrast cystography only permit
classification of the tumor as invasive and non-invasive; this is not
sufficient to determine the prognosis or to decide upon the therapy.

        Bimanual examination under anesthetic although valid is limited
by the excessive subjectivity of the findings and because the examinations can be falsified or invalidated in obese patients and in those
who have already undergone surgery. This technique is only valid
when in association with TUR.

        Cystoscopy permits a more direct approach to the tumor and can
define the site, volume, morphology and characteristics of the im-

plant base but because the data it provides is limited to the exophytic portion of the lesion it can only indicate that stage of
infiltration very approximately.

When CT scan was first used it was heralded as the most valid
technique for staging bladder tumors. Undoubtedly it is much more
precise and gives more details than the techniques already described
since it can show up at least 3 groups of neoplastic lesions. Although these are not concordant with the pathological stages they
help to make interpretation more precise, though still not enough for
prognostic and therapeutic ends. The first group includes non-infil-
trating neoplasms or tumors where the infiltration is limited to the
superficial muscle layer, roughly described as stages T1 and T2
(Figure 11). The second group (T2, T3A) includes lesions invading

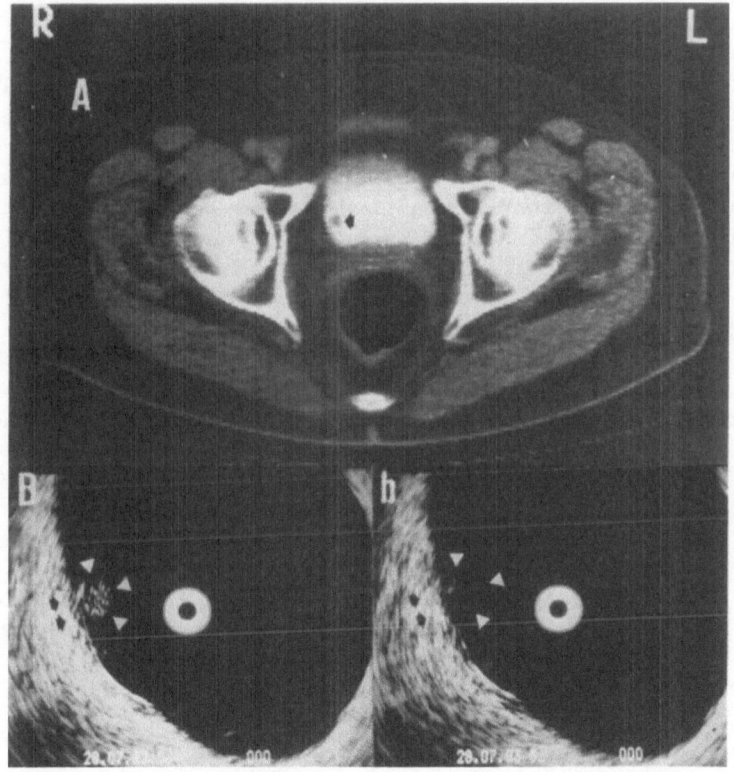

Fig. 11.   CT scan and transurethral scan in the same case - CT scan
           (A): filling defect on right side (arrow).
           Transurethral scan (B): Pedunculated tumor with no muscular
           infiltration (arrows). (b) Tomogram of the same site under
           gain control adjustment: the muscle layer is intact at the
           tumor base (black arrows).

the muscular tissues and the third, lesions invading the perivescical
tissues (T3B) or other nearby structures and organs (T4).

Since the information provided by bloodless methods is not very
useful for a realistic staging of bladder neoplasms recourse to
techniques such as TUR is more than justified since it is the only
staging method that can give reliable data on the degree of infiltration. However, even this technique has its critics because since
the total thickness of the bladder wall must be resected three problems may follow: (i) extra-vesical dissemination of neoplastic cells
by the irrigation fluid after vesical perforation, (ii) possible
infection of the peritoneal cavity following perforation of the
posterior wall of the bladder, (iii) unacceptable multiple perfor-
ations in the case of multiple tumors. There are also technical
difficulties, for example when the lesion involves the anterior wall
or when cellular alterations induced by the hot loop of the resecto-
scope make pathological interpretation difficult.

In view of these problems there is little doubt that transurethral ultrasonography is the best technique for resolving the staging
of bladder neoplasms bloodlessly. Furthermore because it can identify and analyze the anatomical structures of the bladder wall it gives
greater details of the neoplastic alterations. Four groups of
lesions can be defined by this method (Figure 12). The first (U1)
includes all those lesions recognized by the transurethral echo as
superficial irregularities with a completely normal underlying parietal structure, a sure indication of the absence of infiltration of
the muscle layer. These lesions therefore come in the Ta/T1 clinical
stage (Figure 13).

In the group U2, parietal irregularities are more extensive but
the continuity of the bladder wall is not interrupted and the uniform
thickness suggests only superficial involvement of the muscle layer.

Two easily carried out technical expedients confirm whether the
stage has been correctly classified or not.

(i)   The image subtraction technique – since the echogenic level of a
      neoplasm is lower than that of the underlying bladder wall it is
      possible to make the tumor echoes disappear by varying the
      instrument's sensitivity, leaving those of the underlying wall
      unaltered if the tumor has not invaded it.
(ii) Because ultrasonography is, in all its phases, a dynamic
      examination parietal elasticity can be verified by varying the
      volume of vesical filling. In superficial invasive forms this
      is always well preserved. This type of lesion is compared to
      stage T2 of the TNM classification (Figure 14).

Group U3 includes those lesions which ultrasonically are characterized by a loss in the continuity and a thickening of the under-

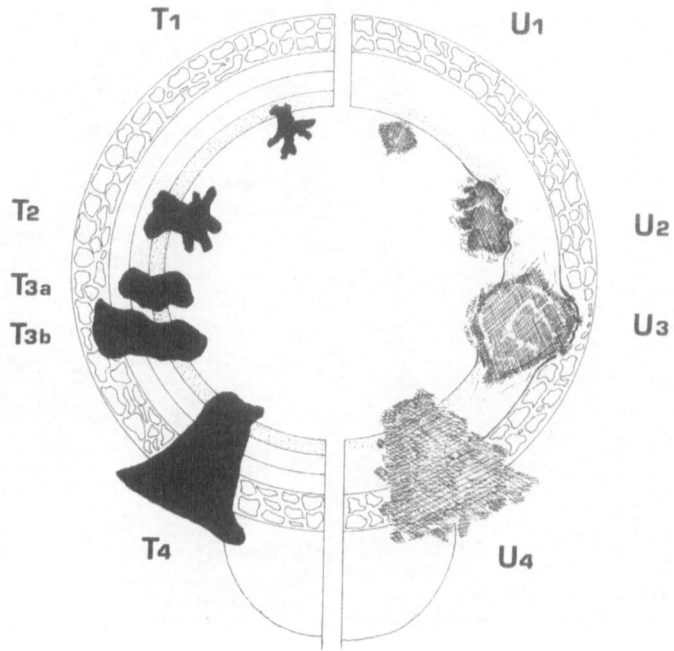

Fig. 12.   Comparison of histopathological classification to
           ultrasonic criteria.

lying bladder wall. When the image is subtracted the echoes are
disordered and broken and it is not difficult to demonstrate a
notable loss of elasticity in the regions with lesions. This is
comparable with stage T3, ie neoplasia infiltrating the deepest
muscle layer (Figure 15-16).

        The last group, U4, includes lesions with extra-vesical con-
tours, recognizable because of the total confusion and disarrangement
in the parietal structure of the bladder and the extra-vesical confusion of the neoplastic echoes (Figure 17). With transurethral
ultrasonography it is therefore possible to define a fairly well
standardized model for every stage of neoplasia.

        Transurethral ultrasonography is without doubt a great advance
over the more conventional techniques for obtaining more reliable
staging even though it is not always easy to distinguish clearly
between stage T2 and T3. Further improvements in the technique and
more experience in using it will result in better agreement between
ultrasonic and pathological staging.

        The use of this technique during surgery has excited a lot of
interest. In fact during TUR the operator can at intervals substi-

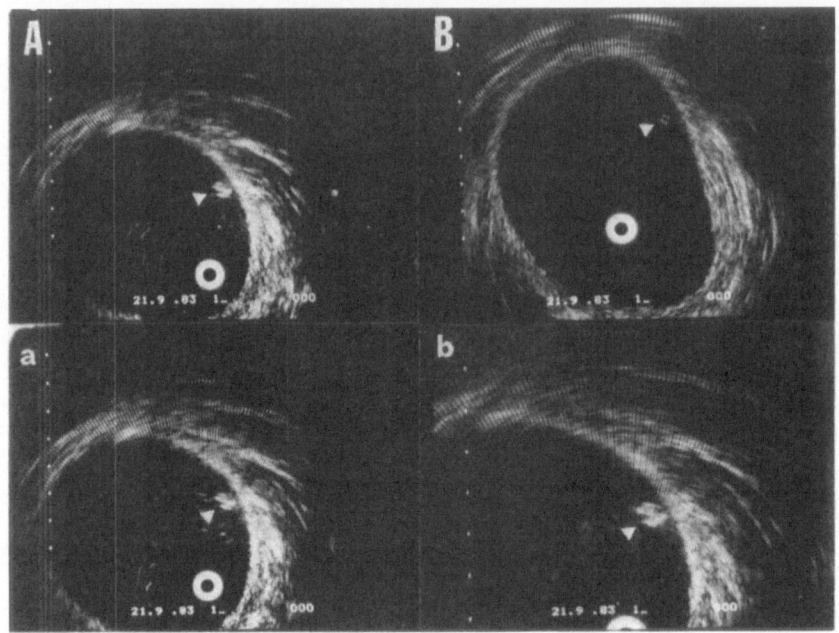

Fig. 13. Transurethral scan: (A-a) small tumor without infiltration
of bladder wall (arrows). (B) Tomogram of same site under
gain control adjustment: image of muscle layer is intact at
tumor base (arrow). (b) Close-up view clearly shows the
normal bladder wall at the tumor base (arrow).

tute the endoscope with an ultrasonic probe and easily evaluate the
extension in depth of the resected tissue and, more important, the
amount of residual neoplastic tissue. This residue is subsequently
removed by endoscopic surgery so assuring the optimal outcome.

The limits of transurethral ultrasonography, still apparently
insurmountable, are the impossibility of distinguishing the in situ
cancers or very small (less than 0.5 mm in our experience) lesions.
Some regions of the bladder, the neck and anterior wall, are difficult to reach and exploration can only be investigative and therefore less reliable. The new probes will overcome this disadvantage.

Some images can even deceive an attentive eye by mimicking a
neoplasm. This happens when irregularities are generated by the
ureteral ostia and interureteric bar, by hyperplastic inflamed mucosa, following the results of heat treatment and especially by
pronounced trabeculation. It is not always easy to decode images
from bladders which have already undergone surgery since scar tissue
may appear similar to neoplastic formations.

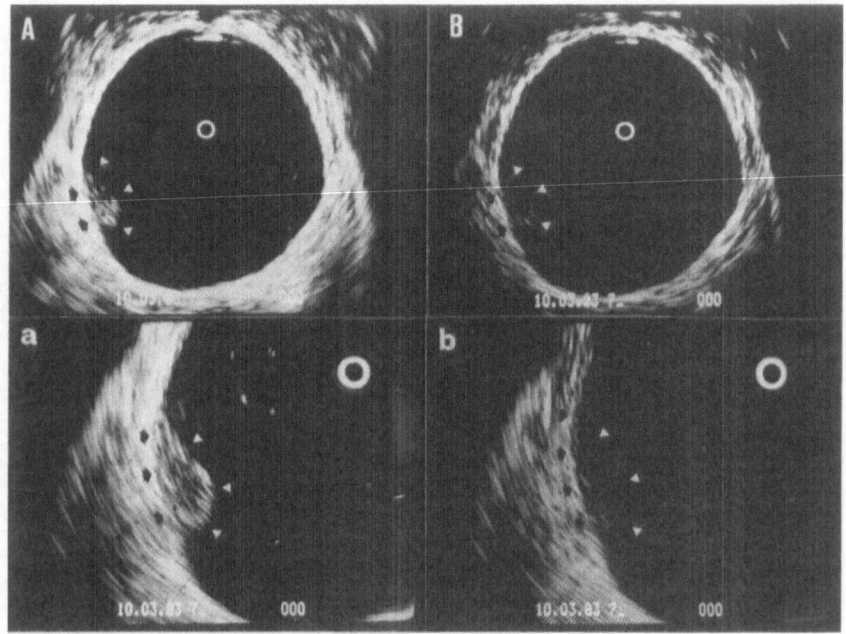

Fig. 14.   Transurethral scan: (A) Tumor with infiltration in the
           superficial muscular layer (arrows). (a) Close-up view
           shows bladder wall slightly indented at tumor base (black
           arrows). (B-b) Tomograms under gain control adjustment:
           black arrows indicate muscle layer superficially indented
           at tumor base, continuity of outer bladder wall
           uninterrupted.

     Another disadvantage is the advisability, often necessity of
using anesthetics for this examination since it is still not considered prudent to rely only on the ultrasonic diagnosis and so the
usual TUR staging techniques are done at the same time.

     For two years we have been using the procedure of a cystoscopic
examination with an ultrasonic probe in the sheath and the patient
under anesthetics to map the bladder wall and take multiple biopsy
sampling of the neoplasm. Cytology, bimanual examination and in the
more dubious cases staging resection complete the protocol. In this
way we propose to resolve both the staging and grading of the neoplasm keeping the number of resections to a minimum for the reasons
mentioned above.

CONCLUSIONS

     In patients with bladder tumors the histological stage of the
tumor and above all the degree of infiltration of the bladder wall

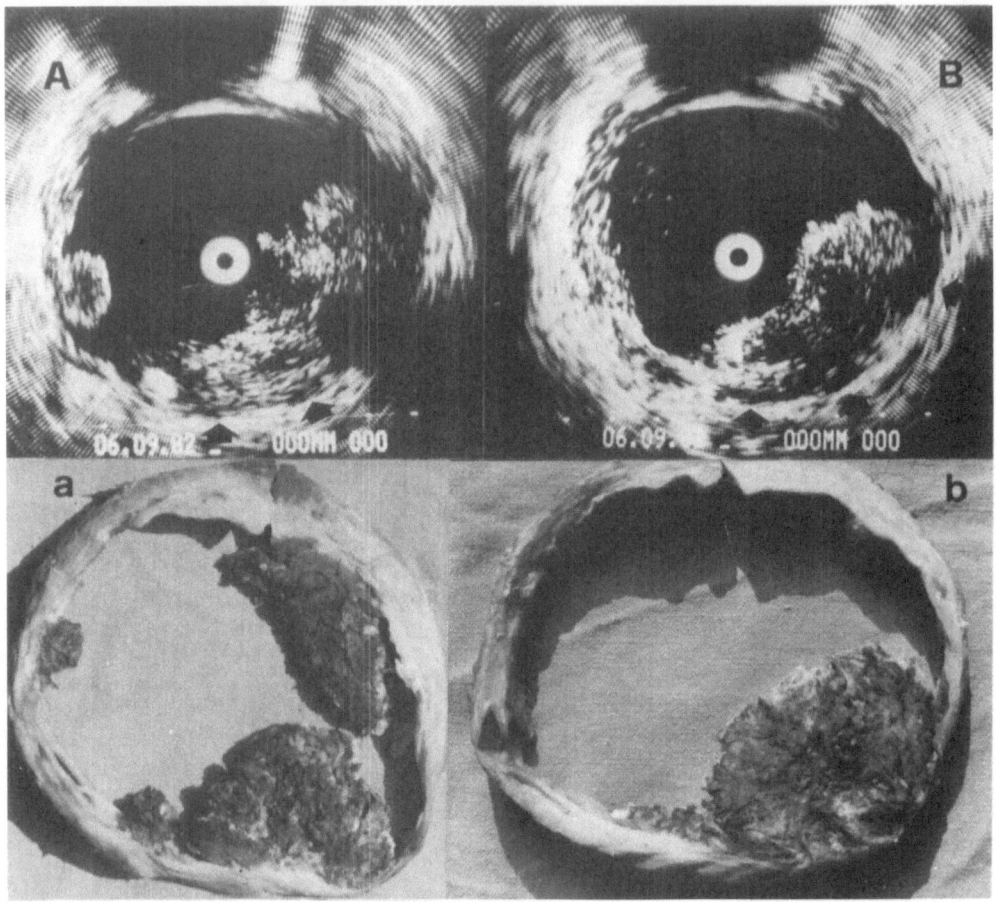

Fig. 15. (A-B) Transurethral scan of deeply invasive bladder tumors
          with interrupted continuity of bladder wall. Muscle layer
          is destroyed by tumors (arrows). (B-b) Deep infiltration
          was confirmed after total cystectomy.

always dictates the choice of therapeutic procedure. Therefore a
method capable of demonstrating without doubt the extent of neoplastic invasion of the bladder wall is of prime importance.

    Transurethral ultrasonography offers real advantages in this
field. The propagation of ultrasound is optimal because it does not
interfere with other tissues or organs located between the bladder
wall and the transducer - this makes it preferable to the transabdominal and transcrectal techniques. Ultrasonography can accurately define the limits of the neoplasia, study the bladder surface
and evaluate the degree of infiltration.

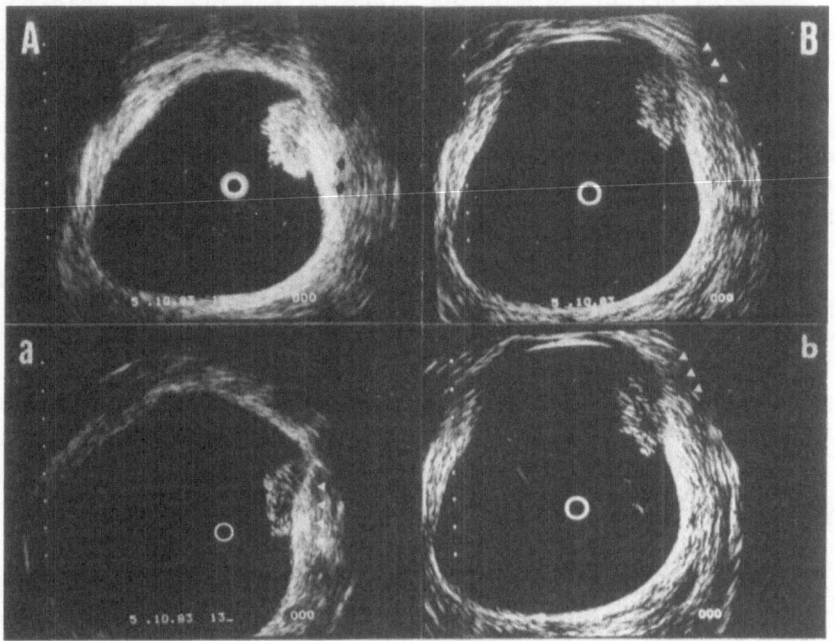

Fig. 16.   Transurethral scan: (A-a) Bladder wall muscle layer
           interrupted by tumor. Thickening of the muscular layer is
           evident at the lateral margin of the tumor (arrows). (B-b)
           Tomogram under gain control adjustment: continuity of the
           bladder wall is interrupted (arrows).

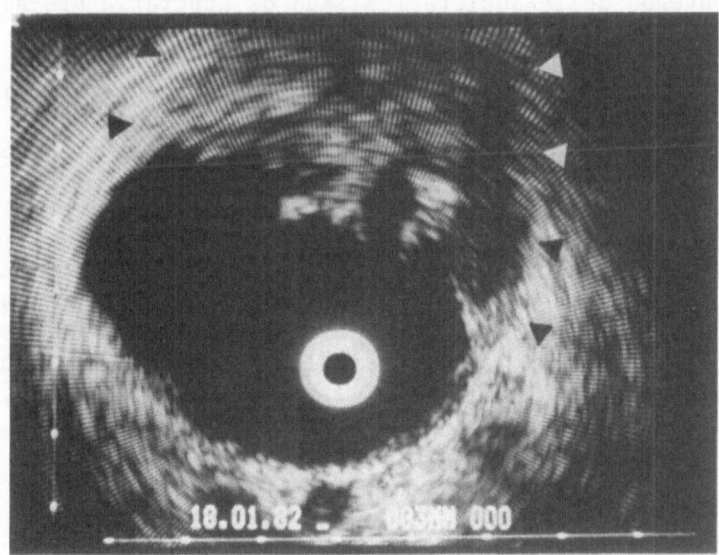

Fig. 17.   Transurethral scan shows bladder tumor with perivesical
           extension (arrows).

Special attention must be given to the overall thickness of the
bladder wall, which in the absence of mucous edema corresponds exactly to that of the muscle layer. Noting the extent of infiltration in
the wall it seems possible to distinguish between category T2 and T3
lesions. The invasive tumor gives a slight echogenecity, usually
reduced, accompanied by a superficial interruption of the parietal
echoes or the disappearance of the stronger echoes of the normal
bladder wall.

The transurethral method can give a more detailed image of the
tumor and the underlying bladder wall and therefore correct staging
can be given. Ultrasonographic examination is always dynamic and
only the observations of the operator throughout the entire scan can
supply the results mentioned above since observations on the basis of
a few pathological images can be greatly misleading.

REFERENCES

1. J. Gammelgaard and H. H. Holm, Transurethral and transrectal
       ultrasonic scanning in urology, J.Urol., 124:863 (1980).
2. H. H. Holm and A. Northeved, A transurethral ultrasonic scanner,
       J.Urol., 111:238 (1974).
3. F. Micali, M. Porena, and G. Virgili, L'indagine ecografica
       della vescica. Proceeding del 1° Convegno della Societá
       Italiana di Ecografia Uro-Nefrologica, Roma (1982) (in
       press).
4. F. Micali, M. Porena, and G. Virgili, Ecografia transuretrale
       nello staging dei tumori della vescica. Proceeding del 1°
       Convegno della Societá Italiana di Ecografia Uro-Nefrologica,
       Roma (1982) (in press).
5. S. Nakamura and T. Niijima, Staging of bladder cancer by
       ultrasonography: a new technique by transurethral intra-
       vesical scanning. J.Urol., 124: 341 (1980).
6. S. Nakamura and T. Niijima, Transurethral real time scanner,
       J.Urol., 125:781 (1981).
7. T. Niijima and S. Nakamura, Transurethral ultrasonography
       bladder cancer staging and other clinical application, in:
       "Clinical bladder cancer," L. Denis, P. H. Smith and M.
       Pavone Macaluso, eds., Plenum press, New York, p. 47 (1982).
8. J. Schuller, V. Walther, E. Schmiedt, G. Staheler, H. W. Bauer
       and A. Schilling, Intravesical ultrasound tomography in
       staging bladder carcinoma, J.Urol., 128:264 (1982).

EXPERIENCES WITH TRANSRECTAL AND

TRANSURETHRAL ECHOGRAPHY

H. J. de Voogt, A. H. J. Miedema and J. F. Felderhof

Department of Urology
Free University Hospital
Amsterdam, The Netherlands

INTRODUCTION

Wild[1] in 1957 was the first to alter echography using a rectal probe which was drawn back slowly and rotated, with the intention of diagnosing lesions of the sigmoid and rectal wall. Several urologists have used this idea to develop echography of the prostate and the bladder.

Takahashi[2] made the first A-scan of the prostate transrectally in 1963 and a year later he developed a transrectal probe which produced bladder and prostate tomograms[3]. However the poor quality of the images prohibited clinical application.

In 1967 Watanabe[4] ostained better images by using more modern techniques. This started expansion of the field of clinical application. Nishi[5] in 1968, was the first to try ultrasound transurethrally. In 1973 the methods were introduced in the USA by King[6].

Soon many investigators followed these pioneers. In Europe Holm and Gammelgaard[7] in close cooperation with the Danish Institute of Biomedical Engineering and the firm Brüel and Kjaer, were the first to develop an apparatus with one probe that could be used transurethrally as well as transrectally.

In 1981 we got the opportunity to try the second generation of the Brüel and Kjaer instrument. Our experiences with it during two years of use in the out-patient department are presented here.

Methods and Materials

The instrument consists of a probe with a built-in motor unit and interchangeable sounds/transducers for use in the bladder[4,5 and 6 Mhz] and in the rectum[3,5 Mhz], the latter one surrounded by a small rubber balloon filled with water. The probe fits in a 24 F cystoscope. The scanner rotates with a velocity of 16 revolutions/min and the ultrasound echoes are reproduced on A- and B- scans. The B mode can be photographed as a semi-real-time scan for recording the images. The third generation of the instrument has a real-time-scanner (gray-scale).

For volumetry of the prostate a special support is available by which the rectal probe can be retracted at 5 mm intervals. The volume is computed by adding up the different surface areas of the prostate using a special formula.

150 patients were examined; their diagnoses are given in Table 1.

Results and Discussion

The Bladder and prostate can be examined ultrasonically using Transabdominal, Transurethral or Transrectal techniques.

In our experience transurethral echography is superior for imaging the bladder wall and its lesions and especially for staging bladder tumors.

The bladder wall is 3-6 mm thick, usually with a smooth inner surface. In the normal bladder the mucosa is not clearly distinguishable from the muscular part of the wall. The air bubble in the dome and the ureteric orifices are easily identified and even the smallest papillomatous lesions are clearly visible. In large papillomas the base is often better seen with ultrasound than with the cystoscope.

The possibility of determining the depth of infiltration of bladder tumors enabled us to stage them before TUR as already described by Niijima, et al.,[8]. In all our 22 patients with bladder cancer our preliminary staging by echography was confirmed clinically (bimanual palpation during anaesthesia at the time of transurethral reaction) and by the pathologist.

In T_1 lesions the mucosal lining continues uninterrupted under the tumor.

In T_2 tumors the mucosal lining at the tumor base is slightly indented or has a minimal interruption.

Table 1. Diagnosis in 150 Patients
Examined by Transrectal or
Transurethral Echography

Benign prostatic hypertrophy	58
Cancer of the prostate	48
Bladder tumors	22
Other lesions	22

In T_3 growths the mucosal lining is clearly interrupted by irregular tumor echoes penetrating the bladder wall.

In T_4 lesions the limits of the tumor cannot be seen outside the bladder wall or there is penetration of prostate and seminal vesicles, or a dilated ureter is visualized.

Other possibilities for transurethral echography suggested by Harada are the visualization of stones, ulceration, chronic cystitis cystica, ureterocele, pseudo diverticula, dilated ureters and ureteric tumors[9]. Also follow-up of tumors after TUR, radiation or chemotherapy can be controlled this way in addition to cystocopy. For the prostate transabdominal and transrectal echography are giving equally good results. Both methods have their advantages and limitations.

Transcretal ultrasound is also invasive and takes time. The volume of the prostate can be assessed accurately and prostatic sizes at different times can be compared because the scanner can be placed directly under the prostate at the same place.

Transabdominal sonography is of course non-invasive and takes less time. The patient has to have a full bladder to act as a window for the ultrasound. It is especially good at detecting intravesical enlargement of the middle lobe of the prostate. In adipose patients however this method of imaging the prostate can be very difficult or impossible.

The image of the normal prostate is crescent-shaped and symmetrical. The capsule provides a clear boundary and the internal structure should be echogenically homogeneous. In benign prostatic hypertrophy the antero-posterior diameter increases in the first place, which causes the shape to become more spherical. Sometimes the capsule is thicker and often the adenomatous part or the middle lobe can be seen clearly. However the echoes of the inner structure can become very non-homogeneous and thereby cause confusion with cancerous nodules as well as with benign structures. Cysts and adenomas are usually regular; peri-urethral glands may be echodense, whilst the urethral wall is poorly echogenic; uncalcified stones often give echodense areas in the subcapsular zone near the rectal wall.

Table 2.   Various Diagnoses, Confirmed
by Transrectal or Trans-
urethral Echography

Bladder stone	1
Prostatic stones	3
Normal bladder	1
Irradiation ulcer of the bladder	1
Trigonitis cystica	1
Prostatitis	3
Prostatic cyst or abscess	2
Distal ureteric tumor	1
Normal prostate	8
Ectopic kidney emptying in vesicula	1

All this makes the diagnosis of T_1 and T_2 prostatic carcinoma very difficult if at all possible, despite the optimistic view of Watanabe[10].   However in T_3 and T_4 lesions the asymmetric aspect of the prostate and the discontinuity of the capsular boundaries clearly indicate the diagnosis.

In these cases measuring the volume of the prostate is of prime importance, as follow-up after irradiation or hormonal therapy can thereby be monitored.   In 3 patients we could demonstrate a good response by diminution in size of the prostate and in one patient we saw progression (the primary tumor growing in size despite hormonal therapy).

Finally in 22 patients various other diseases, as specified in Table 2, were diagnosed and/or confirmed by transurethral or trans-rectal echography and confirm the earlier findings of Tanahashi[11] in this respect.   Bladder deformities especially, whether congenital or acquired can be diagnosed or visualized by transrectal echography with minimal discomfort for the patient.

Conclusions

The indications for transurethral echography are:

1. Diagnosis and staging of bladder tumors;
2. Visualizing atypical lesions, such as tumor in a diverticulum, a distal ureteric tumor and subtrigonal lesions;
3. Follow-up of bladder tumors after irradiation or chemotherapy.

The indications for transrectal echography are:

1. Follow-up of prostate cancer after irradiation or hormonal therapy (response-monitoring of primary tumor);

2. Diagnosis of deformities of the bladder and prostatic cysts or abscess.

REFERENCES

1. J. J. Wild and J. M. Reid, Progress in techniques of soft tissue examination by 15 MC pulsed ultrasound, in: "Ultrasound in Biology and Medicine," E. Kelly, ed., Amer.Inst.Biol.Sci., p.38-45 (1957)
2. H. Takahashi and T. Ouchi, The ultrasonic diagnosis in the field of Urology (1st report), Proc.Jpn.Soc.Ultrasonics Med., 3:7 (1963).
3. H. Takahashi and T. Ouchi, The ultrasonic diagnosis in the field of Urology (2nd report), Proc.Jpn.Soc.Ultrasonics Med., 4:35 (1964).
4. H. Watanabe, H. Katoh, and T. Katoh, Diagnostic application of the ultrasonotomography for the prostate, Jpn.J.Urol., 59:273 (1968).
5. M. Nishi, Diagnosis of prostatic disease by application of ultrasonic A-scope indication, Acta Urologica Jpn., 14:3 (1968).
6. W. W. King, R. M. Wilkiemeyer, and W. H. Boyce, Current status of prostatic echography, JAMMA, 226:444 (1973).
7. H. H. Holm and J. Gammelgaard, Transurethral scanning in: "Diagnostic Ultrasound in Urology and Nephrology," H. Watanabe, ed., Igaku-Shoin, Tokyo, p.84-89 (1981).
8. T. Niijima, S. Nakamura, and T. Shiraishi, Transurethral scanning and scanning in abdominal wall, in: "Diagnostic Ultrasound in Urology and Nephrology," H. Watanabe, ed., Igaku-Shoin Tokyo p.96-104 (1981).
9. K. Harada, Cystitis and other diseases, in: "Diagnostic Ultrasound in Urology and Nephrology", H. Watanabe, ed., Igaku-Shoin, Tokyo, p.104-108 (1981).
10. H. Watanabe, Prostatic Cancer, in: "Diagnostic Ultrasound in Urology and Nephrology", H. Watanabe, ed., Igaku-Shoin, Tokyo, p.130-135 (1981).
11. Y. Tanahashi, Seminal vessels, in: "Diagnostic Ultrasound in Urology and Nephrology", H. Watanabe ed., Igaku-Shoin, Tokyo, p.141-147, (1981).

COMPUTED TOMOGRAPHY IN STAGING OF BLADDER CARCINOMA

T. Norlindh*, S. Hellsten**, U Nyman*,
and I. Andersson*

Departments of Diagnostic Radiology,* and Urology,**
Malmö General Hospital, Malmö, Sweden

MATERIAL AND METHODS

Eighty-two consecutive patients (65 men and 17 women, mean age 69 years, range 45-91 years) with a bladder carcinoma verified at cystoscopy were referred for computed tomography prior to any biopsies or transurethral resections. A Siemens Somatom 2 with a slice thickness of 4 or 8 mm, scanning time 4.8 seconds and a 256 x 256 matrix was used. The urinary bladder was scanned with contiguous 8 mm thick slices with the patient in the supine position. Depending on the size and location of the tumor 4 or 8 mm thick scans were performed with reconstructive enlargement, angulation of the gantry, and/or with the patient in various positions. Intravesical media offering different degrees of contrast (Hounsfield units - HU) including urine, saline, Intralipid® (Kabi-Vitrum) (1-15 HU), peanut oil (100 HU) and carbon dioxide were used. The following criteria of CT staging were used: CT 0, no visible tumor: CT 1-3a, tumor without extramural extension: CT 3b, tumor extending into perivesical fat; CT 4a, tumor involving adjacent organs; CT 4b, tumor involvement of the pelvic and/or abdominal wall. Clinical stage (T) was based upon clinical examination, urography, cystoscopy, bimanual examination under anaesthesia and histopathological examination of the tumor following transurethral resection.

The final (postsurgical) stage (pT) was based upon histopathological findings from adequate transurethral resections, a partial or total cystectomy, and autopsy as well as the clinical course during 2-3 years follow-up. Two patients with artefacts from metal prostheses of the hip and one without adequate histo-pathological verification were excluded from the study which thus comprised 79 patients.

RESULTS

In sixty five patients the final tumor stage was pT3a or less.
In ten of these patients having a tumor sized less than one cm or
flat carcinoma in situ the tumors could not be visualized at CT.
In the remaining 55 patients correct CT staging was made in five and
overstaging in four. The overstaging was mainly due to partial
volume effect in the cranial or caudal part of the bladder where cuts
at right angles to the bladder wall could not be obtained. This
phenomenon led to haziness of the tumor contour simulating perivesical spread. Moreover when peanut oil or gas was used there were
difficulties in identifying the relation of the tumor mass to the
bladder wall. In one patient a 5 cm large tumor was clinically
staged as extramural (T3) whereas CT showed no signs of perivesical
spread. At histo-pathological examination after segmental resection
of the bladder tumor stage was shown to be PTa.

Fourteen patients were considered to have perivesical extension
of their tumors (pT3b or more). CT identified 10 or these tumors
although one tumor with microscopic invasion of the prostate (pT4a)
was incorrectly staged CT3b. Four patients were understaged at CT
(CT3a or less). In two of these patients tumor growth was found in
the perivesical fat at fractional transurethral resections (pT3b).
In the remaining two patients the tumors were shown to infiltrate
into the prostate (pT 4a). Tumor infiltration of the prostate is
difficult to detect with CT essentially because of lack of fat planes
between the prostate and the bladder and since the tumor and the
prostatic tissue have about the same density. In six patients the
tumor was understaged (T3a or less) at clinical examination although
perivesical growth was revealed at CT in all of them.

CONCLUSIONS

In low-stage bladder tumors the accuracy of the staging procedure was generally not improved by CT in contrast to patients with
high-stage tumors in whom the addition of CT to previous clinical
methods seemed to improve the accuracy essentially by visualization
of extravesical tumor growth with or without extension to adjacent
organs or the pelvic wall. Low density urine, saline and fat emul-
sions seemed advantageous over peanut oil and gas as intravesical
contrast medium.

INTRAVESICAL ADRIAMYCIN IN PATIENTS WITH

CARCINOMA IN SITU OF THE URINARY BLADDER

S. Hellsten[1], A. Ek[1], H. Henriksson[3], I. Idvall[4],
C-E. Lindholm[2], K. Lindholm[4], P. Mikulowski[3],
and W. Månsson[1],

Departments of Urology[1], Oncology[2], Pathology[3]
and Cytodiagnosis[4], Lund and Malmö
University of Lund, Sweden

Flat carcinoma in situ of the urinary bladder is identified with increasing frequency as a result of the growing use of urine cytology and the increasing practice of endoscopic mucosal biopsy from unstable looking areas of the bladder epithelium[1]. Since the natural history of carcinoma in situ is most variable the choice of treatment may be extremely difficult and is still under trial. Once progress to invasion is proven cystourethrectomy is to be considered as external irradiation has been shown to be inefficient[2]. Less mutilating alternatives to removal of the bladder have been searched for such as intracavitary treatment using different chemotherapeutic agents. This report concerns our experience with intravesical Adriamycin therapy in patients with carcinoma in situ of the bladder with respect to efficacy, tolerance and side effects.

Material

The present series comprises 22 patients. The age range was 52-77 years (mean 68 years). 18 patients were male and 4 female. 7 patients had no history of bladder carcinoma (primary carcinoma in situ) whereas 15 patients had been treated previously for bladder tumors by surgery or radiotherapy (secondary carcinoma in situ).

Adriamycin was given monthly by the technique of Eksborg et al.,[3] i.e. the dose of the drug and volume to be instilled was calculated in relation to bladder capacity and the solution was retained for one hour. Endoscopic control with multiple bladder biopsies as well as urine cytology was performed before treatment and

after every third instillation. A treatment course of at least 12
instillations was aimed at, unless tumor progress or complications of
treatment occurred.

The total dose of Adriamycin given was 240 - 2990 mg (mean 964
mg) and the drug was administered on 3 - 29 occasions (mean 12
occasions).

Results

In 2 patients histology as well as urine cytology showed no
evidence of malignancy at follow-up after 6 and 9 instillations,
respectively. Ten more patients possibly had a beneficial effect of
the treatment. Three patients became negative or showed atypia only
in both histology and cytology at 2 consecutive controls, but later
became positive during continuous treatment (2 patients) or after
cessation of treatment (1 patient). An additional 7 patients became
temporarily negative or showed atypia only, in either histology (4
patients) or cytology (3 patients) on 2 consecutive occasions.
Except for occasional absence of malignant cells or atypia in histology or cytology, the remaining 10 patients showed persistent malignancy. Six patients with tumor progression underwent cystourethrectomy (5 patients) or external radio-therapy (1 patient). In 3
more patients transurethral resection of papillary low grade tumors
occurring during the intracavitary treatment with Adriamycin was
performed.

The Adriamycin treatment was discontinued in 4 patients due to
severe cystitis. In another patient an acute allergic (anaphylactic)
reaction was observed probably related to the administration of
Adriamycin. No renal or hepatic side effects were observed and there
were no instances of myelosuppression.

Summary

Our study indicates that intracavitary Adriamycin therapy may be
of value in some patients with carcinoma in situ of the urinary
bladder as no evidence of malignancy was found at follow-up in 2/22
patients and a possible beneficial effect was seen in a further 10
patients. The treatment was generally well tolerated even among old
patients and the frequency of side effects was low, mainly in terms
of chemical cystitis. Our study also indicates that a close follow-
up of the patients should be performed to permit early detection of
progressive disease necessitating more aggressive treatment.

REFERENCES

1. P. R. Riddle, G. D. Chisholm, P. A. Trott, and R. C. B. Pugh,
    Flat carcinoma in situ of bladder, Brit.J.Urol., 47:160
    (1976).
2. M. R. Melamed, N. G. Voutsa, and H. Grabstald, Natural history
    and clinical behavior of in situ carcinoma of the human
    urinary bladder, Cancer, 17:1533 (1964).
3. S. Eksborg, S.-O Nilsson, and F. Edsmyr, Intravesical
    instillation of Adriamycin , A model for standardization of
    the chemotherapy, Eur.Urol., 6:218 (1980).

PLASMA ABSORPTION AND TISSUE CONCENTRATIONS OF

ADRIAMYCIN DURING INTRAVESICAL THERAPY

F. Garofalo,* C. M. Camaggi,** M. G. Lalanne,*
P. De Santis,* G. Nanni,* E. Tisatto,** S. Secchiero,**
B. Angelelli,** and F. Pannuti **

The 2nd Division of Urology* and the Division of
Oncology,** Malpighi Hospital, Bologna, Italy

Adriamycin, an anthracycline derived from Daunomycin, is widely used for intravesical instillation in the treatment of bladder tumors[1,2,3] and in the prophylaxis of relapses[4,5,6].

Many clinical studies have shown the preventive efficiency and the minimal absorption of the drug through the bladder wall, which makes adriamycin particularly suitable for topical use.

The aim of the present study is to clarify some of the pharmacokinetic aspects of adriamycin including:

1) the concentrations of the drug in healthy tissue and in various types of bladder tumors at different time intervals after topical instillation;
2) the extent and duration of the tissue impregnation by the drug;
3) the extent and duration of measurable blood levels;
4) possible correlations between 1, 2 and 3 above.

Although this study is still at a preliminary stage we would like to propose it as a research model and if possible draw conclusions.

MATERIALS AND METHODS

All patients had previously had routine laboratory examinations including EKG, chest X-ray, and urography and were then admitted to the 2nd Urology Division of "M. Malpighi" hospital prior to surgical or endoscopic resection of their carcinoma of the bladder.

The staging of each patient was defined with the TNM classification according to the UICC recommendations.

A few days prior to the operation, the plasma uptake of adriamycin was evaluated in some of the patients after intravesical instillation of 50 mg adriamycin dissolved in 50 ml of normal saline solution using a suitable urethral catheter. The solution was eliminated by micturition 45 mins after instillation.

Blood samples were taken before instillation, after 20, 45, 60, 90 mins and after 3, 4, 6, 10, 24 36, 48 and 72 hours and were later analysed for serum levels of adriamycin and adriamycinol. The blood samples (5ml) were immediately centrifuged and 1-2 ml of plasma were taken from each and stored at -20°C until analysis.

Tissue samples were obtained by endoscopy or surgery after 30 mins, and 1, 3 and 6 hours after micturition. Samples were taken from various parts of the neoplastic mass and from healthy parts of the bladder wall; half of each specimen was used to analyse the concentration of adriamycin and adriamycinol; the other half, appropriately labelled with alphanumeric codes, was sent for histological examination and "grading".

The outline of the laboratory technique[7] was as follows:

(i) to the tissue pulverized in a microdissector for 2-3 minutes, was added 2 µg of a stock solution of daunorubicin hydrochloride, 1 ml phosphate buffer and 10 ml of chloroform: heptanol (9:1). The mixture was shaken for 20 mins (horizontal shaker, 120 impulses/min) and then centrifuged at 4000 rpm for 10 mins.

(ii) The organic layer was collected and extracted with 0.5 ml of 0.3M phosphoric acid in a vortex mixer.

(iii) The acidic phase, after treatment with 2 ml of n-hexane to remove nonpolar contaminants, was analysed by high pressure liquid chromatography. Chromatographic analysis was performed with a Perkin Elmer series 3B liquid chromatograph equipped with a 650-10LC fluorescent detector. In addition, a Supelcosil LC-18 column (15 cm x 4 mm internal diameter) was used, and quantitation was performed with a Perkin Elmer Sigma-10 data system.

The mobile phase consisted of methanol:water (3:1) plus 0.5% acetic acid and 2.5 mM sodium pentane sulphonate.

The method was calibrated in the usual way, with standards made by supplementing blood bank plasma with stock solutions of the drug.

RESULTS

We found no evidence of adriamycinol in any plasma or tissue samples.

The average plasma concentration of Adriamycin was between a maximum of 3.15 ng/ml 20 mins after voiding and a minimum of 0.18 ng/ml after 10 hours. After 24 hours the drug could no longer be found in the blood stream (Figure 1).

In the neoplastic tissues the average concentrations of adriamycin ranged from a maximum of 7.85 μg/g after 30 mins to a minimum of 0.43 μg/g after 6 hours (Figure 2).

In the neoplastic papillary lesions the concentrations measured in the different samples were higher than those in the solid forms at the same time (Figure 3). While the median concentration in papillary lesions was 6.85 μg/g 30 mins after voiding, we found the value to be 1.05 μg/g in solid tumors. One hour after the average concentration of the chemotherapeutic agent in the papillary lesions decreases to 4.39 μg/g while the value is 1 μg/g in solid forms.

In the samples taken at 3 hours the concentration of adriamycin in the papillary lesions was 3.55 μg/g and 0.68 μg/g in the solid tumors.

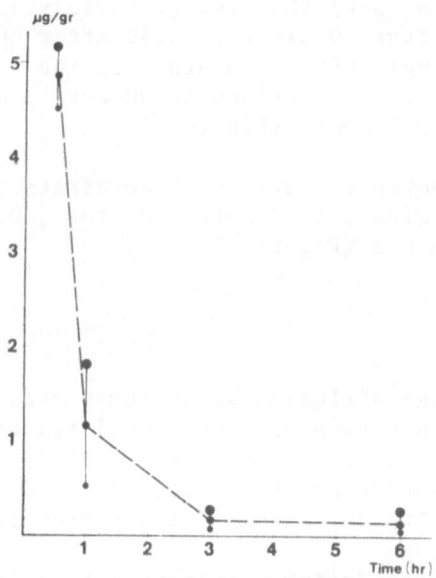

Fig. 1. Plasma concentration following Intravesical
         instillation of Adriamycin.

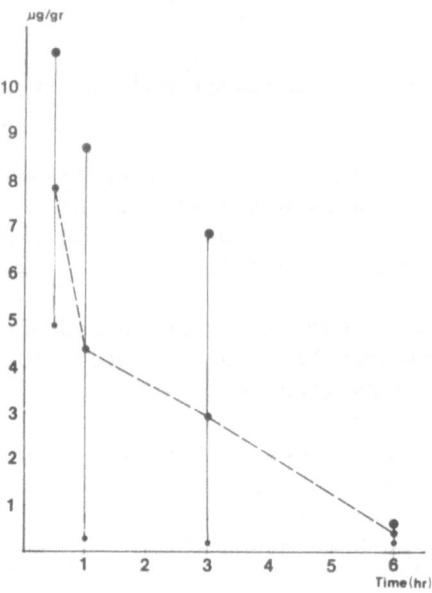

Fig. 2. Tumor tissue concentration after Intravesical
         instillation of Adriamycin.

Subdividing the analytical data on the basis of "grading" of the
tumor a different pattern was revealed in the decrease in tissue
concentrations of adriamycin in the hours following instillation. In
the differentiated forms (a-b) the average adriamycin concentration
ranged from 6.32 µg/g after 30 mins, to 4.39 after 60 mins, 3.57
after 3 hours and 0.49 µg/g after 6 hours. In the undifferentiated
tumors (c-d) the median concentrations found were 2 µg/g after 60
mins and 0.68 µg/g after 3 hours (Figure 4).

In the healthy tissues the average concentrations of adriamycin
are 4.85 µg/g after 30 mins., 1.20 after 60 mins., 0.20 after 3 hours
and 0.19 µg/g after 6 hours (Figure 5).

DISCUSSION

It is known that the diffusion of adriamycin[8,9,10,11] through
the tissues depends on a series of dimensional factors amongst which
the concentration of the solution used, the period of exposure, the
extension of surface contact and the anatomical and biological
characteristics of the tissues are all of importance.

Since the first three factors are constant in this study, any
eventual differences in the permeability and in the capacity to become impregnated by the drug (in a more or less stable manner) must
be attributed to tissue characteristics. As Jacobi (1980) observed

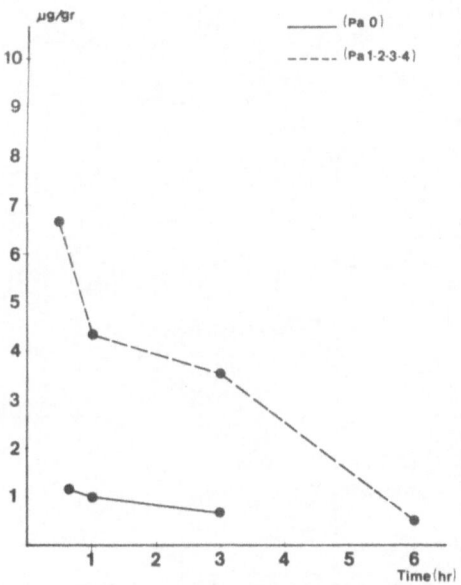

Fig. 3.   Tissue concentration in papillary (---)
and solid (——) tumor.

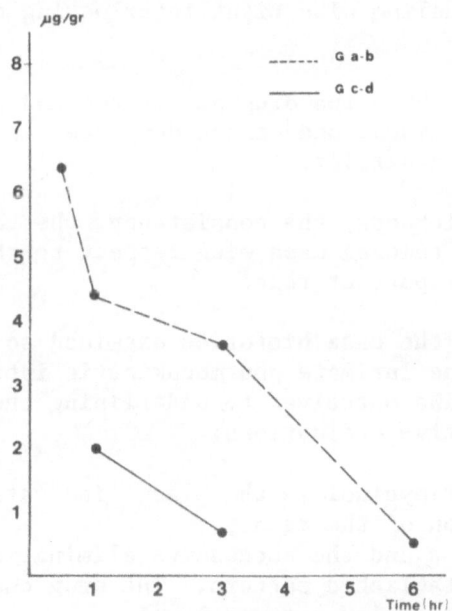

Fig. 4.   Tissue concentration related to grade of tumor.

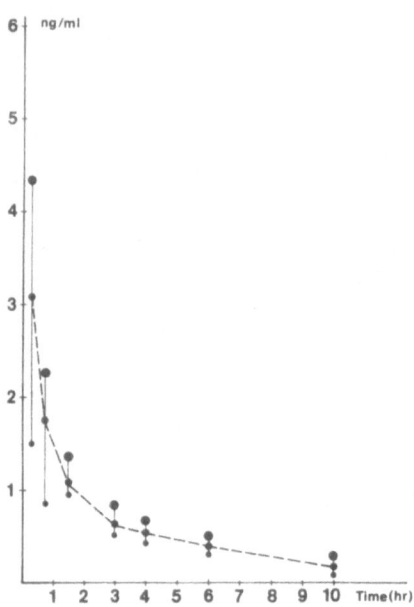

Fig. 5.   Tissue concentration in normal tissue.

the plasma uptake of adriamycin is blocked by the urothelium when
provided with its three normal cellular layers, an integral basal
membrane and an epithelium with tight interlocking cellular junc-
tions.

From the way in which the drug penetrates and persists in tumor
tissue or in healthy tissue one cannot deny the importance of the
urothelial lining as a barrier.

However, the thickness, the consistency, the vascularization,
and the depth of the tumoral mass with respect to the epithelial
surface play a very important role.

On the basis of the case histories examined so far it is not
possible to define the intimate pharmacokinetic intra-tissue mechanism, so we will confine ourselves to underlining the general concepts
arising from comparative evaluations:

1) the absence of adrimycinol in the plasma indicates a minimal
   systemic absorption of the agent;
2) systemic absorption and the successive elimination of adriamycin
   follows a well established pattern. The drug concentration in the
   plasma remains well below a toxic level;
3) absorption of adriamycin by neoplastic tissue indicates the
   possibility of a more efficient therapeutic action in the low

grade forms as opposed to the anaplastic forms and in the
papillary lesions as opposed to the solid tumors;
4) concentration of the chemotherapeutic agent in the healthy tissues
is such as to ensure a therapeutic effect on possible degenerative
and preneoplastic conditions at the preclinical stage.

REFERENCES

1. C. Merrin, R. Cartagena, Z. Wajsman, G. Baumgartner, and G. P.
    Murphy, Chemotherapy of bladder carcinoma with cyclophospha-
    mide and adriamycin, J.Urol., 114:884 (1975).

2. F. Edsmyr, T. Berlin, J. Boman, M. Duchek, P. L. Esposti, H.
    Gustafson, and H. Wikstromm, Intravescical therapy with
    Adriamycin in patients with superficial bladder tumors, in:
    "Diagnostics and Treatment of Superficial Urinary Bladder
    Tumors", F. Edsmyr, ed., Montedison, Lakemedal AB, Stockholm,
    (1979).

3. M. Nakazono and S. Iwata, A preliminary study of chemothera-
    peutic treatment for bladder tumors, J.Urol., 119:598 (1978).

4. M. S. Soloway and C. Martino, Prophylaxis of bladder tumor im-
    plantation: intravescical and systemic chemotherapy, Urology,
    7:29 (1976).

5. M. D. Banks, J. E. Pontes, R. M. Izbicki, and J. M. Pierce Jr.,
    Topical instillation of adriamycin in the treatment of recur-
    ring superficial transitional cell carcinoma of the urinary
    bladder, J.Urol., 118:757 (1977).

6. C. C. Schulman, Quimioterapia intravesical en tumores super-
    ficiales de vejiga, Arch.Esp.Urol., 36:32 (1983).

7. C. M. Camaggi, E. Strocchi, V. Tamassia, A. Martoni, M.
    Giovannini, G. Iafelice, N. Canova, D. Marraro, A. Martini,
    and F. Pannuti, Pharmacokinetic studies of 4'-Epi-Doxorubicin
    in cancer patients with normal and impaired renal function
    and with hepatic metastases, Cancer Treatment Reports,
    66:1819 (1982).

8. R. S. Benjamin, C. E. Riggs, and N. R. Bachur, Pharmacokinetics
    and metabolism of Adriamycin in man, Clin.Pharmacol.Ther.,
    14:592 (1973).

9. M. Pavone-Macaluso and N. Gebbia, Absorption of drugs from the
    bladder and intravesical chemotherapy, Urol.Res., 4:1 (1976).

10. M. Pavone-Macaluso, N. Gebbia, F. Biondo, S. Bertolini, G.
    Caramia, and F. P. Rizzo, Permeability of the bladder mucosa
    to thiotepa, adriamycin and daunomycin in men and rabbits,
    Urol.Res., 4:9 (1976).

11. S. Eksborg, Measurements of plasma levels of Adriamycin and
    adriamycinol after intravesical instillation of adriamycin,
    in: "Diagnostics and Treatment of Superficial Urinary Bladder
    Tumors", F. Edsmyr, ed., Montedison, Lakemedal AB, Stockholm,
    (1979).

COMBINATION CHEMOTHERAPY WITH CISPLATIN AND VM-26 IN ADVANCED TRANSITIONAL CELL CARCINOMA OF THE BLADDER, AN EORTC STUDY

G. Stoter[1], A. T. van Oosterom[2], J. H. Mulder[3]
M. de Pauw[4] and S. D. Fossa[5]
[1]Free University Hospital, Amsterdam, The Netherlands
[2]University Hospital, Leiden, The Netherlands, [3]Radio-
therapy Institute, Rotterdam, The Netherlands, [4]EORTC
Data Center, Institut Jules Bordet, Brussels, Belguim
[5]Norwegian Radium Hospital, Oslo, Norway

Presently, cisplatin is the most active single agent in the treatment of advanced bladder cancer. Dosages of 70-120 mg/m^2 i.v. every 3 weeks, or equivalents thereof, have yielded response rates of 30-40% in pretreated patients and 40-50% in non-pretreated subjects with a median response duration of 6 months[1-5]. Various combinations of cisplatin with agents such as adriamycin, cyclophosphamide and 5-fluorouracil have failed to improve these results[6-10].

In an attempt to improve the therapeutic results, the EORTC Urological Group initiated a phase II study (protocol 30802) with a combination of cisplatin and VM-26 based on the following rationale. First, a phase II study in the early 1970's conducted by this group with VM-26 suggested that the drug may be effective in advanced bladder cancer. Two CRs and 3 PRs were observed in 30 heavily pretreated patients[11]. Second, Burchenal has demonstrated synergism between platinum- and epipodophyllotoxin derivates[12]. Finally, the side effects of the two drugs differ. Myelosuppression is the main adverse effect of VM-26 whereas cisplatin affects the bone marrow only mildly. The objectives of the study were to determine the response rate and the duration of response of this combination.

Patients were eligible for the study if they had histologically proven transitional cell cancer of the urinary tract with bidimensionally measurable distant metastases (lung, liver, lymph nodes, skin) or a pelvic tumor mass measurable by CT scan. Further requirements included a performance status on the WHO scale of 0-2, creatinine clearance $\geqslant$40ml/min and normal bone marrow function. Patients with previous treatment consisting of cisplatin or VM-26

were not accepted. Bone metastases, hepatomegaly and serous ef-
fusions were not accepted as measurable lesions, and patients with a
history of congestive heart failure were excluded because of the
vigorous hydration scheme.

Pretreatment evaluation included history and physical examination, a routine hematological and biochemical screen, creatinine
clearance, IVP and chest X-ray. Except for IVP, these studies were
repeated before each treatment cycle. CT-scan was performed prior to
treatment and repeated after every second treatment cycle if necessary to measure indicator lesions.

The treatment regimen consisted of cisplatin 70 mg/m^2 i.v. on
day 1 and VM-26 100 mg/m^2 i.v. on days 1 and 2, the course being
repeated every 3 weeks. Cisplatin was given over a 4 hour period
with hydration. VM-26 was administered in 250 ml of saline over 30
minutes. For cases of severe bone marrow depression, dose reductions
of 25-50% of VM-26 were required.

The definitions of response were according to the WHO criteria[13]. CR is defined as the disappearance of all known disease
for at least 4 weeks and PR required a 50% or more decrease in the
product of the two largest perpendicular diameters of all measurable
lesions for no less than 4 weeks, with no new lesions developing.
Stable disease was considered less than a 50% decrease or less than a
25% increase in measurable disease, with no new lesions developing
and progression was more than a 25% increase in measurable disease,
or the appearance of new lesions.

Patients were considered evaluable for response if they had
completed a minimum of 2 cycles of therapy.

Fifty-eight patients were entered by 12 institutions. Eight
patients were ineligible for the following reasons: histology failed
to show transitional cell carcinoma in 3 patients; performance status
was >2 in another 3 patients; 1 patient had hepatomegaly only, and 1
had a second malignancy.

Nine patients were inevaluable for treatment response. In 4
patients inadequate dosages were given, 2 patients were treated with
incorrect intervals. After one cycle of therapy, 1 patient refused
further treatment, 1 patient developed renal failure due to obstructive uropathy, and a third patient developed urosepsis without leukopenia.

The average age of the patients was 61 years and males dominated
the subject population 6:1. Thirteen patients underwent cystectomy
with ileal conduit and 18 patients received radiotherapy to the
bladder. Topical chemotherapy was administered to 6 patients; 5
intravesically and 1 intra-arterially. Of six patients with tumor

limited to the pelvis, 2 had received prior radiotherapy. No patient had received previous systemic chemotherapy.

Forty-one evaluable patients were treated with an average number of 4 cycles (range 2-9). Four achieved CR (10%) and 17 PR (41%) for an overall response rate of 51%. Eleven patients had stable disease, 7 progressed, and 2 died secondary to their malignancy. Of the 6 patients in whom the tumor was confined to the pelvis, 2 received prior radiotherapy; one had stable disease and the other had progression of disease after 2 cycles. In the 4 non-irradiated cases, 2 achieved PR, one had stable disease and one had progression. These numbers are too small to evaluate the influence of previous radiotherapy on the chemotherapy.

The median duration of response was 6 months. The median survival of patients with CR was 12 months. Patients with PR and no change had a median survival of 6 months and patients with progressive disease had a median survival of only 3 months.

The major toxity was nausea and vomiting, which led to cessation of treatment in 6 patients. Leukocytopenia (1400-3800/mm^3) was observed in 10 patients and required dose reductions in 7. Thrombocytopenia (<100,000/mm^3) was observed in one patient (85,000/mm^3). No sepsis or bleeding complications occurred.

The overall response rate of 51% (95% confidence limits: 36-67%) with a median response duration of 6 months in this patient population without prior systemic chemotherapy does not differ from cisplatin as a single agent. Therefore, this study fails to confirm a synergistic effect between cisplatin and the epipodophyllotoxin derivate, and it appears that VM-26 has only weak activity in advanced bladder cancer. VM-26, when used as a single agent at a dose of 20-30 mg/m^2 i.v. daily, every 3 weeks, has recently been reported as ineffective, although this study can be criticized for suboptimal doses and extensively pretreated patients[14].

The results of the present study support the view that cisplatin combination chemotherapy is not yet superior to single agent cisplatin in bladder cancer[5-10], and since multidrug treatment may also increase the toxicity, combination chemotherapy with cisplatin is not recommended for the routine treatment of transitional cell cancer of the bladder.

REFERENCES

1. H. W. Herr, Cis-diamminedichloroplatinum II in the treatment of advanced bladder cancer, J.Urol., 123:853 (1980).
2. S. Ostrow, M. J. Egorin, D. Hahn, S. Markus, A. Leroy, P. Chang, M. Klein, N. R. Bachur, and P. H. Wiernik, Cis-dichlorodiam-

402   mineplatinum and adriamycin therapy for advanced gynaecological and genitourinary neoplasms, Cancer, 46:1715 (1980).

3. P. C. Peters and M. R. O'Neill, Cis-diamminedichloroplatinum as a therapeutic agent in metastatic transitional cell carcinoma, J.Urol., 123:375 (1980).

4. A. Yagoda, Chemotherapy of metastatic bladder cancer, Cancer, 45:1879 (1980).

5. G. Stoter, S. D. Williams, and L. H. Einhorn, Genitourinary tumors, in: "Cancer Chemotherapy Annual 3", H. M. Pinedo, ed., Excerpta Medica, Amsterdam, p.322 (1981).

6. A. Yagoda, R. C. Watson, and W. F. Whitmore, Cisplatinum (DDP) regimens in bladder cancer, Proc.Amer.Soc.Clin.Oncol., 20:347 (1979).

7. M. B. Troner, Cyclophosphamide, adriamycin and platinum (CAP) in the treatment of urothelial malignancy, Proc.Amer.Ass.Cancer Res., 20:117 (1979).

8. M. L. Samuel, CISCA combination chemotherapy, in: "Cancer of the Genitourinary Tract", D. E. Johnson and M. L. Samuels, eds., Raven Press, New York, p.97 (1979).

9. S. D. Williams, L. H. Einhorn, J. P. Donohue, Cisplatinum combination chemotherapy of bladder cancer, Cancer Clin.Trials, 2:335 (1979).

10. J. H. Mulder, S. D. Fossa, M. de Pauw, and A. T. van Oosterom, Cyclophosphamide, adriamycin and cisplatin combination chemotherapy in advanced bladder carcinoma: an EORTC phase II study, Eur.J.Cancer Clin.Oncol., 18:111 (1982).

11. M. Pavone-Macaluso and the EORTC Genito-Urinary Cooperative Group: Single drug chemotherapy of bladder cancer with adriamycin, VM-26 or bleomycin, Eur.Urol., 2:138 (1976).

12. J. H. Burchenal, K. Kalaher, L. Lokys, and G. Gale, Studies of cross resistance, synergistic combinations and blocking of activity of platinum derivatives, Biochemie, 60:961 (1978).

13. WHO handbook for reporting results of cancer treatment. WHO offet publication 48, 23, WHO Geneva (1979).

14. R. Qazi, P. Elson, and J. D. Khandehar, Phase II evaluation of VM-26 in patients with metastatic transitional cell carcinoma of the urinary tract: an Eastern Cooperative Oncology Group Study, Cancer Treat Rep., 66:405 (1982).

CHEMOTHERAPY FOR G3-pTis CARCINOMA OF THE BLADDER

H. R. England, R. C. Tiptaft, R. T. D. Oliver
A. M. I. Paris and J. P. Blandy

Department of Urology, The London Hospital
London, E1, England

In recent years perhaps one of the main contributions towards rational management of urothelial malignancy has been the increasing recognition of carcinoma insitu and the acceptance of its significance as an early phase in the development of invasive tumor. The disease is a progression from epithelial atypia and although bladder is the commonest site the process is really a field change with any part of the urothelium liable to be affected and of particular significance is the epithelium lining the prostatic ducts. The nature of the lesions is against successful control by endoscopic resection or diathermy and usually, although not always, radiotherapy is ineffective whilst radical surgery cannot seriously be advanced as routine primary treatment.

At the course on bladder cancer in this center in 1978 papers were presented on treatment by chemotherapy and we read a preliminary report on 9 patients treated by intravenous cyclophosphamide[1,2]. Some 5 years further on, we now report results in a small group treated by a combination of agents and each followed for 3 years or longer.

PATIENTS AND METHODS

Patients

24 patients are included, 22 men and 2 women. At the time when urothelial malignancy was first diagnosed, the ages ranged from 40 to 76 and almost half were over 65.

Diagnosis

All patients had symptoms of "malignant cystitis" consisting of frequency, usually with urgency and often urge incontinence in addition to suprapubic and genital pain. Although the severity was variable, symptoms alone demanded treatment in all.

Diagnosis entailed urine cytology, positive in all but 1, and 4 to 6 representative cup-forceps biopsies of which a minimum of 3 had to show G3-pTis carcinoma to allow inclusion in the series. Resection biopsies of prostatic urethra and prostate itself are now taken in addition.

9 had not been previously treated for bladder cancer, the in situ changes being their primary lesion. The other 15 had already had therapy for overt tumor, 12 confined to epithelium with grades G1 x 1, G2 x 4 and G3 x 7, and 3 invading muscle, all grade G3. In two at presentation, there was widespread associated G3-pTis cancer which persisted after radiotherapy had successfully dealt with G3-pT1 tumors. The remaining 13 had insitu carcinoma diagnosed at intervals ranging from 6 months to 14 years (average 3 years) after treatment of primary tumor.

Treatment

Initially intravenous cyclophosphamide 1.0 G/m^2 every 3 weeks for 8 doses was given but experience soon showed that if a response was not obtained after 4 injections further cyclophosphamide conferred no benefit; 4 injections subsequently became standard therapy.

As expected, when treatment was stopped and the patient had responded, the great majority subsequently again developed positive cytology and/or biopsies and required further treatment. As cyclophosphamide itself had come under suspicion as possibly being carcinogenic, intravesical Adriamycin or Mitomycin C were used in these recurrent cases, and for the non-responders. Most patients therefore have been treated in sequence with more than 1 of these 3 agents.

In the early stages after a response had been obtained we used Adriamycin prophylactically but soon abandoned this policy in favour of giving treatment only when proven disease was present. Doses of agents used are shown in Table 1, and Table 2 lists the combinations given and the numbers of patients receiving them.

Follow-up

Cytology together with cystoscopy and multiple representative biopsies were performed every 3 months, a period extended to 6 months when patients remained clear for long periods.

Table 1.   Dose Schedule of Agents used

Cyclophosphamide:	1 G/M^2 3 weekly x 4
Adriamycin:	80 mgm in 100ml saline intravesically monthly
Mitomycin C:	30 mgm in 30ml saline intravesically weekly x 6

Table 2.   Therapeutic Agents Used and Numbers of Patients

Agents	Numbers		
	Primary	Secondary	Total
Cyclophosphamide alone	2	6	8
" + Adriamycin	2	3	5
" + Mitomycin C	5	5	10
Mitomycin C alone	-	1	1
TOTALS	9	15	24

RESULTS

The most dramatic effect of treatment was the relief of symptoms of malignant cystitis which usually began within 48 hours of starting cyclophosphamide.

Primary G3-PTis carcinoma

All nine patients received cylcophosphamide as initial treatment. Six responded and two have remained clear since for 3 and 6 years respectively. The other four recurred after 6 and 9 months and 2 and 3 years when three responded again, either to Mitomycin C or to Adriamycin. The fourth failed Mitomycin but responded again to cyclophosphamide. Two of three cyclophosphamide non-responders became clear after subsequent Mitomycin C but one failed this agent too to become to date the only total non-responder in the primary group.

Except for the non-responder (2 years) follow up ranges from 3 to 6 years (average 4+) and during this time one has died from a proven primary pancreatic carcinoma whilst four others have grown pT1 tumors. Three were successfully dealt with by endoscopic resection and one by cystectomy.

At the last examination six of eight survivors were apparently
free of disease while two others were found to have recurrent in situ
cancer requiring further treatment.

Secondary G3-pTis Carcinoma

14 of 15 received cyclophosphamide and 10 responded. One remains without recurrence after $5\frac{1}{2}$ years and a second died after 6
months from primary bronchogenic carcinoma. Among the remaining 13
the pattern of responding a second time and/or to another agent was
the same as seen in the primary group.

However, only six have escaped major disease progression or
incident. As a result of therapy one required cystectomy for intolerable symptoms following instillation of Adriamycin into a previously irradiated bladder - there was no malignancy on section. One
developed a G3-pT1 tumor, and refused further treatment for this
potentially curable lesion and subsequently died. Two others grew G3
-pT3 carcinomas and had cystectomies - one is dead from metastases
and the other is well after 3 years. The remaining four all developed G3-pT4a tumors on the basis of in situ changes within prostatic
ducts and two are already dead.

At the last examination and after an average follow up of over 4
years, of 10 survivors 8 were free of disease including two after
cystectomy. Two others are very likely to have residual disease
after pT4a tumors.

Table 3 lists the state of all patients at last assessment.

Toxicity

Vomiting coming on about 15 hours after injection and lasting
some 24 hours was a constant feature of cyclophosphamide therapy but
there were no other untoward sequelae. Instilling Adriamycin into
previously irradiated bladders proved capable of causing crippling
symptoms and limited the use of this agent as 11 patients in the
series had been irradiated.

The use of Mitomycin C was trouble free except for a few minor
skin rashes and complaints of bladder discomfort.

Discussion

In situ cancer will respond to a variety of agents and 95% of
this group responded at least once to one of the agents used. Only
three, however, have remained clear of disease since their initial

Table 3.   State of Patients at Last Assessment.

Assessment		Numbers	
	Primary	Secondary	Total
Free of disease	5	6	11
Free of + cystectomy	1	2	3
Being treated for pTis cancer	2	–	2
Died of bladder tumor	–	2	2
Died pT4a tumor	–	2	2
Likely to have residual tumor	–	2	2
Died of other causes when clear	1	1	2
TOTALS	9	15	24

treatment, all following cyclophophamide, and for 3, 5½ and 6 years, while three others had periods of freedom before recurrence of 2½ to 3 years also after that agent.   Response rates for the different agents used were: Mitomycin C 8/11 (73%), cyclophosphamide 16/23 (70%) and adriamycin 3/5 (60%).

If the whole spectrum of in situ cancer is considered, then there is a wide variation in the time taken for progression to occur but experience suggests that when "cystitis" symptoms are present tumor formation is not far behind.

In 1970 Utz et al., reported on 62 patients with symptomatic primary carcinoma in situ.   Initial treatment was largely conservative so the outcome reflects the natural history of this segment of the disease[3].   When reviewed in 1980 it was reported that within 5 years 73% had developed invasive cancer and that 57% of them (42% of the series) had died as a result[4].   On chemotherapy our tiny group of comparable primary patients has fared better.   So far none has died from the disease or looks like doing so, and only one patient has had a cystectomy.

Patients with secondary in situ disease are at greater risk which is reflected by the poor outcome.   This should not be unexpected as all of them have already demonstrated a high malignant potential by growing high grade tumors.   The main hazard was the liability to develop invasive tumor within prostatic ducts.   20% of men with G3 bladder tumors have been shown to have associated in situ changes within prostatic ducts or in the prostatic urethra[5], when there was extensive in situ bladder disease.   Farrow et al., found the same changes in periurethral prostatic ducts in 7 of 19 men (37%)[6].

These findings emphasise the field-change nature of the disease and indicate the need to take representative samples of prostate for proper assessment.

If prostatic ducts or urethra are involved, intravesical agents will not come into contact with them, which is why we favor systemic cyclophosphamide. In theory it could influence the whole urothelium and we have been encouraged by the experience in one patient with easily demonstrated carcinoma in situ within prostatic ducts. After a course of cyclophosphamide 3 years ago the changes can no longer be demonstrated.

In 3 cases outside this series Methotrexate has had no effect on the disease.

CONCLUSIONS

From our experience with this small group the following conclusions suggest themselves:-

1. If chemotherapy is to be the main line of treatment for these patients follow-up must be impeccable and frequent.
2. The results in primary in situ disease are encouraging, and this group has been well controlled.
3. Secondary carcinoma-in situ is a much higher risk disease and less favorable for treatment by chemotherapy.
4. The main danger area has been the prostatic ducts, and if significant in situ changes remain after a course of cyclophosphamide the patient is best served by radical surgery. This also holds good for those who escape control to the extent of developing a G3 bladder tumor as they are a high risk group whose tumors will at least have breached the lamina propria.

REFERENCES

1. G. Jakse and F. Hofstadter, Adriamycin therapy in Carcinoma in situ: A preliminary report, in: "Bladder Tumors and Other Topics in Urological Oncology," M. Pavone-Macaluso, P. H. Smith and F. Edsmyr, eds., Plenum Press, London, pp.327-8 (1980).
2. H. R. England, E. A. Molland, R. T. D. Oliver and J. P. Blandy, Flat carcinoma in situ of the Bladder Treated by Systemic Cyclophosphamide - A Preliminary Report: ibid pp.371-5.
3. D. C. Utz, K. A. Hanash, and G. M. Farrow, The plight of the patient with carcinoma in situ of the bladder, J.Urol., 37:93-99 (1970).
4. D. C. Utz, G. M. Farrow, C. C. Rife, J. W. Segura and H. Zincke, Carcinoma in situ of the bladder. Cancer, 45:1842-8 (1980).

5. N. Gowing, Urethral carcinoma associated with cancer of the bladder, Brit.J.Urol., 32:428-38 (1960).
6. G. M. Farrow, D. C. Utz and C. C. Rife, Morphological and Clinical Observations of Patients with Early Bladder Cancer treated with total cystectomy. Cancer Res., 36:2495-501 (1976).

TREATMENT OF ADVANCED BLADDER CANCER

Mark S. Soloway
Department of Urology
Baptist Memorial Hospital
Memphis
Tennessee, USA

As we analyze our "progress" in the management of patients with bladder cancer with particular emphasis on patients with invasive tumors (T2+), we are keenly aware of our limited ability to cure the disease at this stage. Since we seem to be altering the natural history of the disease for patients identified with superficial bladder cancer we must focus some of our effort on earlier diagnosis and initiation of therapy. A recent report by Kaye et al.,[1] indicated that a high percentage of patients with invasive bladder cancer were first seen by a physician when tumors had already infiltrated. Since this data has important implications I sought to see if our patient population at the University of Tennessee Center for the Health Sciences and associated hospitals was similar. The primary question we sought to ask was "When do patients with advanced bladder cancer first seek medical care? How many with invasive tumors were initially diagnosed with superficial tumors?"

The charts of 297 patients with bladder cancer were reviewed. Ninety (30%) patients had tumors which invaded into the bladder wall or had metastasized either regionally or to distant sites when first seen. To our surprise, 82 of these 90 patients, or 91%, had an advanced stage when they were first seen by a clinician. Thus they never had a diagnosis of superficial bladder cancer. 51 (62%) of these patients had tumors which were clinically localized to the bladder (stages B-C). 45 of the 51 patients were treated with radiation therapy and surgery, while 9 had urinary diversion only because of extensive local disease. 36 of the 45 patients had a cystectomy and 9 had lymph node metastases.

The message is clear. Few patients with advanced bladder cancer are seen at a stage when there is a high likelihood of cure. A

screening program entailing methods such as urinary cytology will
have to be instituted in high risk populations with the hope of
detecting the disease when the potential for cure is high. The
expertise and the techniques are available. The cost of such a
program is the primary deterrent.

MANAGEMENT OF MUSCLE INVASIVE BLADDER CANCER

There is a great deal of controversy regarding the most appropriate management of patients with locally extensive bladder cancer,
stages (B-D1 or T2-T4). There are advocates of pre-operative radiation therapy with the hope of downstaging the tumor and thus improving survival. There is evidence that downstaging occurs in a
high percentage of these patients and if the tumor is downstaged, the
survival rate is improved[2-3]. It is unresolved, however, that
giving this entire group of patients preoperative radiation plus
cystectomy improves the survival rate when compared to cystectomy
alone[4,5]. An alternative which has many advocates is the full dose
of radiation therapy for initial therapy with salvage cystectomy if
the tumor is not eradicated or there are subsequent tumor occurrences[6].

Radical cystectomy involves the removal of the bladder with the
surrounding fibro-fatty tissue and the peritoneal covering. In the
male it includes the removal of the prostate and seminal vesicles and
in the female it includes the fallopian tubes, ovaries and uterus as
well as the anterior vaginal wall. In 1973 Pearse et al.,[7] reported on the results of 52 patients having cystectomy either for
lesions of category T2 or greater, or for multifocal superficial
tumors with an increasing tumor grade. Fourteen of the 52 patients
had tumors of stage 0, A or B-1. All patients had a pelvic lymph
node dissection as part of the procedure. The ten year survival of
the entire group was 27%. The operative mortality was 19%. Analysis
of survival by stage indicated that those with 0, A or B-1 tumors had
a 50% ten year survival; 42% for those in the B-2 category; and 13%
for C lesions. There were 11 patients with regional lymph node
metastases (D1) and none survived.

Roswell Park Memorial Institute reported on 92 patients[8].
Pelvic node dissection was not performed. Fifteen had preoperative
radiation therapy of 4000R. The five year survival for those with 0,
A or B-1 disease was 50% and it was 32% for those with B-2, C or D
disease. The operative mortality was 9%.

I recently reviewed my results of 56 consecutive radical cystectomies performed at the University of Tennessee. There were 41
males and 15 females. Their average age was 62 with a range of
40-76. The preoperative stages were CIS 5%, A 16%, B 32%, and C 46%.
Thus only 21% of the patients did not have invasion into the muscle

layer. It has been my policy to avoid extirpative surgery unless muscle invasion has occurred. Those with superficial disease had failed intensive intravesical chemotherapy. 77% of the patients had transitional cell carcinoma with a Grade 2-3 tumor in 70%. The other 7% of those with TCC had Grade 4 lesions. 12% had squamous cell carcinoma, 5% had adenocarcinoma, 2 patients (3%) had leiomyosarcoma and one patient had a small cell sarcoma of the bladder.

All patients had a three day preparation of the bowel with both mechanical cleansing agents and oral antibiotics. A urethral catheter or hemovac drain was used in all patients. If a pelvic Foley catheter was used this was always removed within 48 hours. The method of division was uretero-ileal cutaneous diversion. The ureters were stented with silastic stents which were removed at one week. The ileo-ileal anastomosis was performed with staples in most cases.

Almost all patients with invasive tumors received 4000R preoperative radiation therapy in accordance with a National Bladder Clinical Collaborative Group protocol. Surgery was carried out within 7-14 days of completing the radiation therapy. Forty of the 56 patients had preoperative radiation therapy. The average hospital stay was 14 days with a range of 8-53 days. The average blood loss was 1260 ml with a range of 500-2500 ml. 32% of the patients had early complications and 20% had late complications (Table 1). Metastases to the lymph nodes were documented in 11 (20%) patients.

Thirty five (62.5%) of the patients have no evidence of disease with a follow-up of 1-84 months. Nineteen patients have been followed for over three years and 8 (42%) are alive.

A review of several recent cystectomy series indicated that there has been a gradual decline in operative mortality over the last few years[9]. In my series of 56 cystectomies there was no operative mortality. Recent reports by Freiha[10], Skinner, et al.,[11] and Johnson and Lamy[12] indicate an operative mortality of 2-3%. This improved mortality must be considered when reviewing prior studies with an attempt to use these as a baseline for determining the impact of a new modality, e.g., adjuvant chemotherapy, since there may be as much as 10% improvement in survival simply related to the improved pre and postoperative care.

CHEMOTHERAPY FOR ADVANCED BLADDER CANCER

Only a few years ago, it was unclear whether it was appropriate for a patient to receive chemotherapy for advanced bladder cancer, either metastatic or locally extensive, since there was little evidence that any drug would make an impact on survival. Currently approximately 30-50% of patients with advanced transitional cell

Table 1.   Complications following Cystectomy

Early		Late	
None	38	None	45
Ileus	5	Ureteral Stenosis	9
Wound Dehiscence	5	Bowel Obstruction	2
Wound Infection	2	Stomal Stenosis	1
Cardiovascular	3	Draining Sinus	1
Confusion	2		
Pelvic Abscess	2		
Bowel Obstruction	1		
UGI Bleeding	1		
Pneumonia	1		
Drainage	1		
Sepsis	1		

carcinoma of the urinary tract which includes tumors originating in the renal pelvis, ureter, bladder, urethra, or prostatic ducts can achieve a complete or partial objective response as a result of single or combination chemotherapy. Four chemotherapeutic agents have been shown to have activity. These include cis-platin (DDP), methotrexate (MTX), doxorubicin (ADM), and vinblastine sulfate (VLB). The first two drugs appear to have more activity than the latter two. Combinations of these agents may provide response rates higher than can be achieved by the single agents alone, and studies to document this are currently being initiated.

Unfortunately most of the responses to chemotherapy have been partial, i.e. a reduction in an objectively measurable lesion of greater than 50% of the summed products of the longest and its perpendicular diameter. Complete remissions are increasingly being documented, but still represent the minority of the objective responses. Stabilization of disease may or may not be biologically significant. If often represents the inability of the observer to document a slowly progressing metastatic lesion in sites such as bone or in the pelvis which are difficult to measure. Alternatively, stabilization may simply represent the natural history of a slow growing tumor.

Cis-Platin

Cis-platin (DDP) is a non-cell-cycle specific drug which appears to function as an alkylating agent by inhibiting DNA by intrastrand cross linkage[13]. Pharmacokinetic studies indicate that there is an active component which is nonprotein bound while the protein bound component may in part relate to the toxicity of this agent. The half life is approximately one hour. Twenty-five percent of the drug is excreted in the urine within 24 hours.

The dose used in most of the studies evaluating this drug in the treatment of urothelial tract tumors is 70 mg/M^2 intravenously every 3-4 weeks. ˙Other schedules and doses have been used. Unfortunately, there is no good data indicating a dose-response relationship. In some of the combination regimens DDP doses up to 100 mg/M^2 have been used with an improved response rate but it is unclear whether this is related to the addition of the other drugs or partially relates to the higher dose of DDP. Adequate hydration prior to cis-platin administration is necessary to reduce the chance of nephrotoxicity. The protocols utilized by the National Bladder Cancer Collaborative Group (NBCCGA) have incorporated a mannitol-induced diuresis administering 12.5 gm of mannitol a few minutes prior to cis-platin. This is one of the commonly employed methods to lessen the likelihood of nephrotoxicity. In addition to mannitol, 500 ml of D5-½ normal saline is infused one hour before and again one hour after DDP. Potassium and magnesium should also be added to the infusion. It has been our practice to administer DDP in an outpatient setting if feasible. If patients are travelling long distances, however, this may be difficult and an overnight admission may be required to ensure continued hydration and the administration of appropriate anti-emetics. Recently there has been a report that 3% NaCl infusion will dramatically reduce the chance of nephrotoxicity and that a higher dose of cis-platin can be safely used[14]. This report requires confirmation by a larger series of patients.

In addition to nephrotoxicity and nausea and vomiting, other toxicities which are related to DDP include: hypomagnesemia, peripheral neuropathy, myelosuppression, anemia, and hypersensitivity reaction.

Several platinum analogs are undergoing clinical trials. Although they do not appear to be more active than DDP, the toxicity is different. In particular many have substituted nephrotoxicity for myelosuppression.

In a recent review Yagoda[15], indicated that, regardless of dose or schedule, 30% of patients receiving DDP for urothelial tract tumors achieve a complete or partial response. The 95% confidence limits are 25-30%. The NBCCGA performed a prospective randomized study comparing DDP to DDP + cyclophosphamide[16]. In 105 evaluable patients there were 10 responders (20%) among the 49 evaluable patients who received DDP alone. This includes four complete and six partial responses. An additional 16 patients (32%) were considered stable. Among the 56 who received DDP + cyclophosphamide, there were two complete and five partial responders (12.5%). Thirty-four percent were stabilized as a result of the combination treatment. There was no statistical difference between the differences in the response rates for these two treatment arms.

Not unexpectedly, patients with a better performance status were more likely to respond than patients with impaired performance. Sixty-six percent of the patients who responded had progressed at the six month interval. In the entire group 27.5% of the patients had their creatinine rise above 1.5 mg%.

In a non-randomized trial, Yagoda[17] found that cyclophospham-ide + DDP provided a similar response rate to that of DDP alone although the average duration of response was two months longer for those receiving the combination regimen. In a later study, he evalu-ated the combination of DDP + ADM and a 48% partial response rate was observed[18].

An important facet of DDP therapy is that responses are usually seen within one or two courses of treatment. Thus if the patient has not responded by the 9-12 week interval, further therapy with DDP is unlikely to be of benefit.

All tumor sites may not respond identically. Pulmonary metasta-ses are more likely to regress as a result of DDP. In some cases, DDP may be continued after radiation therapy or surgery eradicates a local pelvic recurrence.

When ADM and/or cytoxan are combined with DDP, the response rates approximate 45%[19-10]. The lower limits of the 95% confi-dences exceed the upper limit for DDP alone (35%)[21]. There is only one study in which DDP was randomized to the three-drug combination of DDP + ADM + cyclophosphamide and the response rates were 24% vs. 37% respectively[22].

Yagoda recently reported a response rate of 46% in patients receiving the three-drug combination[21]. DDP, 70 mg/M^2, was given on day 1, cyclophosphamide on day 2, and ADM on day 3. The doses of cyclophosphamide and adriamycin were 250 and 45 mg/M^2 respectively. The median duration of response was eight months and five patients remained in remission for over one year. The response rate was greater for those with transitional cell compared to squamous or adenocarcinoma. Myelosuppression, of course, does occur with the addition of other agents like cyclophosphamide and ADM.

A report from M.D. Anderson Hospital[23] reported 52% response rate with the same three-drug combination. In this study, higher doses were used: DDP 100 mg/M^2, cyclophosphamide 600 mg/M^2, and ADM 50 mg/M^2.

The Southeastern Cancer Study Group treated 43 patients with metastatic urothelial cancer with the same three-drug regimen al-though a lower dose of DDP was used[24]. There was a 38% response rate which included three complete responses.

Yagoda[21] reviewed the accumulated reports in the literature of 174 cases who received this three-drug combination; 45% responded compared to an overall response rate of 31% in 273 cases given DDP alone. Thus the three-drug combination seems to induce a longer duration of response and survival and to be more active than DDP alone.

Methotrexate

Methotrexate (MTX) is a cell-cycle-phase-specific agent which binds to the enzyme dihydrofolate reductase. Since this reduces the intracellular levels of tetrahydrofolic acid, the production of DNA and RNA is inhibited. Methotrexate can be delivered in a standard or a high dose form. The high dose method requires rescue with citrovorum factor (CF). This inhibits methotrexate and rescues cells from damage. There is no conclusive evidence that the high dose methotrexate administration will result in a higher response rate for urothelial tract tumors.

Most therapeutic schedules begin at a dose of 30-40 mg/M^2 weekly with an escalation to 60 mg/M^2. Patient tolerance is variable and it depends upon pre-treatment leukocyte and platelet counts. The drug can also be given in a dose of 50 mg I.M. weekly or in a range of 100-200 I.M. with CF rescue, 10 mg I.M. 6 hourly for 48 hours. Salicylates should not be given at the same time as MTX. Renal function must be adequate for drug administration or toxic levels may rapidly occur.

The major toxicities of MTX include myelosuppression, stomatitis, and gastrointestinal ulceration. Because of the importance of renal clearance, adequate hydration must be ensured. Some have also advocated alkalinization with sodium bicarbonate prior to administration of the drug.

Studies performed in England have used three different schedules for MTX administration with a 13% response rate in 23 patients using a dose of 50 mg every two weeks compared to a 53% (20/38) response rate when using a high dose regimen with CF rescue[25]. A majority of the patients had only pelvic disease and monitoring is difficult in this group of patients. In patients with measurable disease, Oliver[26] noted three responses in nine patients. Natale et al.,[27] used a dose of 40 mg/M^2 weekly and found an objective response rate of 26%. Previously untreated patients had a 36% response rate.

Yagoda[21] reviewed the 236 cases reported in the literature using MTX and found a 29% CR + PR rate with 95% confidence limits of 23-35%.

The combination of methotrexate + DDP is attractive since the two most active drugs are combined. Such trials are being initiated by the NBCCGA and EORTC. Nephrotoxicity will be a prime factor in determining the feasibility of this particular regimen since with any significant nephrotoxicity, MTX levels may be increased with associated severe mucositis or myelosuppression.

Adriamycin

Adriamycin (ADM) acts by intercalating between DNA strands thus affecting transcription. It can probably affect all phases of the cell cycle but is primarily a cell-cycle-phase specific agent. The most commonly used dose is 60-75 mg/M^2 intravenously every three weeks. For patients with prior chemotherapy or radiation therapy, the usual initial dose is 30-45 mg/M^2 with a nadir leukocyte or platelet count at 8-12 days. An elevated bilirubin requires dose reduction.

The major toxicities of ADM are nausea, vomiting, alopecia, myelosuppression, and cardiomyopathy.

There is a paucity of studies that have used adriamycin as a single agent and many were done several years ago. Yagoda reviewed the literature and utilizing strict criteria for complete or partial responses found the rate to be 18% in 223 cases[21]. The Southwestern Oncology Group performed a randomized trial comparing adriamycin to the combination of ADM + DDP and indicated that the single agent provided a response rate of 20% in 40 patients compared to 36% response rate in 36 patients receiving the combination[28]. Complete responses have been uncommon using ADM as a single agent.

The EORTC[29] evaluated the combination of ADM + 5FU in 52 evaluable cases and found a response rate of 40% which included four complete responses. Many of the patients had disease confined to the pelvis and thus the criteria of response were relatively soft and it is unclear whether the response rate in fact is so high.

Vinblastine

Vinblastine sulfate (VLB) is a vinca alkaloid which inhibits mitosis by arresting the cell cyle in metaphase. It is thus considered a cell-cycle-phase-specific agent. The dose used in patients with urothelial cancer has been 3-4 mg/M^2 on a weekly basis. The primary side effect is myelosuppression. Neurotoxicity can also occur but is much less than with the other vinca alkaloid, vincristine.

Blumenreich, et al.,[30] noted an 18% objective response rate in a heavily pretreated group of patients receiving VLB in a dose of 4-6 mg/M^2 weekly. The average duration of response was four months with minimal toxicity. The EORTC completed a trial evaluating vincristine for urothelial tract tumors and found minimal activity - a response rate of 8% in 37 cases[31].

Adjuvant Chemotherapy

There have been few studies which have carefully and systematically analyzed the effect of adjuvant chemotherapy following radiation therapy or radical surgery for bladder cancer. The few that have been done do not provide any optimism utilizing single agent chemotherapy. The NBCCGA has recently elected to stop a study which was randomizing patients who had tumor remaining following treatment of invasive bladder cancer with preoperative radiation therapy (4000 R) plus radical cystectomy. They were receiving either DDP or no chemotherapy. Although the results have not been analyzed, insufficient numbers of patients elected to continue the prescribed eight doses of chemotherapy. Patients having no tumor in the pathologic specimen (PO) were not randomized. In addition many patients elected not to be randomized following cystectomy. Thus although a large number of patients were entered, the number finally receiving the intended course was extremely limited.

312 patients were considered for this protocol but 181 were not entered for a variety of reasons. 131 patients were entered and 114 received the appropriate course of preoperative radiation therapy. 99 patients subsequently underwent radical cystectomy and pelvic lymph node dissection. Of these, 43 were ineligible for randomization to adjuvant chemotherapy either because they had no tumor in the bladder and thus were expected to have an excellent survival and should not have the morbidity of chemotherapy or because they were too sick or refused to enter the randomization. Thus only 54 patients were eligible for randomization. 24 served as controls and 30 received DDP. Only 6 of these had at least 8 courses as initially described in the protocol. Despite a high initial entry there will be too few patients to make any meaningful comment regarding the efficacy of adjuvant chemotherapy.

I have reviewed the patients I entered into this protocol to gain some idea of the potential effect of DDP. Seven patients were P3NO and received DDP and 4 are living from 12-29 months. Three have died of liver metastases from 14-21 months after surgery. Seven patients had one or more lymph nodes involved with tumor. All of these patients have died despite DDP with the longest survivor being 30 months following his surgery. Thus there does not appear to be any major impact from the addition of cis-platin.

The number of patients is small and these conclusions are only anecdotal. Recently, Skinner et al.,[32] reported their results of adjuvant chemotherapy over the last several years much of which included the use of DDP. Once again there was no advantage to those patients who received the adjuvant chemotherapy regimen.

Oliver[26] has used adjuvant methotrexate in a dose of 100 mg/M^2 with CF rescue on a weekly basis. Methotrexate was delivered as an adjuvant to radiation therapy. Once the radiotherapy was initiated the drug was given every two weeks for one year. There appeared to be some survival benefit early in the trial for patients receiving the MTX compared to a historical group treated only with radiation therapy. Although the results appeared promising at six months after initiation of treatment, at 12 months there was no significant difference when compared to the historical group.

Cis-Platin + Radiation Therapy

Since the use of radiation therapy prior to radical cystectomy has many advocates and has improved patient survival for those patients who are downstaged, it would be attractive to attempt to increase downstaging with the use of a radiosensitizer. In addition, since many patients with extensive local disease have micrometastases, a systemic agent with local radiosensitizing properties would be of particular benefit.

There is evidence from animal and _in vitro_ systems that cis-platin provides an additive antitumor effect when combined with irradiation[33-36]. The animal experience has stimulated pilot studies in patients with advanced local or metastatic disease when radiation therapy was being used as the primary mode of therapy. Soloway et al.,[37] used DDP plus radiation therapy in eight patients with advanced bladder cancer. The tumor was locally recurrent or metastatic in five patients and this was used as primary therapy in three patients. The treatment was well-tolerated and there was no additional toxicity from the DDP radiation regimen. The mean survival from initiation of chemotherapy for the four patients with persistent pelvic tumor, local recurrence or metastatic disease was 24 months (range 13-48). This compares with a median survival of three months from detection of metastases as reported by Babaian et al.,[38] in a group of patients who did not receive chemotherapy. Jakse et al.,[39] used a similar regimen in eight patients who were not candidates for cystectomy for advanced local disease. Six of these patients were free of tumor by clinical evaluation with a mean follow-up of 7.7 months. It should be mentioned that following the conclusion of radiation therapy, chemotherapy was continued with VM-26 every three weeks until the end of the sixth month.

Although these two clinical studies were quite preliminary, the potential advantages are evident. The NBCCGA has recently initiated a trial of DDP plus radiation therapy for patients with advanced local disease. Early reports are encouraging.

Intra-Arterial Infusion Chemotherapy

Intra-arterial infusion of chemotherapeutic agents for disease localized to the pelvis theoretically presents a higher concentration of drug to the neoplasm without increasing toxicity. Wallace et al.,[40] recently reported their experience in 15 patients receiving intra-arterial DDP, 80-120 mg/M^2 over a 24 hour period combined with intravenous ADM + cyclophosphamide. Nine of the patients had an objective response. The median survival in this group of patients was indicated to be 52 weeks. All the patients in this series were thought to have metastatic disease and many had failed previous radiation therapy.

Acknowledgement

Supported in part by PHS Grant CA 18643 by the National Cancer Institute, National Institute of Health and by Research Funds from the Veterans Administration.

REFERENCES

1. K. W. Kaye and P. H. Lange, Mode of presentation of invasive bladder cancer: Reassessment of the problem, J.Urol., 128:31 (1982).
2. W. F. Whitmore, Jr., Management of invasive bladder neoplasms, Sem.in Urology, 1:34 (1983).
3. B. Van der Werf-Messing, Preoperative irradiation followed by cystectomy to treat carcinoma of the urinary bladder category T3 Nx, MO, Int.J.Rad.Oncol.Biol.Phys., 5:395 (1979).
4. M. J. Droller, The controversial role of radiation therapy as adjunctive treatment of bladder cancer, J.Urol., 129:897 (1983).
5. H. M. Radwin, Radiotherapy in bladder cancer: A critical review, J.Urol., 124:43 (1980).
6. J. P. Blandy, H. R. England, S. J. Evans, H. F. Hope-Stone, G. M. Mair, B. S. Mantell, R. T. Oliver, A. M. Pierce, and R. A. Resdon, T3 bladder cancer - the case for salvage cystectomy, Brit.J.Urol., 52:506 (1980).
7. H. D. Pearse, J. T. Pappas, and C. V. Hodges, Radical cystectomy of bladder cancer: 10 year survival, J.Urol., 109:623 (1973).
8. Z. Wajsman, C. Merrin, R. Bore, and G. P. Murphy, Current results in treatment of bladder tumors with total cystectomy at Roswell Park Memorial Institute, J.Urol., 113:806 (1975).

9. J. E. Montie, Technique of radical cystectomy, Sem.in Urol.,
    1:42 (1983).
10. F. S. Freiha, Complications of cystectomy, J.Urol., 123:168
    (1980).
11. D. G. Skinner, E. D. Crawford, and J. J. Kaufman, Complications
    of radical cystectomy for carcinoma of the bladder, J.Urol.,
    123:640 (1980).
12. D. E. Johnson and S. M. Lamy, Complications of a single stage
    radical cystectomy and ileal conduit diversion: Review of 214
    cases, J.Urol., 117:171 (1977).
13. J. J. Roberts and J. M. Pascoe, Cross-linking of complimentary
    strands of DNA in malignant cells by anti-tumor platinum com-
    pounds, Nature, 235:282 (1972).
14. C. L. Litterest, Alterations in the toxicity of cis-diamminedi-
    choloplatium (II) and in tissue localization of platinum as a
    function of NaC concentration in the vehicle of admini-
    stration, Toxicol Appl.Pharmacol., 61:99-108 (1981).
15. A. Yagoda, Chemotherapy for advanced urothelial cancer, Sem.in
    Urol., 1:60 (1983).
16. M. S. Soloway, A. Einstein, J. Coombs, M. P. Corder, G. R. Prout
    Jr., and W. Bonney, A comparison of cis-platin and the com-
    bination of cis-platin and cyclophosphamide in advanced
    urothelial cancer - A National Bladder Cancer Collaborative
    Group A study, Cancer, (in press).
17. A. Yagoda, R. C. Watson, N. Kemeny, W. Barzell, H. Grabstald,
    and W. F. Whitmore, Diamminedichloride platinum II and cyclo-
    phosphamide in the treatment of advanced urothelial cancer,
    Cancer, 42:2121-2130 (1978).
18. A. Yagoda, Phase II trials with cis-diamminedichloride platinum
    II in the treatment of urothelial tract tumors, Cancer Treat
    Rep., 63:1565-1572 (1979).
19. A. Yagoda, Chemotherapy of metastatic bladder cancer, Cancer,
    45:1879-1888 (1980).
20. P. H. Smith, Chemotherapy of bladder cancer. A review, Cancer
    Treat.Rep., 65:165-173 (1981).
21. A. Yagoda, G. Bosl, and H. Scher, Advances in chemotherapy of
    bladder cancer, in: "Recent Advances in Urologic Cancer",
    N. Javadpour ed., Williams and Wilkin Co., Baltimore, pp.
    211-254 (1982).
22. J. D. Khandekar, P. J. Elson, and W. D. Dewys, Comparative
    activity and toxicity of cis-diamminedichloroplatinum vs.
    cyclophosphamide, adriamycin and DDP (CAD) in disseminated
    transitional cell carcinoma of the urinary tract, Proc.Am.
    Ass.Cancer Res., 22:461 (1981).
23. M. L. Samuels, C. Logothetis, A. Trindade, and D. E. Johnson,
    Cytoxan, adriamycin and cis-platinum (CISCA) in metastatic
    bladder cancer. (Abstract #C415) Proc.Am.Assoc.Cancer Res.,
    21:137 (1980).
24. M. Troner and G. Hemstreet, Cyclophosphamide, adriamycin and
    cis-platinum (CAP) chemotherapy of metastatic transitional

cell carcinoma of the bladder. (Abstract #642) Proc.Am.Assoc. Cancer Res., 19:161 (1978).

25. A. G. Turner, Methotrexate, in: "Bladder Cancer - Principles of Combination Therapy", R. T. D. Oliver, W. T. Hendry, and H. J. G. Bloom, eds., Kent, Butterworth, pp.219-229 (1981).

26. R. T. D. Oliver, Methotrexate as salvage or adjunctive therapy for primary invasive carcinoma of the bladder, Cancer Treat Rep., 65:179-181 (1981).

27. R. B. Natale, A. Yagoda, R. C. Watson, and W. F. Whitmore, Methotrexate: An active drug in bladder cancer, Cancer, 47: 1246-1250 (1981).

28. O. R. Gagliano, Adriamycin vs. adriamycin plus cisplatinum in transitional cell bladder carcinoma: A SWOG study (Abstract #C-110), Proc.Am.Assoc.Cancer Res., 21:347 (1980).

29. EORTC: The treatment of advanced carcinoma of the bladder with a combination of adriamycin and 5-fluorouracil, Eur.Urol., 3:276-278 (1977).

30. M. S. Blumenreich, A. Yagoda, R. B. Natale, and W. F. Whitmore, Phase II trial of vinblastine sulfate for metastatic urothelial tract tumors, Cancer, 50:435-438 (1982).

31. P. H. Smith, Presented at the Second International Conference - A treatment of urinary tract tumors with adriamycin, San Francisco, CA, September 4, (1982).

32. D. G. Skinner, G. Leskovsky, and J. R. Daniels, Adjuvant chemo-therapy following cystectomy for deeply invasive bladder cancer: Current status, Proc.of the AUA, p.130 (1983).

33. K. H. Luk, G. Y. Ross, T. L. Phillips, and L. S. Goldstein, The interaction of radiation and cis-diamminedichloroplatinum (II) in intestinal crypt cells, Int.J.Rad.Oncol.Biol.Phys., 5:1471 (1979).

34. E. B. Douple, W. L. Eaton, Jr., and M. C. Tulloh, Skin radio-sensitization studies using combined cis-dichlorodiammed-platinum (II) and radiation, Int.J.Rad.Oncol.Biol.Phys., 5:1382 (1979).

35. D. R. Burholt, L. L. Schenken, C. J. Kovacs, and R. F. Hageman, Response of the murine gastrointestinal epithelium to cis-dichlorodiammineplatinum (II) radiation combinations, Int.J. Rad.Oncol.Biol.Phys., 5:1377 (1979).

36. F. M. Maggia and E. Glatstein, Summary of investigations on platinum compounds and radiation interactions, Int.J.Rad. Oncol.Biol.Phys., 5:1407 (1979).

37. M. S. Soloway, C. R. Morris, and B. Sudderth, Radiation therapy in cis-diamminedichloroplatinum in transplantable and primary murine bladder cancer, Int.J.Rad.Oncol.Biol.Phys., 5:1355 (1979).

38. R. J. Babaian, D. E. Johnson, L. Llamas, and A. G. Ayala, Metasteses from transitional cell carcinoma of urinary bladder, Urology, 16:142 (1980).

39. G. Jakse, H. Frommhold, and H. Marberger, Combined cis-platinum and radiation therapy in patients with stage pT3 and pT4 bladder cancer: A pilot study, J.Urol., 129:1502 (1983).

40.   S. Wallace, V. P. Chuang, M. Sanuels, and D. Johnson, Trans-
        catheter intra-arterial infusion of chemotherapy in advanced
        bladder cancer, Cancer, 419:640 (1982).

COMBINED RADIO-CHEMOTHERAPY IN

ADVANCED BLADDER CANCER

Gerhard Jakse and Hermann Frommhold

Departments of Urology and Institute of Radiology
University of Innsbruck
Innsbruck, Austria

Patients with locally advanced transitional cell carcinoma of the bladder treated by irradiation alone or in combination with transurethral resection of the exophytic tumor fail mostly because of local tumor persistence or development of new tumors elsewhere in the bladder. Only a few patients keep the bladder after definitive radiotherapy without the risk of future tumor recurrence. Therefore for further improvement of irradiation results additional treatment modalities are warranted.

There are recent data on FANFT-induced animal bladder tumors demonstrating an increased effect of irradiation if doxorubicin, cyclophosphamide or cis-platinum were administered before the initiation of radiotherapy[1,2]. These experimental data were supported by Herr et al.,[3] in a prospective study in which an increased downstaging was achieved in patients treated by cis-platinum and 20 gray before cystectomy versus patients receiving radiotherapy alone. There also is an anecdotal report by Soloway on 4 patients with advanced local tumor who experienced favorable results after combined treatment with cis-platinum and radiotherapy[1].

Since an integrated therapy might result in an increased cure rate in patients with stage T3 bladder cancer, and therefore might be considered as a definitive procedure in those not being candidates for cystectomy, we initiated a study to evaluate the systemic and local toxicity of such a treatment modality[4].

Material and Methods

We studied 22 patients with advanced transitional cell carcinoma of the bladder of whom 4 had had previous treatment. In these

patients transurethral resection was done initially for low stage
tumor, but tumor progression was noted within 6 months or more. The
clinical data are given in Table 1.

The local tumor extent was estimated by bimanual examination
under anesthesia, percutaneous sonography and CT, and transurethral
resection performed in 2 consecutive sessions with a separate biopsy
of the tumor base.

Transurethral resection was followed by a combined chemotherapy-
radiotherapy regimen after an interval of 4 weeks. Thereafter cis-
platinum (1.6mg/kg) was given intravenously on day 1 and 3 weekly
until the end of radiotherapy (i.e. 4 cycles). Radiotherapy consisted of a tumor dose of 60 gray administered using a split course
technique with either 60-cobalt or 18mev photons. The first irradiation was applied 8 to 10 hours after the termination of the
cis-platinum infusion.

Evaluation of toxicity. Before the integrated treatment bladder
capacity was estimated by spontaneous voided urine volume and maximal
filling of the bladder by normal saline. Stool and voiding frequency, serum creatinine, complete blood count and body weight were
monitored throughout therapy.

Results

All but 6 patients received the planned treatment schedule. The
reasons for incomplete treatment are given in Table 2. The problem
related only to the chemotherapy as all patients save the one who
died of a myocardial infarct completed the radiotherapy. The results
in relation to stages are shown in Table 3 and the causes of death in
Table 4. Recurrent tumors were observed in 2 patients (pT1).
Follow-up ranged from 6 to 36 months.

Table 1. Clinical Data at Initiation
of Combined Chemoradiotherapy.

Mean Age: 67 years (58–80)
male: 17, female: 5
Pathology
Grade: II=1, III=5, IV=16
Stage: pT3a=10, pT3B=8,
pT4a=3, T4=1
Prior therapy
Transurethral resection – 4

Table 2. Reasons for incomplete therapy. Radiotherapy
         was completed in all but 1 patient who died of
         myocardial infarction, but cis-platinum was
         stopped after 2 cycles in the patients listed

pT3a	urinary infection	persistence
	refused	–
pT3b	refused	persistence
	myelosuppression	–
	refused	–
	myocardial infarct	–

Toxicity. The hematological effects were not severe. Only 2
patients experienced a white blood count below 3.500 due once to
Cis-platinum therapy and once to VM-26 which was given as maintenance
therapy in 4 patients after termination of the above mentioned treatment schedule. An increase in serum creatinine was noted in 3
patients, 2 of whom had tumor persistence and progressive dilation of
the upper urinary tract. One patient with a solitary kidney experienced an increase of creatinine to 3,8 mg%, which finally decreased
to 2,5 mg%. All patients suffered some form of gastrointestinal
symptoms; because of this 3 patients refused further chemotherapy
after the second cycle of cis-platinum. Voiding discomfort was mild
and transient in all but three patients, who experienced a decrease
in bladder capacity. 13 patients needed antibiotics and/or antispas-
molytic medication. Diarrhoea was observed in 10 patients; one of
these needed local treatment with steroids. A decrease in body
weight was noted in all patients, with an average loss of 1.5 kg.
Those patients without evidence of metastatic disease regained their
pretreatment body weight within 4 to 6 weeks after termination of
cis-platinum treatment and irradiation.

Discussion

The main purpose of this study was to determine whether the
toxicity of a combination of definitive radiotherapy and chemotherapy
with cis-platinum is acceptable. Only one patient had permanent
objective damage due to a cumulative effect of the combined treatment; decrease of renal function in association with a solitary
kidney. In 2 other patients an increased serum creatinine level
evidently was a sequel of progressive urinary tract obstruction and
could not be attributed to the possible nephrotoxicity of cis-
platinum. Contracted bladders may be found in 4 to 12% of patients
following definite radiotherapy. Therefore, of our 19 patients
followed for 6 months the presence of 3 with a contracted bladder
cannot be considered as due to increased local toxicity, especially
since two had had several transurethral resections and local chemotherapy before or had an extensive local tumor mass.

Table 3. Clinical results in relation to tumor stage

	pT3a	pT3b	pT4a	T4b
alive NED	6×	4	1×	–
alive with tumor	2×	1	2×	–
dead NED	1	1	–	–
dead with tumor	1	2	–	1

× tumor recurrence elsewhere in the bladder, stage pT^1

Table 4. Cause of death in 6 patients in relation to tumor
          stage and eradication of local tumor

Stage	Time	Local	Cause of Death
pT3a	5 mos	–	cardiac failure
	8 mos	–	metastatic disease
pT3b	6 mos	–	metastatic disease
	4 mos	–	myocardial infarction
	10 mos	+	unknown
T4b	7 mos	+	metastatic disease

Conclusion

The systemic and local toxicity of the integrated therapy is
acceptable, especially when the mean age of these patients (67 years)
is considered. The value of such treatment in respect of tumor
eradication cannot be estimated because of the small number of
patients and the short follow-up. However the number of bladders
free of tumor is surprisingly high. Integrated therapy should be
applied in a prospective randomized trial on patients not fit or not
wanting radical cystectomy for invasive transitional cell carcinoma
of the bladder.

REFERENCES

1. M. S. Soloway, C. R. Morris, and B. Sudderth, Radiation therapy
    and cis-diammine-dichloroplatinum (II) in transplantable and
    primary murine bladder cancer, Int.J.Rad.Oncol.Biol.Phys., 5:
    355 (1979).
2. T. E. Weldon, E. Kursh, L. J. Novak, and L. Persky, Combination
    radiotherapy and chemotherapy in murine bladder cancer,
    Urology, 14:47 (1979).
3. H. W. Herr, A. Yagoda, M. Batata, P. Sogani, and W. F. Whitmore,
    Preoperative cis-platinum and radiation therapy in bladder

cancer: a pilot study, Read at the annual meeting of American Urological Association, abstract 584 Boston, Massachusetts, May 10-14 (1981).

4. G. Jakse, H. Frommhold, and H. Marberger, Combined cis-platinum and radiation therapy in patients with stages pT3 and pT4 bladder cancer: a pilot study, J.Urol., 129:502 (1983).

IRRADIATION OF ADVANCED BLADDER CANCER (T4)

S.D. Fosså

General Department
The Norwegian Radium Hospital
Oslo, Norway

INTRODUCTION

The object of treatment of T4 bladder cancer is to achieve cure
or at least to prolong survival and to relieve the patient's symptoms
and improve his/her performance status.

The present retrospective study was performed with regard to
these criteria especially in order to analyse the results of radiotherapy of T4 bladder carcinoma, to define prognostic parameters and
to identify a group of patients who have the greatest benefit from
radiotherapy.

PATIENTS AND METHODS

Two hundred and thirty-four patients with histologically proven
category T4[1] bladder carcinoma (T4a:40;T4b: 194) were admitted to
the Norwegian Radium Hospital between 1972 and 1978. Forty patients
had distant metastases (M1 category). Table 1 gives further patient
characteristics.

One hundred and eighty patients (159 without distant metastases)
received radiotherapy, in most cases given to an anterior pelvic
field (16 x 16 cm) by a Betatron 33 MV (Table 2, Figure 1). The
intention was to treat patients without distant metastases by a total
dose according to a cumulated radiation equivalent (CRE) value > 1700
– (1800) radiations equivalent units (reu), adequate radiotherapy.
Of 104 M 0 patients + 5 patients with distant metastases which were
not discovered at start of treatment 71 patients received pelvic
irradiation with a CRE value below 1700 reu due to acute intolerable

Table 1. Patient Characteristics

Parameter		Number of patients	
Sex:	males	169	
	females	65	
Age (mean, years)	males		66,8
	females		69,8
T-category	T4a	40	
	T4b	194	
Metastases	present	40	
	absent	194	
Localization:	non-regional lymph nodes	13	
	lung	7	
	cutis/subcutis	6	
	liver	6	
	skeleton	4	
	multiple sites	4	
Performance status (at start)	0	105	
	1	70	
	2	29	
	3	17	
	4	13	
Histology	Transitional cell carcinoma	162	
	WHO2	32	
	WHO3	130	
	Undifferentiated	54	
	Adenocarcinoma	1	
	Squamous cell carcinoma	6	
	Non-classifiable	3	
Small vessel invasion	Present	117	
	Absent	117	
Urography (at start)	Hydronephrosis and/or delayed excretion	62	
	Non-functioning kidney	81	
	Normal urography*	83	
	Unknown	8	
Serum creatinine	> 150 µmol/l	55	
	≤ 150 µmol/l	179	

* Including patients with a normally functioning remaining kidney
  after previous unilateral nephrectomy.

Table 2.   Radiotherapy Characteristics (180 patients)
           a) Irradiation source
           b) CRE value

a) Irradiation Source	Number of Patients
Betatron 33 M V	144
Linear accellerator	26
Co^{60}	5
Combination	5

b) CRE value [reu]	Number of Patients
> 1700	109*
1601–1700	16
1501–1600	17
1401–1500	16
$\leq$ 1400	22

*$1800 \leq CRE > 1750$:   29 patients.

gastrointestinal side effects and/or rapidly decreasing performance status during irradiation, or because of widespread disease at start of treatment. As a rule, the daily radiation dose was 2 Gray (Gy) given 5 times weekly. Fifty-four patients with distant metastases and/or an extremely bad performance status at admission were not irradiated at all.

In 1974 a prospective trial compared the effect of irradiation + adjuvant 5-fluorouracil (5-FU)-12 mg/kg i.v. weekly during radio-therapy and for at least 3 months afterwards (27 patients) with the efficacy of irradiation alone (33 patients) (Table 3).

All patients were followed to December 31st 1981 by regular cystoscopies and clinical examinations and the differences between the curves for crude survival[2] were evaluated by the log rank test[3].

RESULTS

All Patients (234)

The overall 5 year survival was 10% (Figure 2), 18 patients with no evidence of disease (NED) at the end of the observation time. Patients with M 0 disease survived significantly longer than those of Ml category.

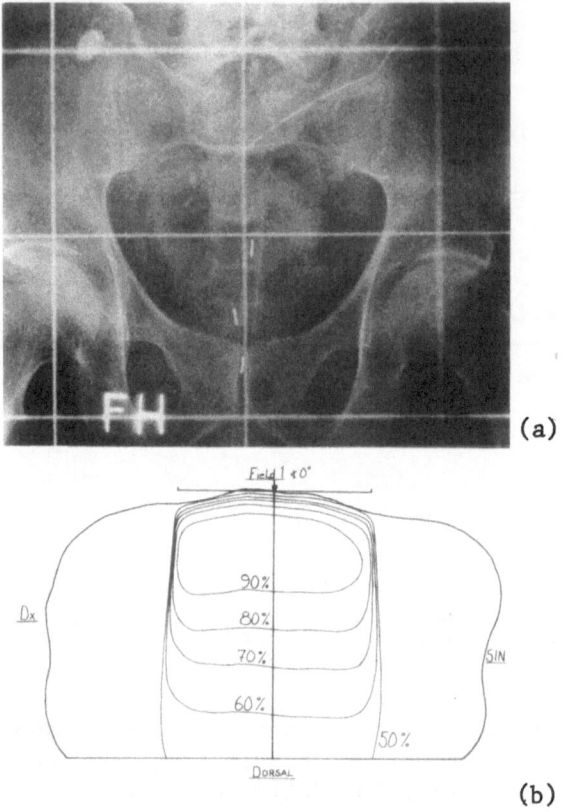

(a)

(b)

Fig. 1.   (a) Anterior irradiation field (16 x 16 cm) for T4 bladder
          cancer. (b) Dose distribution for radiotherapy of T4
          bladder cancer using a Betatron 33 M V.

Patients without metastases (194) (Table 4a)

The initial performance status (Figure 3), the primary tumor's
subcategory (Figure 4) and the CRE value of radiotherapy (Figure 5)
had a significant influence on the patient's survival.

Patients of M 0 Category, Treated by Adequate Radiotherapy (104 (Table 4b)

The only prognostic parameter was the improvement of the urographic findings (Figure 6) 3 months after treatment whereas the
initial urographic observations and the different histological criteria appeared to be without prognostic impact.

Table 3. Details of Patients With T4 Bladder Cancer Treated by
Radiotherapy With and Without Adjuvant 5-FU (prospective
randomized trial).

Parameter	Number of patients		
	with 5-FU 27	Without 5-FU 33	Total 60
Sex:			
males	19	23	42
females	8	10	18
T-category			
T4a	7	3	10
T4b	20	30	50
Performance status			
0-1	26	28	54
2-4	1	5	6
Histology			
Grade 2	3	7	10
Grade 3	16	18	34
Undifferentiated	6	7	13
Others	2	1	3
Urography			
Normal	14	11	25
Hydronephrosis/delayed excretion	4	8	12
Non-functioning kidney	8	14	22
Unknown	1	0	1
CRE-value			
> 1700	23	28	51
1601 - 1700	1	2	3
≤ 1600	3	3	6

Irradiation + 5-FU Versus Irradiation Alone (Figure 7)

The administration of 5-FU together with radiotherapy had an
effect on 5 years crude survival which was almost significant
(p:0.06).

Palliation Effect

In this retrospective study it was not possible to evaluate the
palliative effect of radiotherapy as regards the urinary symptoms.
Three months after radiotherapy an increased performance status was
observed in only 8 of 103 evaluable adequately treated patients,
whereas the general condition was worse in 41.

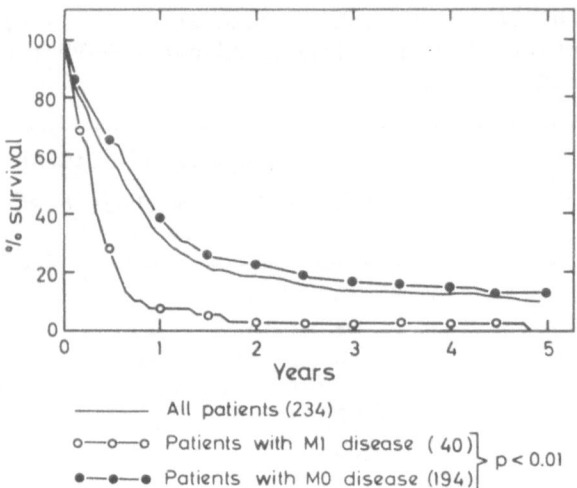

Fig. 2.   Crude survival in 234 patients with T4NX bladder carcinoma
          (NRH, 1972-1978).

Complications

Acute gastrointestinal and urinary side effects worsened during
radiotherapy with increasing total dose. Fifty-five M O patients
could not go through the complete course of planned radiotherapy due
to acute irradiation toxicity.

Late serious complication after radiotherapy, demanding major
surgery occurred in 7 patients. (Gastrointestinal side effects: 5;
bladder complications: 1; combined gastrointestinal/urinary side
effects:1).

DISCUSSION

As also shown by other authors, the prognosis of T4 bladder
cancer patients is unfavorable (4-10). Only 10% of all patients
survived for 5 years. However, certain subgroups may have more
favorable prognoses, such as patients with a good initial performance
status and without detectable distant metastases, and those with a
T4a primary tumor. In the latter group crude survival after radiotherapy is at least as good as that obtained by total cystectomy[9].

On the background of the above poor results one may ask: Does
radiotherapy favorably influence the "natural history" of T4 bladder
cancer? The relatively few detailed reports about virtually untreated T4 bladder cancer[11,10,14] show an even worse survival than
demonstrated in the present study. One of the most important con-

Table 4. Prognostic Parameters and Crude Survival in Patients with T4 Bladder Cancer

Parameter		No. of patients	Crude survival (%) 1 year	Crude survival (%) 5 years	p-value Significance of differences (for the total observation period (5 years)
(in 194 MO Patients)					
Sex	males	135	38	12 ⎫	
	females	59	26	6 ⎭	0.03
Performance status (Figure 3)	0-1	150	46	16 ⎫	
	2-4	44	11	0 ⎭	<0.01
T-category (Figure 4)	T4a	36	58	27 ⎫	
	T4b	158	34	9 ⎭	<0.01
Irradiation (Figure 5)	>1700	104	51	16	
CRE [reu]	<1700	55	33	13 ⎫	0.04
	no irradiation	35	9	0 ⎭	
(in 104 MO Patients who Received CRE >1700 reu)					
Grade	2	14	57	21 ⎫	
	3	52	48	12 ⎬	0.48
	4	32	50	14 ⎭	0.52
Urography* (at start)	pathol.	63	48	22 ⎫	
	normal	39	59	9 ⎭	0.81
Changes in urography** (3 months) (Figure 6)	improved	26	69	33 ⎫	
	unchanged	34	68	8 ⎬	0.02
	worsened	10	20	0 ⎭	
Serum alkaline phosphatase (U/1)	>300	10	40	10 ⎫	
	<300	94	52	17 ⎭	0.91
Serum creatinine µmol/1	≤150	81	54	17 ⎫	
	>150	23	39	14 ⎭	0.66

* 2 patients not evaluable
**Only 70 patients evaluable

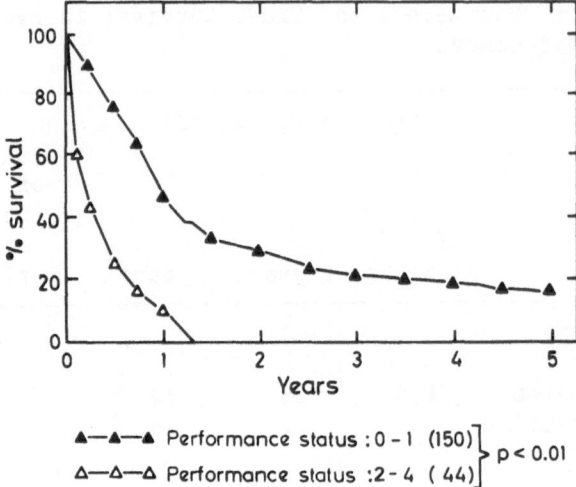

Fig. 3. Crude survival in 194 patients with T4NXMO bladder carcinoma
according to the performance status (WHO) at treatment
start.

ditions for a good result of radiotherapy is, however, that high
total doses are applied[15]. Probably the total dose could have been
increased even more than in the present study, with the object of
improving the cure rate.

The increase of the CRE values of the total irradiation dose,
has, however, to be balanced against the frequency and severity of

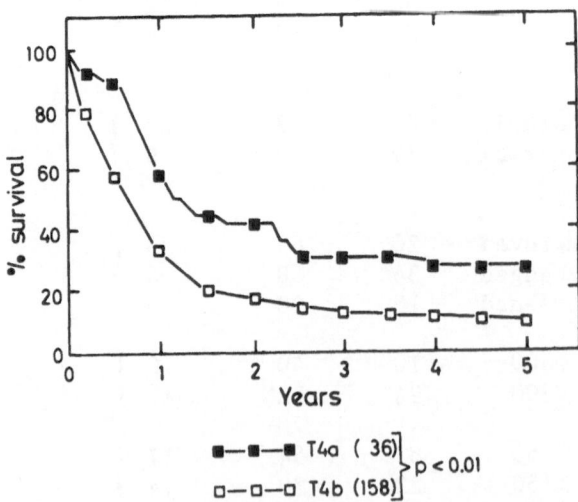

Fig. 4. Crude survival in 194 patients with T4NXMO bladder carcinoma
according to the T4 subcategories.

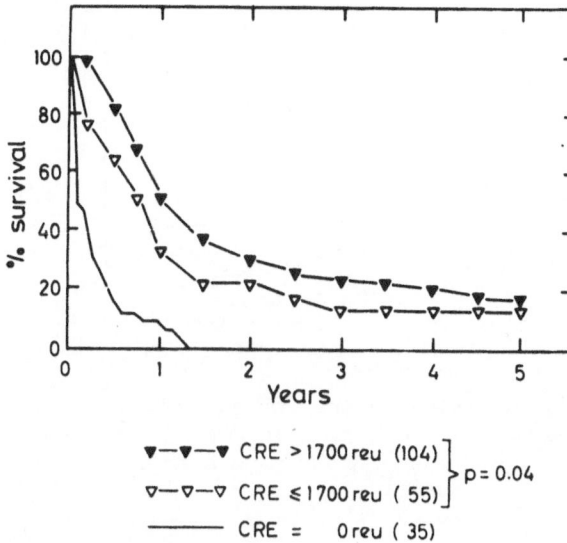

Fig. 5. Crude survival in 194 patients with T4NXMO bladder carcinoma
         according to CRE value.

side effects combined with high dose radiotherapy. After pelvic
radiotherapy with CRE values of 1700-1750 reu, the rate of serious
complications is 2-17%[7,15,16]. These figures are comparable with
the 3% observed in the present study. An increase of CRE to > 1800
reu will raise the complication rate to 20% or more [15,16] which we
consider to be unacceptable in most of the patients.

     Our results of the trial comparing irradiation + 5-FU versus
irradiation alone are in contrast to previous results of combined
chemoradiotherapy of advanced bladder carcinoma [17,18]. The present
results therefore should be confirmed or disproved in a larger
series. Any possible survival benefit of the combination of radiotherapy and 5-FU may be explained by a radio-sensitising effect of
5-FU [19,20] and/or by the fact that 5-FU is a drug with cytostatic
activity in bladder cancer[21].

     This retrospective study gives only limited information about
the palliative effect of radiotherapy in T4 bladder cancer patients.
The important question remains as to whether adequate palliation in
patients with T4 bladder cancer can be achieved by less resource-
demanding treatment. For example, hematuria, can be relieved by a
single dose of 10 Gy[5] and bladder symptoms may be relieved by a
urinary diversion without cystectomy[11,13]. The superiority of
these methods to radical radiotherapy might be expected in patients
with metastatic disease, in older patients or in those with a poor
performance status. However, one has to remember that high doses of
radiotherapy may occasionally be curative in some patients with T4
bladder cancer. This treatment should be applied if there is any
realistic chance of cure in an individual patient.

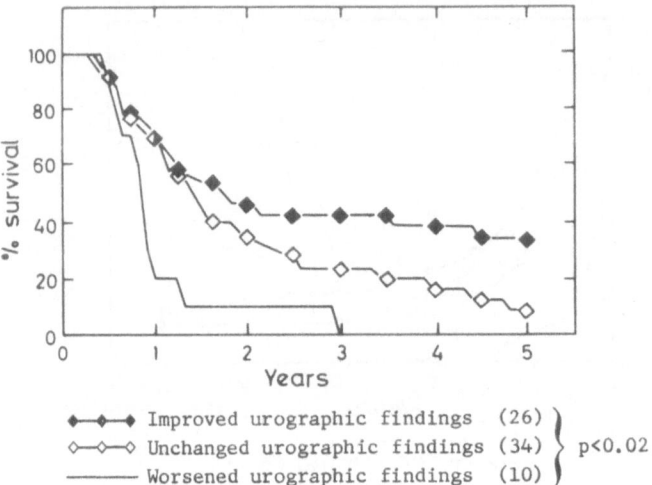

Fig. 6. Crude survival in 70 evaluable adequately irradiated (CRE
         <1700 reu) patients with T4NXMO bladder carcinoma according
         to <u>changes of the urographic findings</u> 3 months after radio-
         therapy.

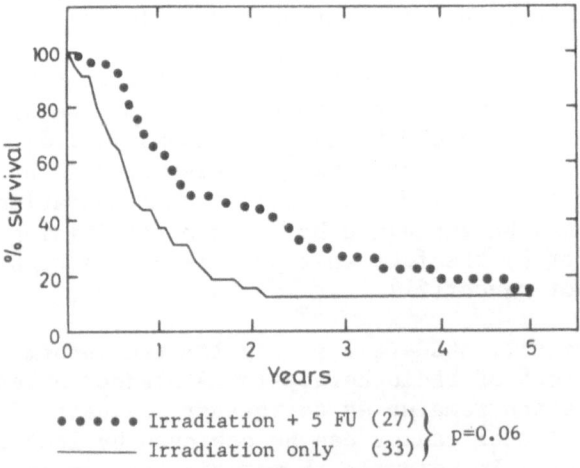

Fig. 7. Crude survival in 60 adequately irradiated (CRE >1700 reu)
         patients with T4NXMO bladder carcinoma according to <u>adjuvant
         treatment with 5-Fluoro-Uracil</u>.

CONCLUSIONS

1. High dose radiotherapy improves the survival in patients with T4 bladder carcinoma and may be curative in some patients.
2. Good risk factors are: Performance status 0-1 (WHO); a primary tumor of T4a category and the absence of M1 disease.
3. Combined chemo-/radiotherapy seems to improve the survival. The combination of radiotherapy with potent cytostatic drugs should be further investigated.
4. The palliation effect of irradiation in patients with T4 bladder cancer should be evaluated in a prospective study, preferably in comparison with other methods of palliative treatment.

REFERENCES

1. F. K. Mostofi, L. H. Sobin, and H. Torloni, Histological typing of urinary bladder tumors, International histological classification of tumors, Wld.Hlth.Org., Geneva, 10, p.17 (1973).
2. S. J. Cutler and F. Ederer, Maximum utilization of the life table method in analyzing survival, J.Chron.Dis., 8:699 (1958).
3. R. Peto and J. Peto, Asymptotically efficient rank invariant test procedures, J.roy.statist,soc., A 135 (Part 2):185 (1972).
4. B. M. Birkhead, J. G. Conley, and R. M. Scott, Intensive radiotherapy of locally advanced bladder cancer, Cancer, 37:2746 (1976).
5. R. C. Chan, R. B. Bracken, and D. E. Johnson, Single dose whole pelvis megavoltage irradiation for palliative control of hematuria or ureteral obstruction, J.Urol., 122:750 (1979).
6. R. W. Edlund, J. B. Wear, and F. J. Ansfield, Advanced cancer of the urinary bladder. An analysis of the results of radiotherapy alone versus radiotherapy and concomitant 5-Fluoro-Uracil, a prospective randomized study of 36 cases, Amer.J.-Roentgenol., 108:124 (1970).
7. F. Edsmyr, P. L. Esposti, G. Giertz, and B. Littbrand, Radiation treatment of urinary bladder carcinoma, Urol.Res., 6:229 (1978).
8. N. Green and F. W. George III, Radiotherapy of advanced localized bladder cancer, J.Urol., 111:611 (1974).
9. M. A. Batata, F. C. H Chu, B. S. Hilaris, M. Z. Lee, R. W. Varesko, H. S. Lee, E. Visetsiri, Y. S. Kim, R. Ong, and W. F. Whitmore Jr., Preoperative whole pelvis versus true pelvis irradiation and/or cystectomy for bladder cancer, Int.J. Radiation Oncol.Biol.Phys., 7:1349 (1981).
10. S. B. Wassif, and B. van der Werf-Messing, Treatment of bladder cancer at the Rotterdam Radiotherapy Institute (R.R.T.I). With special reference to bladder radium implantation and preoperative radiotherapy followed by cystectomy, Urol.Res., 6:241 (1978).

11. J. E. Meyer, M. Yatsuhashi, and T. H. Green Jr., Palliative
    urinary diversion in patients with advanced pelvic mal-
    ignancy, Cancer, 45:2698 (1980).
12. H. R. Sauer, M. S. Blick, and D. J. Meehan, A study of untreated
    bladder cancer, J.Urol., 63:124 (1950).
13. I. Silber, W. T. Bowles, and J. J. Cordonnier, Palliative
    treatment of carcinoma of the urinary bladder, Cancer, 23:586
    (1969).
14. G. R. Prout and V. F. Marshall, The prognosis with untreated
    bladder tumors, Cancer, 9:551 (1956).
15. R. Morrison, The results of treatment of cancer of the bladder -
    a clinical contribution to radiobiology, Clin.Radiol., 26:67
    (1975).
16. J. Kirk, G. W. H. Wingate, and E. R. Watson, High-dose effects
    in the treatment of carcinoma of the bladder under air and
    hyperbaric oxygen conditions, Clin.Radiol., 27:137 (1976).
17. G. R. Prout, N. H. Slack, and J. D. J. Bross, Irradiation and
    5-Fluoro-uracil as adjuvant in the management of invasive
    bladder carcinoma. A cooperative group report after 4 years,
    J.Urol., 104:116 (1970).
18. B. Richard, Yorkshire Urology Cooperative Research Group, The
    combination of chemotherapy and radiotherapy in invasive
    carcinoma of the bladder, in: "Bladder Cancer. Principles of
    Combination Therapy," R. T. D. Oliver, W. F. Hendry, and H.
    J. G. Bloom eds., Butterworths, London, pp.247 (1981).
19. M. A. Bagshaw, Possible role of potentiators in radiation
    therapy, Am.J.Roentgenol., 85:822 (1961).
20. Y. Nakajima, T. Miyamoto, M. Tanabe, I. Watanabe, and T.
    Terasima, Enhancement of mammalian cell killing by 5-Fluoro-
    uracil in combination with X-rays, Cancer Res., 39:3763
    (1979).
21. S. D. Fossa, and T. E. Gudmundsen, Single-drug chemotherapy with
    5-FU and adriamycin in metastatic bladder carcinoma, Brit.J.
    Urol., 53:320 (1981).

THE NATURAL HISTORY OF PROSTATIC CANCER -

THE ARGUMENT FOR A NO TREATMENT POLICY

D. W. W. Newling[1], R. R. Hall[2], B. Richards[3],
M. R. G. Robinson[4], and J. W. Hetherington[5]

Department of Urology,
[1]Princess Royal Hospital, Hull, [2]Freeman Hospital,
Newcastle, [3]York District Hospital, York, [4]Pontefract
General Infirmary, Pontefract, [5]St. James Hospital
Leeds, UK

INTRODUCTION

In patients with prostatic obstruction certain clinical features
of their presentation suggest a malignant pathology. In a recent
analysis of 166 cases of carcinoma of the prostate 36 patients with
no objective evidence of metastases who had none of the presenting
symptoms and signs associated with a poor prognosis received no
initial treatment of their cancer, apart from relief of obstructed
micturition. Five Surgeons in the North of England have now collected 130 cases of carcinoma of the prostate in whom urethral obstruction was relieved by transurethral resection after which further
therapy was deferred.

PRESENT SERIES (HULL)

In an analysis of 166 cases of prostatic cancer in Hull who have
now been followed up for a period of one to eight years - an average
follow-up of 4.3 years, 74 have died (44.5%), 41 (25%) from prostatic
cancer and 33 (20%) from other causes (Table 1). The clinical staging and pathological grading of these patient's tumors is shown in
Table 2.

The objective assessment of patients with prostatic cancer has
always included Serum Acid Phosphatase estimation, by an enzymatic
method and a radionucleotide bone scan. In this series 43 of 166
(26%) had an elevated Serum Acid Phosphatase at presentation. Eight-

Table 1. Causes of Death in 166 Patients with Prostate
          Cancer.

Prostatic cancer	41
Cardio-vascular	16
Bronchopneumonia	13
Carcinoma of colon	1
Pulmonary fibrosis	1
Carcinoma of stomach	1
Septicaemia	1

Of the 33 patients who died of causes other than
carcinoma of the prostate 5 had metastases at the time
of death.

Table 2. Stage and Grade at Presentation.

Stage		
T_0	21	(13%)
T_1	34	(22%)
T_2	55	(35%)
T_3	37	(24%)
T_4	10	(6%)

3 patients had previous abdominoperineal resection.

6 other patients were not properly staged.

Grade		
G_1	38	(23%)
G_2	87	(52%)
G_3	41	(25%)

een of these 43 (42%) had a positive bone scan, supported by conventional x-rays at presentation and 10 others subsequently developed
bony metastases. 22 of these 43 (51%) have died of prostatic cancer.
Of the 20 patients with a positive bone scan at presentation 15 (75%)
had symptoms from their metastases and all have died of disseminated
prostatic cancer.

The symptoms of these patients at presentation have been analyzed to establish those which should alert the clinician to the
diagnosis of malignant disease and those which indicate a poor prognosis (Table 3). Of these the following are of importance:

Anorexia and weight loss: 26 patients (16%) had a weight loss
of 10% or more during the twelve months prior to diagnosis. Of these

Table 3. Symptoms of Prognostic Importance.

1.	Anorexia + weight loss (>10%)
2.	Anaemia
3.	Pain
4.	Duration of symptoms

17 died of prostatic cancer, of whom 10 had no evidence of dissemination at presentation.

Anaemia was common. 75 (46%) having a haemoglobin less than 13 gms% at presentation of whom 25 had proven metastases and 10 an elevated Serum Acid Phosphatase level. In 32 patients anaemia was found at presentation for no obvious reason and 19 of those have died of prostatic cancer.

Pain. Usually perineal or suprapubic, not associated with infection, retention or metastases, was often associated with high stage, high grade malignant disease. 48% of the patients in whom local pain was a dominant presenting symptom, have died of prostatic cancer.

The duration of symptoms seem to be an indication of the grade of tumor with the higher grades (G3) having the shortest history.

Although aematuria was more common in patients whose pathology was malignant it carried no prognostic importance. Oedema was not indicative of malignancy or of a poor prognosis when found in patients with prostatic cancer at the time of presentation.

From this study it seems that anaemia, anorexia and weight loss, pain and haematuria, suggest a malignant pathology in patients presenting with prostatism and that all these features, except haematuria, are of prognostic importance.

Among the patients in this series, 36 had no poor prognostic features and were not given any additional treatment, apart from relief of their obstruction by transurethral resection. These appeared to do no worse than similar patients treated with hormone manipulation.

It was concluded from this study that patients with evidence of dissemination, particularly if symptomatic, require systemic treatment. Patients without objective evidence of dissemination but who are anaemic, have lost weight or have pain, should probably be treated. There are, however, some patients with clinically detectable carcinoma of the prostate without any of the above negative prognostic factors in whom treatment may be deferred or withheld altogether.

UNTREATED PROSTATIC CANCER

Five Surgeons in the North of England have now collected 130 cases of carcinoma of the prostate in whom treatment was deferred, to look into this last suggestion in more detail. As very few Urologists now treat focal TO pT_1 tumors these have been excluded from this series which comprises TO pT_2 or T2 - T4 local tumors. The length of follow up is from 6 - 141 months and the average age at presentation was 73.5 years. The distribution by stage of these cases is shown in Table 4.

In this series there are 11 patients with symptomless metastases. Three main parameters have been studied, viz, the number of patients who have progressed wither locally or by metastases; the time to that progression; and patient survival, particularly those patients dying of prostatic cancer.

Of the 51 who progressed, the progression was local in 30 (59%). 25 patients developed metastases of whom 11 progressed locally and systemically. 35 of 51 progressions was seen during the first two years, the average time to progression being approximately 14 months. (Table 5). Of the 11 patients with metastases at presentation 7 (64%) have progressed.

Table 4. Stage at Presentation of 130 Cases of Untreated Prostatic Cancer

T_0	pT_2	M_0	37
T_2	M_0		36
T_3	M_0		34
T_4	M_0		12
$T_0 - T_4$		M1	11

Table 5. Progression in 130 Patients With untreated Prostatic Cancer

Total number progressing	51
Local progression	30/51
Metastases	25/51
Progression without symptoms	11/51

Table 6.   Deaths in the 130 Treated Patients (of 166 patients studied).

	From Carcinoma prostate	Non-Carcinoma deaths
T_1	Nil	Nil
T_2	10	14
T_3	14	2
T_4	9	1
	33	17

Thirty-seven of the 130 untreated patients (28.5%) have died. 15.3% have died non-cancer deaths and 13.2% have died from carcinoma of the prostate. The length of survival in those dying of carcinoma of the prostate was on average 18.4 months while those who died of non-cancer causes survived, on average, 23 months.

From the study of the 166 cases looked at for prognostic factors there were 130 who did receive treatment and although not strictly comparable, since many had negative prognostic factors, their survival statistics may be of interest (Table 6).

CONCLUSIONS

From this series it can be said that patients with clinical, non metastatic carcinoma of the prostate progress relatively slowly. Those that progress do so within three years. Untreated, patients with no symptoms and signs other than urethral obstruction are marginally more likely to die of causes other than prostatic cancer. There is, therefore, probably a place for the study of standard treatment, for example, bilateral subcapsular orchidectomy versus standard treatment deferred until symptoms or signs or progression occur in patients with prostatic cancer whose only symptoms are those of outflow obstruction.

Such a study should be designed to answer two questions:

1)   Do patients with clinically obvious carcinoma of the prostate who have no negative prognostic factors progress and, if so, how quickly?
2)   If treatment is withheld until progression occurs, do these patients respond as well as those treated as soon as the diagnosis is made?

REFERENCES

1.   W. F. Whitmore, The natural history of prostatic cancer, Cancer
       32:1104-1112 (1973).
2.   L. M. Franks, etiology, epidemiology and pathology of prostatic
       cancer, Cancer 32:1092-1995 (1973).
3.   J. N. Corriere, J. L. Cornog, J. J. Murphy, Prognosis in
       patients with carcinoma of the prostate, Cancer 25:911-918
       (1970).
4.   Scott et al., Hormone Therapy of Prostatic Cancer, Cancer
       45:1929-1935 (1980).
5.   C. E. Blackard, G. T. Mellinger, D. F. Gleason, The treatment of
       stage 1 cancer of the prostate, J.Urol. 106:729-733 (1971).

125-I-IMPLANTATION OF PROSTATE CANCER

H. J. de Voogt, J. Battermann and T. A. Boon

Department of Urology and Radiotherapy of
Antoni van Leeuwenhoek Hospital
Amsterdam, Holland

INTRODUCTION

Hilaris and Whitmore[1] introduced interstitial irradiation of
the prostate with 125-I seeds in 1970. Long before that time other
forms of interstitial irradiation had been used[2] of which the
transurethral injections with radio-active gold (198 Au) by Flocks[3]
are probably the best known. For unknown reasons however his work
never was replicated or continued. However if we look at qualities
of irradiation and half life of some of the X-ray omitting isotopes,
it is clear that the choice of 125-I is more realistic. With a half
life of 60 days emitting X-rays in a very limited space, the seeds
can stay in the prostate. They give a total dose of 10 - 20,000 rad
to the prostate, but the surrounding tissues such as bladder and
rectum receive only very little irradiation, with little or no effect
observed by the patient. In addition there is usually no impairment
of sexual function and micturition remains normal if there are no
obstructive symptoms. One advantage of retropubic implantation is
that it can be preceded by regional lymphnode dissection, allowing
for an accurate N-staging. Though it is possible to implant the
seeds transperineally (without surgery) under ultrasound guidance[4],
the advantage of N-staging is thereby lost, an opportunity which in
our opinion should not be missed.

As it took a long time to get permission from the authorities to
use this kind of therapy in Amsterdam, we could only start in 1980.
So far we have treated 18 patients and it is our intention to present
the early results and complications. However it must be clear that
survival rates cannot be given and that we only want to point out
some of the difficulties of the technique, as well as the limitations
and indications for this treatment procedure.

Method

Patients not older than 70 years with histologically proven prostatic carcinoma in stage T_1 and T_2 and with a prostate of not more than 40 g. are selected. PAP, bone scan and pedal lymph-angiography are done to exclude the possibility of distant metastases.

Preoperative subcutaneous minidose heparin is given and continued until the patient is completely mobilized.

The patient is put on the operating table in a modified lithotomy position with elevation of the pelvis. An O'Connor sheet is applied and a Foley catheter 18 Fr. is inserted. The space of Retzius is opened using a midline incision and [after ligating and dividing the vasa deferentia] an extraperitoneal staging lymphadenec-tomy is performed, including external iliac, hypogastric and obtur-ator nodes on both sides. The uppermost chain of the iliac lymph vessels is spared to prevent post operative lymphoedema of legs.

The prostate is then mobilized by incision of the endopelvic fascia. The radiotherapist inserts 6 - 10 hollow needles in the prostate, guided by a finger in the rectum. When they are placed correctly, the 125 I-seeds are inserted through the needles which are then withdrawn.

After careful haemostasis a drain is left in the space of Retzius and the wound is closed. A postoperative pelvic X-ray is carried out for localization of the seeds and computerized dosimetry.

The catheter can be removed after 2 days and the drain after 4 - 5 days. Hospital stay is 7 - 10 days when no complications occur.

Results and Discussion

From 18 patients treated in the last two years the data as to age, T-stage and grade, volume of prostate are given in Table 1. In 5 patients lymphnodes were positive, which was expected from lymph-angiography in 3. In 12 patients aspiration cytology of the prostate was repeated after six months and in 6 malignant cells could no longer be found whilst 5 still had malignant cells; of these 1 developed a local recurrence, and the other 4 became negative after one year. The postoperative complications are given in Table 2.

As yet in only one patient whose lymphnodes were negative has a local recurrence been noted after one year. Presumably this was caused by extension of the primary tumor in one of the seminal ves-icles, which was not appreciated as such at the time of 125-I-implan-tation. This patient is now on hormonal treatment.

Table 1. T and G Categories and Estimated Weight of Prostate in
          18 Patients Treated

Age	No. of Patients	T	No. of Patients	G	No. of Patients	Est. Weight	No. of Patients
50–60	6	T_1	3	G_1	8	15–20g	3
61–65	6	T_2	15	G_2	7	20–25g	6
65–70	3			G_3	3	26–30g	3
71–75	3					>30g	6

Table 2. Postoperative Complications

Edema	3
Wound dehiscence	3
Thrombosis	1
Haematoma	1
Infection	2
Death from myocardial infarction	1
No complications	8

All patients are on regular follow-up, every 3 months for the
first year and thereafter every 6 months. Those patients who had
positive nodes are followed with particular care. As soon as signs
of progressive disease appear, we intend to put them on secondary
hormonal treatment.

As it is still far too early for survival rates or comparison
with other series, we will only discuss here some of the results of
Hilaris and Whitmore who have the largest experience and compare
these with the results of external irradiation.

In 1979 this group at the Memorial Hospital in New York presented a review on 5 year survival rates on more than 200 patients
[5]. Their patient selection was rather liberal as they included
also patients with stage C (T_3) lesions. More than 50% of the T_3
cases appeared to have positive nodes and the crude survival of these
was 65% with half the patients having recurrent disease or distal
metastasis. However for T_1 and T_2 cases the crude 5-year survival
was 100%. Actuarial 5-year survival for node-negative patients was
92% and for node positive 46%.

They found that tumor regression was extremely slow, but it was
based on rectal examination and pelvic X-rays only. They had a 10%
local failure rate and there was evidence that local tumor control
was highly dose dependent: 90% tumor control was achieved with a
matched dose of 18000 rads.

Patients with positive lymphnodes have poor prognosis, as expected (only 46% 5-year survival) which was not improved by additional external irradiation.

Complications could be described as minimal and were mainly related to the lymphadenectomy.

Recently Whitmore[6] presented the 10-year survival rates of the first 100 patients and pointed out that several circumstances make comparison with other treatment series difficult as bone metastases were monitored only by skeletal X-rays and and elevated PAP was not a criterion for exclusion of patients. Overall survival was 82% 5-year and 52% 10-year, but stratifying according to grade, node staging and stage of local tumor revealed that these were all strongly related to survival. Positive nodes indicated the worst prognosis (87% 10-year survival if node negative versus 14% if node positive). He also pointed out that patients with positive lymphnodes and/or local recurrence should be given additional therapy as this certainly influenced survival.

Conclusions

Patients with T_1 and T_2 tumors, whose prostates were within 20-40g limits were selected for 125-I implantation. Patients with smaller or larger prostates and with locally confined T_3 tumors were offered external irradiation as alternative. In prostates smaller than 20 g. it is not possible to implant seeds without considerable loss of seeds and/or too small a radiation dose. Probably these rare patients would benefit more by prostatectomy.

A limited lymphadenectomy was done as a staging procedure in the hope of minimising complications. With these limitations we believe that for patients in the younger age groups with T_{1-2} disease 125-I-implantation offers a good treatment with preservation of potency less post operative complications and a survival rate which equals that of radical prostatectomy and is better than external radiation[7,8,9,19].

For those patients that have positive lymphnodes a trial of immediate or delayed additional hormonal or chemotherapy should be seriously considered.

Very recently a study was published on iridium implantation of the prostate[11]. It has the advantage that the implant material is removed at the end of treatment and that no loss of seeds can occur. It is applicable to larger volume prostates (stage C), but it has to be complemented with additional external irradiation, as the local tumor dose is not more than 3500 rads. Time will show which of the techniques is preferable.

REFERENCES

1. B. G. Hilaris, W. F. Whitmore, M. A. Batata, and H. Grabstald, Cancer of the prostate, in: "Handbook of Brachytherapy," B. G. Hilaris, ed., Publ. Sciences Group, Acton, Mass. p.219 (1975).

2. C. L. Deming, Results in one hundred cases of cancer of prostate and seminal vesicles, treated with radium, Ann.Surg., 65:633 (1917).

3. R. H. Flocks, The treatment of stage C prostatic cancer with special reference to combined surgery and radiation therapy. J.Urol., 109:461 (1973).

4. P. P. Kumar and F. F. Bartone, Transperineal percutaneous I-125 implant of prostate, Urology, XVII: 238 (1981).

5. P. C. Sogani, W. F. Whitmore Jr., B. S. Hilaris, and M. A. Batata, Experience with interstitial implantation of Iodine 125 in the treatment of prostatic carcinoma, Scand.J.Urol. Nephrol., Suppl.55:205 (1980).

6. H. B. Grossman, M. Batata, B. Hilaris, and W. F. Whitmore, 125-I-implantation for carcinoma of prostate. Further follow-up of first 100 cases. Urology, XX: 591 (1982).

7. H. Sommerkamp, H. Knüfermann, and M. Wannenmacher, Die interstitielle Strahlentherapie des Prostatakarzinoms mit Jod 125: Klinischer Erfahrungsbericht, Akt.Urol., 13:268 (1982).

8. M. A. Bagshaw, D. A. Pistenma, G. R. Ray, F. S. Freika, and R. L. Kempson, Evaluation of extended field radiotherapy for prostatic neoplasm, Cancer Treat.Rep., 61:297 (1978).

9. F. Edsmyr, P. L. Esposti, and L. Andersson, External irradiation therapy in carcinoma of the prostate, Scand.J.Urol.Nephrol., Suppl. 55:213 (1980).

10. B. Werf-Messing, v.d. Radiation Therapy of Carcinoma of the Prostate, in: "Prostate Cancer," G. H. Jacobi and R. Hohenfellner, eds., Williams and Wilkins, Baltimore, p.195 (1982).

11. L. A. Tansey, A. M. Shanberg, A. M. Nisar Syed, and A. Puthawala, Treatment of prostatic carcinoma by pelvic lymphadenectomy, temporary iridium-192 implant and external irradiation, Urology, XXI-594 (1983).

HORMONE THERAPY

P. H. Smith

Department of Urology
St James's University Hospital
Leeds

INTRODUCTION

Hormonal therapy for carcinoma of the prostate has been firmly established since the pioneering work of Huggins and Hodges[1]. During my own training any patient diagnosed as having carcinoma of the prostate was treated with stilboestrol, usually at a dosage of 5 mg 3 times a day, though many centers used higher dosages. The original report of the Veterans Administration Cooperative Urological Research Group[2] caused a re-evaluation of such a general policy, which had in any case never been uniform, and which would certainly never have been acceptable in many centers in the United States where radical prostatectomy has always been felt to have a role in the management of patients with localized disease[3]. However, for the patient with extra-capsular disease (category T3-4, MO) and for patients with metastases, hormonal therapy, whether by the use of stilboestrol or some other estrogenic compound or by orchiectomy has nearly always been felt to be the correct course of action, though at intervals people have raised the question of deferred therapy, suggesting that this may be at least as effective as immediate treatment in terms of survival, whilst allowing the patient to lead a more normal life for a longer period of time than would otherwise be possible[4].

HORMONES AFFECTING THE PROSTATE

The prostate is influenced by the hypothalamus, the pituitary, the testes and the adrenal. Luteinizing hormone-releasing hormone (L.H.R.H.) stimulates the formation of the gonadotrophins in the pituitary whilst luteinizing hormone (LH) stimulates testosterone

production in the testis. The circulating testosterone is largely
bound to sex hormone binding globulin (SHBG) and only that which is
free is available for uptake by the prostate gland. The adrenal
androgens are less active than testosterone.

The essential aim of hormonal therapy is to eliminate free
circulating testosterone and thus inhibit prostatic metabolism. This
may be achieved by orchiectomy or by the administration of estrogens
which, in addition to suppression of LH release, increase the production of SHBG making testosterone less readily available to the
prostate gland.

In recent years antiandrogens, including cyproterone acetate and
flutamide, which inhibit gonadotrophin release, suppress adrenal
function to a lesser extent, inhibit formation of testosterone in the
testes, its uptake by the prostate cell and the binding of intra-
nuclear dihydrotestosterone (D.H.T) have been increasingly investigated. It seems likely that cyproterone acetate has little effect
either on prolactin secretion or on the production of SHBG. Further
information on this subject is available elsewhere[5,6,7].

HORMONAL MANAGEMENT OF PROSTATIC CANCER

Disease Localized to the Prostate

Since the article by Byar[8] also confirmed by Donoghue[9]
showing that patients with coincidentally diagnosed disease (TO
disease) have a survival similar to that expected for the age group,
irrespective of treatment given, hormonal therapy has been questioned
for these early forms of cancer and it has become increasingly common
to use external or interstitial radiation (or radical prostatectomy)
for such patients and for those with nodules localized to the prostate (category T1 and T2, TNM 1978). Since most are now agreed that
hormone therapy is not the treatment of choice for these patients
these categories of the disease will not be discussed further.

Advanced non-Metastatic Prostatic Cancer

For disease beyond the prostatic capsule local therapy is unlikely to be curative since 50 to 60% of patients with category T3-4
disease already have involved lymph nodes and most will eventually
develop metastases. The conversion from extra-capsular localized to
metastatic disease may however be slow and Byar[4] has demonstrated
that even with deferred therapy up to 50% of patients may show no
progression over a period of up to 5 years.

When this point is considered, together with the known cardio-
toxicity of estrogen therapy, at least in the doses of 5 mg daily

given by the VACURG[2] and 1 mg tds in the more recent study[10] of
the EORTC Urological Group (European Organization for the Research
and Treatment of Cancer) the use of estrogen therapy in patients
without clinical or radiological evidence of metastases bears further
examination.

It will also be remembered that in the VACURG series the majority of cardio-vascular deaths occurred within the first year and that
additional deaths from cardiovascular disorders in the group receiving estrogens outweighed the reduction in deaths from carcinoma
(Table 1).

If these deaths are considered by stage[11], it can be seen that
the excess cardiovascular disease tends to be seen in the patients
with extra-capsular localized disease rather than in those with
metastases, lending support to the concept of avoiding estrogen
therapy in this group of patients and treating, if considered desirable, by means of orchiectomy.

For patients with metastatic disease, especially those with
symptoms, the benefits of estrogen therapy in terms of pain relief
for the majority and objective remission for 30% of the patients[10]
outweight the possible risks.

Metastatic Prostatic Cancer

It is customary to treat all patients who show evidence of
metastases on clinical examination or on radio-isotope bone scan
(subsequently confirmed by X-ray examination). There seems to be
some evidence that the levels of free testosterone are somewhat lower
in patients following estrogen therapy than following orchiectomy
since estrogen therapy stimulates sex hormone binding globulin,
rendering the testosterone unavailable to the cell. That which does
enter the prostate, after conversion to the active dihydrotestosterone (DHT), is combined with an androgen receptor and transported
to the prostatic cell nucleus, increasing its activity. This process
is inhibited by estrogen therapy which:

(a) decreases the plasma levels of luteinizing hormone (LH), follicle
    stimulating hormone (FSH), and testosterone,
(b) increases prolactin secretion which itself stimulates prostatic
    activity and the production of sex hormone binding globulin
    (SHBG), which renders circulating testosterone unavailable to the
    prostate, and
(c) inhibits prostatic metabolism, at least in animal systems, when
    large doses are used.

Table 2 shows the ways in which orchiectomy differs from estrogen therapy. Though its cardio-vascular hazard is less, it may be

Table 1.   Causes of Death in   First VACURG Study[9]

| | Cancer | | Cardiovascular | |
	Stage C	Stage D	Stage C	Stage D
Placebo	18%	47%	33%	24%
Placebo + Orchiectomy	13%	48%	35%	27%
Estrogen	6%	38%	42%	36%
Estrogen + Orchiectomy	9%	38%	42%	27%

Table 2.   Prostatic Cancer.   Effects of Hormone Therapy

	Estrogen	Orchiectomy
T	↓	↓
LH	↓	↑
FSH	↓	↑
Prolactin	↑	−
CVS Hazard	+	−
Breasts	+	−
Impotence	+	±
SHBG	↑	−
Free T	↓↓	↓

marginally less effective, perhaps because the SHBG is not increased.
Of the alternative agents (Table 3) available for the treatment of
prostatic cancer, the antiandrogens or estramustine may offer the
best hope of success.   These drugs include cyproterone acetate and
flutamide but, as mentioned, extensive information is available only
on the former agent.

   Ideal Hormonal Therapy.   In the ideal world one might consider
the combination of treatments outlined in Table 4 which should convey
the maximum benefit and, of course, increase toxicity and add to the
cost of therapy (Table 5).

Contributions of the EORTC Urological Group

   The EORTC Urological Group is now starting its final analysis of
its first two Phase III trials (Table 6).   The conclusions of a preliminary evaluation have been summarized elsewhere[10] and have
demonstrated that stilboestrol and estracyt in the doses used were
equi-effective in the production of objective remissions (approximately 30% of all patients) and were each more effective than cyproterone acetate and medroxyprogesterone acetate.   The cardio-vascular
side effects of stilboestrol were, however, greater than those seen

Table 3. Cancer of the Prostate

Some Possible Alternatives to Stilboestrol Therapy

Other Oestrogens	– Honvan, TACE, Estradurin, Estracyt
Progestogens	– Progesterone, Medroxyprogesterone Acetate
Antiandrogens	– Cyproterone Acetate, Flutamide
Medical Adrenalectomy	– Aminoglutethimide
Cytotoxic Chemotherapy	

Table 4. Treatment of Prostatic Cancer

Androgen Withdrawal	–	Orchiectomy
Adrenal Androgen Suppression	–	Prednisolone
Stimulation S.H.B.G.	–	Estrogens
Prolactin Suppression	–	Dopamine Agonists
Direct Prostatic Inhibition	–	Antiandrogens

Table 5. Cost of Certain Agents used in the Treatment of Prostatic Cancer (1983, U.K. prices)

Drug	Price per Tablet
1. Stilboestrol 1 mg	0.7 p
2. Medroxyprogesterone Acetate 100 mg	37 p
3. Cyproterone Acetate 50 mg	50 p
4. Estracyt 140 mg	£1.14 p

with the other agents and it was clear from further analysis that certain prognostic factors, including performance status, histo-logical grade and the presence or absence of pain, were of greater significance in relation to patient survival than the choice of therapy.

The recognition that a dose of stilboestrol as small as 1 mg tds carried a significant cardio-vascular hazard has reawakened interest in orchiectomy as primary therapy and the present study of the EORTC Urological Group compares low dose stilboestrol 1 mg daily versus orchiectomy versus orchiectomy and cyproterone acetate – this last combination having been reported to be effective by Bracci et al.,[12].

Table 6.

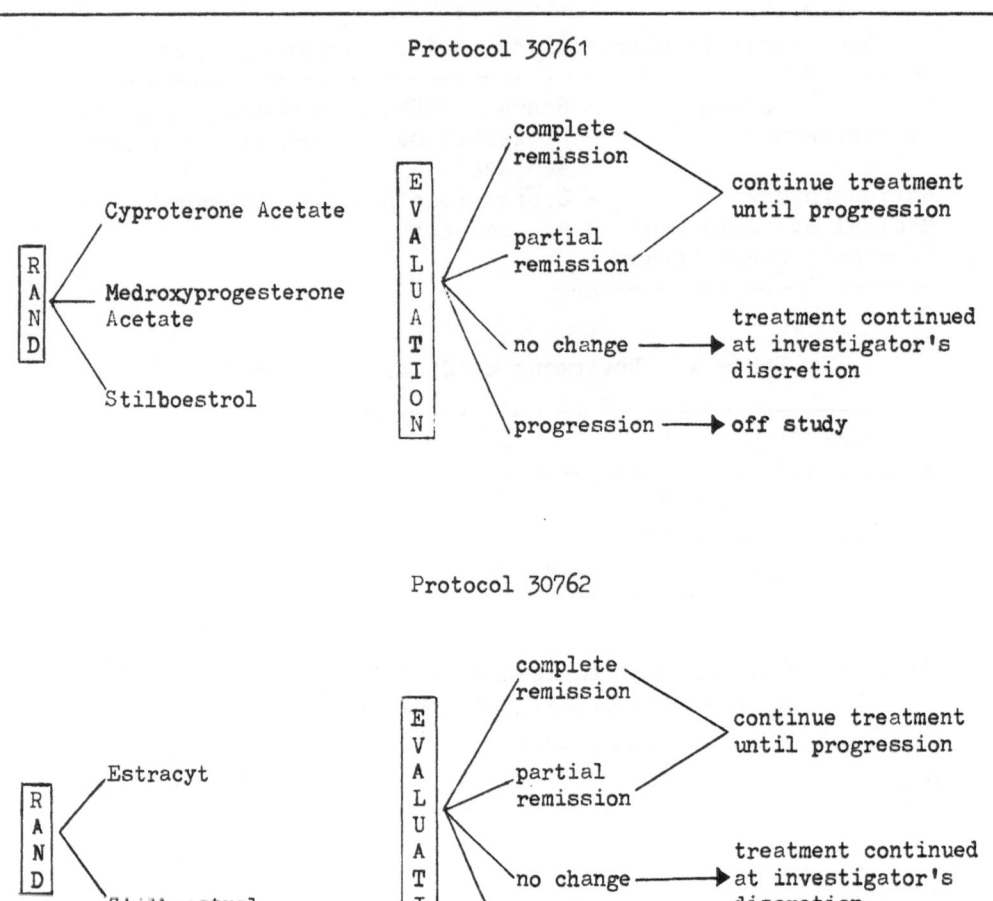

Protocol 30761

Protocol 30762

REFERENCES

1. C. Huggins and C. V. Hodges, The effect of castration, of
    oestrogen and of androgen injection on serum phosphatases in
    metastatic carcinoma of the prostate, Cancer Research, 1:
    293-297 (1941).
2. Veterans Administration Cooperative Urological Research Group
    (VACURG), Treatment and survival of patients with cancer of
    the prostate, Surg.Gync.& Obstet., 124:1011-1017 (1967).
3. F. H. Schröder and E. Belt, Carcinoma of the prostate: a study
    of 213 patients with stage C tumor treated by total perineal
    prostatectomy, J.Urol., 114:257-260 (1975).
4. D. P. Byar, Results of clinical trials of estrogen treatment for
    cancer of the prostate, in: "Proceedings of the 16th Congress

of the International Society of Urology", Volume 2, Doin, Paris (1973).

5. J. D. Altwein, Hormone manipulation for palliative treatment of advanced prostatic carcinoma, in: "Prostate Cancer", G. H. Jacobi and R. Hohenfellner, eds., Williams and Wilkins, Baltimore and London, pp.215-243 (1982).

6. F. Neumann, M. Humpel, Th. Senge, B. Schenck, and U. Tunn, Cyproterone acetate – Biochemical and biological basis for treatment of prostatic cancer, in: "Prostate Cancer", G. H. Jacobi and R. Hohenfellner, eds., Williams and Wilkins, Baltimore and London, pp.269-303 (1982).

7. G. H. Jacobi, U. Tunn, and Th. Senge, Clinical experience with cyproterone acetate for palliation of inoperable prostatic cancer, in: "Prostate Cancer", G. H. Jacobi and R. Hohenfellner, eds., Williams and Wilkins, Baltimore and London, pp.305-319 (1982).

8. D. P. Byar, Veterans Administration Cooperative Urological Research Group, Survival of patients with incidentally found microscopic cancer of the prostate: results of a clinical trial of conservative treatment, J.Urol., 108:908-913 (1972).

9. R. E. Donohue, H. E. Fauver, J. A. Whitesel, and R. R. Pfister, Staging prostatic cancer: a different distribution, J.Urol., 122:327-329 (1979).

10. P. H. Smith, M. Pavone-Macaluso, G. Viggiano, H. De Voogt, B. Lardennois, M. R. G. Robinson, B. Richards, R. W. Glashan, M. De Pauw, R. Sylvester, and The EORTC Urological Group, EORTC protocols in prostatic cancer, in: "Controlled Clinical Trials in Urologic Oncology", p.107-117, L. Denis, G. P. Murphy, G. R. Prout, and F. H. Schröder, eds., Raven Press, New York (1984).

11. J. D. Altwein, Controversial aspects of hormone manipulation in prostatic carcinoma, in: "Cancer of the Prostate and Kidney", M. Pavone-Macaluso and P. H. Smith, eds., Plenum Press, New York and London, pp.305-316 (1983).

12. U. Bracci, Antiandrogens in the treatment of prostatic cancer, Eur.Urol., 5:303 (1979).

LHRH ANALOGUES IN THE MANAGEMENT OF PROSTATIC CARCINOMA

M. R. G. Robinson,* S. Chandrysekran,* D. Dris,*
S. Crispin,* R. Nicholson,** and K. Walker**

* Pontefract General Infirmary, West Yorks, UK
**Tenovus Institute for Cancer Research
Welsh National School of Medicine, Cardiff, UK

In 1971 Schally et al., and Matusuo et al.,[1,2] isolated and characterised hypothalamic neurotransmitters (releasing hormones). Since then synthetic analogues (decapeptides) have been developed which when given in supraphysiological doses, fix pituitary receptors and block endogenous releasing hormones[3]. Such agonistic analogues of lutenising hormone releasing hormone (LHRH) are currently being evaluated clinically in the management of carcinoma of the prostate.

Animal experiments have shown that these agonistic analogues reduce the weight of the testes, seminal vesicles and prostate[4] exhaust and desensitise the pituitary gland[5,6] and inhibit prolactic secretion[7]. They also downgrade gonadal gonadotrophin receptors resulting in decreased steroidogenesis[8] and probably antagonise the action of sex steroids in a different way to cyproterone acetate[9]. The data from clinical trials has shown that in man they first stimulate and then suppress the pituitary gonadal function[10] and in high doses cause a fall of plasma testosterone to castrate levels within 3 weeks[3,11].

Several studies have shown that in patients with carcinoma of the prostate LHRH analogues produce subjective and objective evidence of tumor regression with an associated fall of acid phosphatase levels (Table 1). The two most commonly used analogues have been Buserelin (D-ser[6]-ter-butyl-gonadotrophin-releasing hormone-ethylamine) and ICI 118,630 (p-glu-his-trip-ser-tyrD-ser(buleu-arg-pro-azgly NH_2). Both may be administered subcutaneously and Buserelin is also available as a nasal inhalation. Their reported toxicity has been minimal and includes facial flushing, impotence and irritation at the site of injection.

Table 1.   LHRH Analogues in the Management of Prostatic Carcinoma:
           Clinical Data

Author	No of Patients	Subjective Response	Fall in Acid Phosphatase	Objective Response
Tolis et al.[12]	10	9	5/6	7
Allen et al.[13]	10	8	8/8	8
Waxman et al.[14]	12	9	5/6	5
Walker et al.[15]	8	7	7/8	–
TOTAL	40	33	25/28	20/32

Our current study using ICI 118, 630 has been designed to
determine the effect on endocrine function, the dose required to
reduce plasma testosterone (effectively), the toxicity and subjective
and objective response. Patients are entered if they have histologically proven carcinoma of the prostate and a life expectancy of more
than three months. They are excluded if they have had previous
hormonal therapy including orchidectomy and if they have concurrent
malignancy. It is planned to give 6 patients ICI 118,630 250µg
subcutaneously daily for 3 months, 6 patients 100 µg daily for 3
months and 6 patients 50 µg daily for 3 months. So far data is
available on the first 2 groups of patients.

The preliminary results have already been reported by Walker et
al.,[15]. The current data is summarized in Table 2. Both dosage
regimes reduce plasma testosterone to castrate levels. This was
achieved in under 14 days when 250 µg ICI 118,630 was given daily.
The lower dose (100 µg daily) did not produce castrate levels for 5-6
weeks. Both doses reduce plasma LH but a fall in FSH was only observed in patients receiving 250 µg daily. It can be concluded that
the high doses of LHRH are needed to reduce testosterone levels
rapidly.

In this group of patients receiving 250 µg ICI 118,630 daily,
only 4 of the 6 are evaluable because 2 died during the study. All
had very advanced metastatic disease and the deaths may have been due
to poor patient selection. Allen et al.,[13] however, have reported
an initial rise in testosterone levels in their patients receiving
the same dose of ICI 118,630 and it is possible that the analogue may
initially exacerbate the disease. LHRH analogues should therefore be
used very cautiously in patients with severe advanced disease who are
not receiving other concurrent hormonal therapy.

The evaluable patients in both dosage groups have had the expected subjective response and fall in acid phosphatase. Only one
patient had progressive disease. Toxicity was insignificant.

Table 2.   Preliminary Results

	Dose	
	250 µg daily	100 µg daily
Evaluable patients	4/6*	6/6
Testosterone reduction	4	6
Acid phosphatase reduction	4	5
Subjective response	3	6
"Objective" response**	1	4
Progressive disease	1	0
Major toxicity	0	0

*   2 patients died before the 3 months assessment.
** Digital examination of the primary lesion.

This study confirms the other data that we have reviewed in suggesting that LHRH analogues are of potential value in the management of carcinoma of the prostate. Further Phase II and Phase III Studies of LHRH analogues alone and in combination with other hormonal treatment are needed to evaluate their true value in the management of carcinoma of the prostate.

The lack of toxicity of LHRH analogues makes them an attractive proposition for future endocrine therapy but their use will depend not only on the subjective and objective response they produce but also on their cost and availability as easily administered preparations e.g., intranasal or long term depot injection. In the future they may possibly be used as primary treatment in advanced category carcinoma of the prostate, as an adjuvant to other hormone or cytotoxic therapy, as secondary hormonal therapy or as a test of hormone responsiveness prior to orchidectomy.

REFERENCES

1. A. V. Schally, A. Arimura, Y. Baba, R. M. G. Nair, H. Matsuo, T. W. Redding, L. Debeljuk, and W. F. White, Isolation and properties of the FSH and LH-releasing hormone, Biochem, Biophys.Res.Commun., 43:393 (1971).
2. H. Matsuo, R. M. G. Nair, A. Arimura, and A. V. Schally, Structure of the porcine LH and FSH-releasing hormone, 1. The proposed aminoacid sequence, Biochem.Biophys.Res.Commun., 43:1334 (1971).
3. G. H. Jacobi and U. K. Wenderoth, Gonadotrophin-releasing hormone analogues for prostatic cancer: Untoward side effects of high-dose regimens acquire a Therapeutical Dimension, Eur.Urol., 8:129-134 (1982).
4. C. Auclair, P. A. Kelly, F. Labrie, D. H. Coy, and A. V. Schally, Inhibition of testicular luteninising hormone re-

ceptor level by treatment with a potent LHRH-agonist or human
chorionic gonadotrophin, Biochem,Biophys.Res Commun., 76:855
(1977).

5. J. Sandow, W. von Rechenberg, K. Jerzabak, K. Engelbart, H.
    Kuhl, and H. Fraser, Hypothalamic-pituitary-testicular func-
    tion in rats after supraphysiological doses of a highly
    active LHRH analogue (Buserelin) Acta Endocrinol.(Copenh),
    94:489-497 (1980).

6. A. Corbin, From contraception to cancer: a review of the
    therapeutic applications of LHRH analogues as antitumor
    agents, Yale J. Biol.Med., 55:27 (1982).

7. S. W. J. Lamberts, P. Uitterlinden, J. M. Zuiderwijk-Roest, E.
    G. Bons-von Evelingen, and F. H. de Jong, Effect of a luten-
    ising hormone-releasing analog and tamoxifen on the growth of
    an estrogen-induced prolactin-secreting rat pituitary tumor
    and its influence on pituitary gonadotropins, Endocrinology,
    108:1878 (1981).

8. K. J. Catt, J. P. Harwood, R. N. Clayton, T. F. Davies, V. Chan,
    M. Katikineni, K. Nozu, and M. L. Dufau, Regulation of pep-
    tide hormone receptors and gonadol steroidogenesis, Recent
    Prog.Horm.Res., 36; 557 (1980).

9. P. Lecomte, N. C. Wang, K. Sundaram, J. Rivier, W. Vale, and C.
    Wayne Bardin, The anit-androgen action of gonadotropin-
    releasing hormone and its agonists of the mouse kidney,
    Endocrinology, 110-1 (1982).

10. V. Borgmann, W. Hardt, M. Schmidt-Gollwitzer, H. Adenauer, and
    R. Nagel, Sustained suppression of testosterone production by
    the luteinising hormone releasing-hormone agonist Buserelin
    in patients with advanced prostatic carcinoma. A new thera-
    peutic approach? Lancet, 1:1097 (1982).

11. C. Bergquist, S. J. Nillius, T. Bergh, G. Skarin, and L. Wide,
    Inhibitory effects on gonadotrophin secretion and gonadal
    function in men during chronic treatment with a potent stimu-
    latory luteinizing hormone-releasing hormone analogue, Acta
    Endocrinol.(Copenh), 91:601 (1979).

12. G. Tolis, D. Ackman, A. Stellos, A. Mahta, F. Labrie, A. T. A.
    Fazekas, A. M. Comaru-Schally, and A. V. Schally, Tumour
    growth inhibition in patients with prostatic carcinoma
    treated with luteinising hormone-releasing hormone agonists,
    Proc.Natl.Acad.Sci.USA, 74:1658 (1982).

13. J. M. Allen, J. P. O'Shea, K. Mashiter, G. Williams, and S. R.
    Bloom, Advanced carcinoma of the prostate: Treatment with a
    gonadotrophin releasing-hormone agonist, Brit.Med.J.,
    286:1607 (1983).

14. J. H. Waxman, J. A. H. Wass, W. F. Hendry, H. N. Whitfield,
    G. M. Besser, J. S. Malpass, and R. T. D. Oliver, Treatment
    with gonadotrophin releasing hormone analogue in advanced
    prostatic cancer, Brit.Med.J., 286:1309 (1983).

15. K. J. Walker, R. I. Nicholson, M. Robinson, Z. Crispin, D. Dris,
    A. D. Turkes, and K. Griffiths, Therapeutic potential of the
    LHRH agonist ICI 118,630 in the treatment of advanced
    prostatic cancer, Lancet, 2:413-415 (1983).

CHEMOTHERAPY FOR PROSTATE CANCER - THE EXPERIENCE

OF THE NATIONAL PROSTATIC CANCER PROJECT

M. S. Soloway

Department of Urology
Baptist Memorial Hospital
Memphis, Tennessee, USA

Unfortunately for many patients with advanced prostate cancer the long-term benefit from endocrine therapy is limited and, following relapse to initial hormonal manipulation, further endocrine therapy is only rarely helpful. Thus there is a need to develop effective chemotherapy. The National Prostatic Cancer Project (NPCP) initiated clinical trials utilizing non-hormonal chemotherapy in 1973. Starting with five collaborative institutions, there are now 18 institutions participating in the NPCP. Over 2,000 patients have been accessioned into the clinical trials.

A number of factors have hampered trials in selecting active agents for patients with prostate cancer. This includes the elderly age group, prior radiation therapy which had been given to many of the patients, and the common presentation with advanced disease with multiple areas of the bony skeleton being involved by the time hormonal therapy is discontinued and chemotherapy initiated. Another complicating factor is the difficulty in objectively evaluating response to treatment. Most patients have poorly measurable lesions such as a rectal or a pelvic mass, ureteral deviation and/or bone metastases. The lack of strictly measurable lesions, e.g. lung, lymph nodes, and the use of subjective parameters make it frequently difficult to evaluate response to therapy.

Bone metastases are notoriously difficult to quantitate. The radionucleotide bone scan has become one of the most important parameters for monitoring response to treatment. The objective criteria for monitoring response utilized by the NPCP have been accepted by a number of other groups. The bone scan is an integral part of the response criteria. Response criteria have been published[1], and are indicated in Table 1.

One of the categories utilized to evaluate response is the stable category. This is not the same as a "no change" category but reflects a favorable response to therapy in patients who had documented progression prior to entry onto a protocol. This category has been reviewed in two NPCP papers[2]. Analysis shows a survival advantage for patients indicated as clinically stable compared to those progressing despite therapy.

The first series of protocols initiated by the NPCP were designed for patients who were in progression following hormonal therapy. The first protocol (Protocol 100) evaluated 125 patients randomized to receive either 5-fluorouracil, 600 mg/M^2 I.V. weekly, cyclophosphamide, 1 gm?M^2 I.V. every three weeks, or continuation of any non-chemotherapeutic treatment. There were no complete responders[3]. Partial regressions were observed infrequently and only in patients receiving chemotherapy. Nineteen percent of patients on the standard arm were considered stable compared to 39% receiving cytoxan and 24% on 5-fluorouracil. When the total response rate, partial response plus stable, was evaluated, those receiving cytoxan had a significantly greater chance of responding than those in the standard arm, 46% vs. 19%. Unfortunately the mean survival was not appreciably different from patients randomized to each of the initial therapies.

A concurrent trial entered patients who had received at least 2000 rads of radiation therapy to the pelvis[4]. These separate trials sought to evaluate compounds with reduced risk of myelosuppression. The compound estramustine phosphate, estracyt, was one of the drugs evaluated in this trial. Estracyt is a nitrogen mustard attached to estradiol phosphate. The dose was 600 mg/M^2 given in three divided oral doses. The other two arms in this trial consisted of streptozotocin, 500 mg/M^2 I.V. for the first five days of each six week period or continuation of standard therapy (as in Protocol 100).

The response rate (all stabilization) for those receiving standard therapy was 19%. This was compared to 30% of those receiving estracyt and 32% for the streptozotocin group. Myelosuppression, as expected, was minimal with the most prominent side effect being nausea and vomiting. Once again the mean survival was not significantly different among the three treatment arms although patients receiving chemotherapy survived longer than those receiving standard treatment.

Protocol 300 succeeded the initial protocol for patients who had not received prior radiation therapy[5]. Cytoxan was selected as the standard drug. The other two arms consisted of DTIC, 200 mg/M^2 I.V. days 1-5 every three weeks and procarbazine, 100 mg/M^2 orally daily days 1-22, three weeks rest and treatment on days 44, 65, etc. There were two PRs among the patients receiving DTIC but none in the other treatment arms. The overall response rate (PR + Stable) was similar

for cytoxan, 26% and DTIC, 28%, and less for those receiving procarbazine, 13%.   The toxicity associated with procarbazine and the lower
response rate suggested that this drug has little place in the treatment of prostate cancer.

Protocol 400 succeeded Protocol 200 for those patients who had
received prior radiation therapy[6].   Prednimustine, a combination of
prednisone and chlorambucil, was evaluated both as a single agent and
in combination with estracyt.   Both treatment arms produced a rather
low response rate, 13%.   Toxicity was minimal and was primarily
gastrointestinal.

A subsequent protocol for those who had not received prior
radiation therapy evaluated cytoxan, hydroxyurea, 3 gm/M^2 orally
every three days in three divided doses, and methyl CCNU, 175 mg/M^2
orally every six weeks[7].   The overall response rate (PR + Stable)
was slightly better for those receiving cytoxan, 34%, when compared
to methyl CCNU, 30%.   Only 15% responded to hydroxyurea.   There were
no complete and only a few partial responses among the patients
treated in this protocol.   Approximately 20% of patients on each of
the treatment arms had an improvement in pain.

Patients receiving methyl CCNU had a high incidence of hematologic toxicity.   This can be compared to 36% on the cytoxan arm and
43% for those receiving hydroxyurea.   It should be remembered, however, that many patients with prostate cancer have hematologic depression as a result of tumor infiltration into bone.

In this particular study the mean survival for patients on the
cytoxan arm was superior to the other regimens particularly compared
to hydroxyurea, 40 vs. 20 weeks.   It can be concluded from this study
that hydroxyurea exhibited little activity and was more toxic than
the other two agents.   Methyl CCNU did have moderate activity but
also had appreciable toxicity.

The next in a series of protocols for patients who had received
prior radiation therapy once again compared estracyt to other non-
myelosuppressive drugs[8].   One hundred and twenty-one patients were
randomized to three treatment arms.   Vincristine, 1 mg/M^2 I.V. every
two weeks; estracyt; and the combination of the two agents.   Ninety
patients were evaluable for response.   There was one partial response
in each of the single agent treatment arms.   An additional 22% of
patients receiving estracyt remained stable while 12% receiving
vincristine met the criteria for stabilization.   Twenty-four percent
were stabilized by receiving the combination regimen.   Few patients
had a noticeable improvement in performance status or relief of pain.

Toxicity was not a major problem in this protocol although
nausea and vomiting was recorded in 59% of patients receiving
estracyt and 54% with the combination.   It was also observed in 42%

receiving vincristine which simply indicates the high background
incidence of this side effect in patients with advanced prostate
cancer since vincristine does not usually induce nausea and vomiting
in patients with other malignancies.

The median duration of response was similar for those receiving
estracyt, 20 weeks, and vincristine, 22 weeks. It was 13 weeks for
those receiving the combination. The mean response duration was
greater for estracyt, 41 weeks, than for the other two arms, 27 and
17 weeks. The probability of survival as calculated by the life
table method indicated a slight advantage for estracyt. None of
these differences were significant. Thus is can be concluded from
this protocol that adding vincristine to estracyt did not enhance the
responses seen with either agent alone.

Although the success of chemotherapy in patients with advanced
prostate cancer who have failed hormonal therapy has been limited,
the NPCP did demonstrate some advantage when compared to any other
treatment, e.g. prednisone. Since it is apparent that from the
inception of the tumor there is a heterogeneous cell population, some
of which may be sensitive and others insensitive to hormonal therapy,
a combination approach seemed worthy of trial.

The next protocol (1100) for patients who had not received prior
radiation therapy, compared the use of estracyt, methotrexate, and
cis-platin (DDP)[9]. One hundred and fifty-eight evaluable patients
were entered into this trial. There were few partial regressions and
the overall response rate did not significantly differ in the two
arms: methotrexate 39%, estracyt 34%, and DDP 18%. There were two
PRs in the methotrexate and DDP group and one in the estracyt group.

Methotrexate was associated with moderate toxicity. Seventy-
four percent of patients experienced some mucositis and this was
severe in approximately one half of the involved cases. Most
frequently this resulted in an alternate week by schedule instead of
the originally designed weekly regimen.

Protocol 1200 was the next protocol for patients who had received prior radiation therapy[10]. This protocol was complete in
January 1982. Patients received either estracyt or DDP alone or the
combination of these two agents. One hundred and twenty-four evaluable patients were randomized and no CRs or PRs were observed.
Patients receiving the combination had a 33% incidence of stabilization compared to 21% for DDP alone and 18% for estracyt.

In 1976 the NPCP instituted a trial for patients with newly
diagnosed metastatic prostate cancer[11]. These patients had not
received prior treatment. Patients were randomized to receive either
diethylstilbestrol (DES) 1 mg orally t.i.d., DES + cyclophosphamide,
or estracyt + cyclophosphamide. The initial response was evaluated

at 12 weeks and subsequently at 12 week intervals. The response criteria were the same as in prior NPCP studies. The results of this trial have been recently reported by Murphy, et al.,[11].

Three hundred and on patients were randomized into the study and 246 were eligible for analysis. This represents an exclusion of 18%. The objective response rates were similar among the different treatment arms although there were fewer progressing patients, 12%, in those receiving DES + cytoxan than the group receiving cytoxan + estracyt, 17%, or the hormonal therapy only arm, 19%. There were no statistically significant differences between the treatment arms in the distribution of response categories. Comparisons of survival for the different treatments did not reveal any statistically significant differences either. There also was no difference in response or survival between the two forms of hormonal manipulation, DES or orchiectomy.

The histologic grade of the primary tumor was compared with the objective response rate and once again there was no significant difference for response as compared to tumor grade. The duration of responses was quite similar among the three treatment categories. Two thirds of the patients who entered the protocol are still alive and thus survival experience is certainly subject to subsequent fluctuation. In addition, patients are often given other treatments after progression. The overall survival experience does indicate, however, that the two arms containing a chemotherapeutic agent were similar and consistently better than the curve for the hormonal manipulation only arm. This level reaches a statistical significance of p=0.10 when the hormonal therapy only arm is compared to the cytoxan arm.

Not unexpectedly, the survival curves for the three treatments for patients without pain on entry into the study were better than those for patients who did have pain. For patients with pain, the best overall survival was achieved for patients receiving cytoxan + estracyt. For this subgroup, the response rate of 92% was markedly better than the 73% for hormonal therapy only, p=0.06. The rate of 80% for those receiving cytoxan + estracyt was intermediate between the three arms. Interestingly, when patients entered without pain, suggesting an earlier stage in the natural history of the disease, the best survival was achieved by those receiving hormonal therapy alone.

The current clinical studies in the NPCP now include protocols for all stages of disease except Al and Bl[12]. The current protocol for patients with progression following orchiectomy and who have not received pelvic irradiation includes methotrexate, cytoxan + adriamycin and cytoxan + 5FU + DDP.

Protocol 1600 is the current protocol for those who have received prior radiation therapy and streptozotocin is being compared

TABLE I

COMPLETE RESPONSE

All of the following:

Tumor masses, if present, totally disappeared and no new lesions appeared.

Elevated acid phosphatase returned to normal.

Osteoblastic lesions disappeared with a negative bone scan.

If hepatomegaly is a significant indicator, there must be a complete return in size of the liver to normal and normalization of all pre-treatment abnormalities of liver function.

No significant cancer-related deterioration in weight, symptoms, or performance status.

PARTIAL REGRESSION

Any of the following:

A reduction by 50 percent in the number of increased uptake areas on the bone scan.

Decrease of 50 percent or more in cross-sectional area of any measurable lesions.

There must be at least 30 percent reduction in liver size and at least a 30 percent improvement of all pretreated abnormalities of liver function.

All of the following:

No new sites of disease.

Acid phosphatase returned to normal.

No significant cancer-related deterioration in weight (greater than 10 percent), symptoms or performance status.

OBJECTIVELY STABLE

All of the following:

No new lesions and no measurable lesions increased more than 25 percent.

Osteoblastic lesions remained stable on the bone scan.

Hepatomegaly, if present, did not appear to worsen by more than a 30 percent increase in liver measurements.

No significant cancer-related deterioration in weight (greater than 10 percent), symptoms, or performance status.

OBJECTIVE PROGRESSION

Any of the following:

Significant cancer-related deterioration in weight (greater than 10 percent), symptoms, or performance status.

Appearance of new areas of malignant disease by bone scan or x-ray or in soft tissue.

Increase in any previously measurable lesion by greater than 25 percent in cross-sectional area.

Development of recurring anemia secondary to prostatic cancer.

Development of ureteral obstruction.

NOTE: An increase in acid or alkaline phosphatase alone is not to be considered an indication of progression.

to stilphostrol and two arms in which patients will receive megestrol acetate. In one of these two arms patients will also receive a small dose, 0.1 mg/day, of DES.

A protocol has just been completed which is a successor to Protocol 500 and evaluated patients with newly diagnosed metastatic disease. Standard hormonal therapy with DES or orchiectomy was compared to estracyt alone and the combination of cytoxan + 5FU + DES or orchiectomy. This protocol analysis has not yet been completed.

Supported in part by the National Prostatic Cancer Project (NPCP) Grant 20618, University of Tennessee Center for the Health Sciences, Department of Urology Memphis, Tennessee USA.

REFERENCES

1. R. P. Gibbons, S. Beckley, and M. D. Brady, The addition of chemotherapy to hormonal therapy for treatment with patients with metastatic carcinoma of the prostate, J.Surg.Oncol., 23:133 (1983)
2. N. H. Slack, A. Mittelman, M. D. Brady, and G. P. Murphy, The importance of the stable category for chemotherapy treated patients with advanced and relapsing prostate cancer, Cancer, 45:2393 (1980).
3. W. W. Scott, R. P. Gibbons, D. E. Johnson, and G. R. Prout, Comparison of 5-fluorouracil and cyclophosphamide in patients with advanced carcinoma of the prostate, Cancer Chemother. Rep., 59:195 (1974).
4. G. P. Murphy, R. P. Gibbons, D. E. Johnson, and S. A. Loening, Comparison of estramustine phosphate and streptozotocin in patients with advanced prostatic carcinoma who have had extensive irradiation, J.Urol., 118:288 (1977).
5. J. D. Schmidt, W. W. Scott, R. P. Gibbons, and D. E. Johnson, Comparison of procarbazine, imidazol-carboxamide and cyclophosphamide in relapsing patients with advanced carcinoma of the prostate, J.Urol., 121:185 (1979).
6. G. P. Murphy, R. P. Gibbons, D. E. Johnson, and G. R. Prout, Jr., The use of estramustine and prednimustine vs. prednimustine alone in advanced metastatic prostatic cancer patients who have received prior irradiation, J.Urol., 121:763 (1979).
7. S. A. Loening, S. Beckley, J. P. dekernion, and E. P. Gibbons, Comparison of hydroxyurea, methyl-chloroethyl-cyclohexyl-nitrosourea and cyclophosphamide in patients with advanced carcinoma of the prostate, J.Urol., 125:812 (1981).
8. M. S. Soloway, J. P. deKernion, R. P. Gibbons, and D. E. Johnson, Comparison of estramustine phosphate with vincristine alone or in combination for patients, hormone refactory previously irradiated carcinoma of the prostate, J.Urol., 125:664 (1981).

9. S. A. Loening, S. Beckley, M. F. Brady, and T. M. Chu, Comparison of estramustine phosphate, methotrexate, and cis-platinum in patients with advanced, hormone refactory prostate cancer, J.Urol., 129:1001 (1983).

10. M. S. Soloway, S. Beckley, M. F. Brady, and T. M. Chu, A comparison of estramustine phosphate vs. cis-platinum alone vs. estramustine phosphate plus cis-platinum in patients with advanced hormone refactory prostate cancer who had had extensive irradiation of the pelvic or lumbosacral area, J.Urol., 129:56 (1983).

11. G. P. Murphy, S. Beckley, M. F. Brady, and T. M. Chu, Treatment of newly diagnosed metastatic prostate cancer patients with chemotherapy agents in combination with hormones vs. hormones alone, Cancer, 51:1264 (1983).

12. N. H. Slack, and G. P. Murphy, A decade of experience with chemotherapy for prostate cancer, Urology, 22:1 (1983).

ADVANCES IN DIAGNOSIS AND TREATMENT OF ADRENAL TUMORS

F. Micali, M. Porena, G. Vespasiani and G. Virgili

Department of Urology
University of Perugia
Italy

The diverse embryological derivation, the multiple secretory activities of both the cortex and medulla, the different biological behavior of benign or malignant forms and lastly the primitive or metastatic origin make classification of adrenal tumors extremely complex. The most difficult diagnostic and surgical problems are due to the organ's location which makes a physical approach and normal instrumental or radiological exploration difficult.

Since embryogenetic and biological characteristics concern both the endocrine activity and type of spread they severely condition diagnostic and therapeutic approaches. A brief review of the main adrenal oncotypes (Table 1) and a personal clinical orientation for the most important of these follows.

Tumors metastatic to the adrenal are not discussed although they account for a large percentage of cases. Generally the biggest diagnostic-therapeutic problems derive from their primary source. Primary tumors may be cortical (hyperplasia, adenoma, carcinoma) or medullary (sympathoblastoma, ganglioneuroblastoma, ganglioneuroma, pheocromocytoma and undifferentiated sympathogonioma) in origin.

In clinical practice there is a big difference between endocrine active cortical and medullary forms and non-functional forms. (Table 2). For the first group the characteristic symptomatology and existing laboratory reports lead one to suspect their presence immediately, favoring diagnosis and leaving the definition of site, dimensions, entity and evolutive stage to the instrumental examination.

In non-functional forms the clinical picture is always non specific and connected with localizing symptoms (compression, dis-

Table 1. The Main Adrenal Oncotypes

Metastatic	Tumors	Primary Tumors
Breast	27.3%	Cortex
Lung	12.2%	Hyperplasia
Lymphoma	14.5%	Adenoma
Melanoma	7.3%	Carcinoma
Leukemia	7.3%	Medulla
Ovary	5.4%	Chromaffins-Pheochromocytoma
Rectum	3.6%	Sympathoblastoma
		Sympathetics Ganglioneuroblastoma
		Ganglioneuroma
		Undifferentiated: Sympathogonioma
		Cysts
		Amyloidosis
		Myelolipoma
		Rare Tumors (Teratoma, hemangioma,
		hamartoma)

placement and destruction of adjacent structures). Diagnostic examinations must therefore ascertain both the existence of the tumor and
its anatomical evolutive characteristics.

CORTISOL PRODUCING ADRENAL TUMORS

Hyperplasia, adenoma and carcinoma manifest clinically by pathologically translating the physiological effects of cortisol
(Cushing's syndrome) or aldosterone (Conn's syndrome). In the first
case the alterations are mainly metabolic (protein-catabolic) or
neoglycogenetic, thus making diagnosis easy. In the second case the
symptoms are hypertension, hypokaliuric syndrome with polyuria,
asthenia and sometimes paresis.

When Cushing's syndrome is suspected a series of laboratory
tests are necessary to diagnose the nature and etiology. The etiology may be suggested by suppression with dexamethasone, stimulation
with ACTH and the metyrapone tests. Radiologic, isotopic and instrumental examinations are carried out to confirm the suspicion.

For primary hyperaldosteronism the discriminating tests are
increased aldosterone in the urine, lack of secretory suppression
following measures to increase the plasma volume such as intake of
200 mEq sodium per day for 5 days; administration of 20mg desoxycor-
ticosterone acetate per day for 3 days and determination of renin
activity and/or plasma levels of renin and angiotensin since the

Table 2. Classification of Functional Tumors and Non-Functional Tumors of the Adrenals

	Functional Tumors	Non Functional Tumors
Cortex	Hyperplasia Adenoma Carcinoma	Adenoma Carcinoma
Medulla	Pheochromocytoma	Sympathogonioma Sympathoblastoma Ganglioneuroblastoma Ganglioneuroma Cysts-amyloidosis Myelolipoma-others

angiotensin-renin system in primary hyperaldosteronism is suppressed and cannot be stimulated (Lasix stimulation test). Radiology, histology and instrumental examinations, diagnosing by images must be done to locate the site of this tumor.

Diagnosing adrenal tumors by imaging is the common test for all cortical adrenal tumors (functional and non-functional) and non-secreting medullary adrenal tumors but the pheochromocytoma, because of the special syndromes it often causes cannot be diagnosed by these methods. These tests can also be used to identify and localize adrenal cysts, teratomas and other rarely encountered tumors.

Differential diagnosis is necessary for extra-adrenal tumors and especially expanding renal processes affecting the upper pole, accessory spleens tumors and cysts, of the tail of the tail of the pancreas, omental masses, adenopathies and pedunculated liver cysts. Adrenal cysts are most frequent in the adult population whereas neuroblastoma is characteristic of young people.

The first examination of a patient with a suspected suprarenal mass or clinical signs and symptoms of adrenal dysfunction should be done using the least invasive methods possible, for example, plain abdominal film which will show up any calcification in the pancreas area and displacement of the viscera (gastric air bubble, bowel) or IVP (Figure 1) to interpret anatomical changes in the kidney, displaced and compressed by the adrenal mass.

Since these can only give indirect signs of the presence of tumors, angiography, istope scan, ultrasonography and CT scan must follow since they are the only methods capable of giving direct morphologcial images of the adrenals.

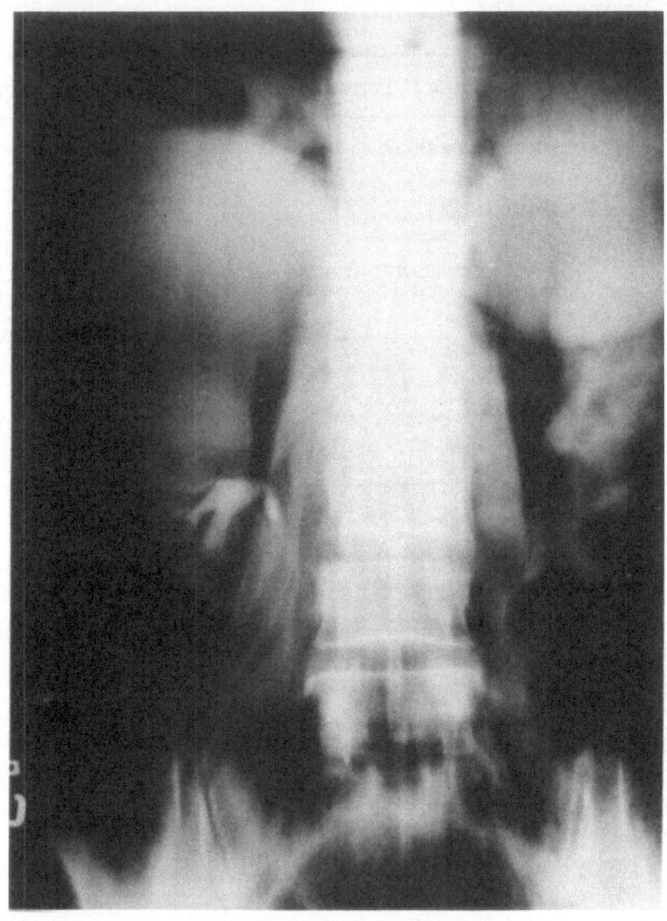

Fig. 1. Compression of kidneys by bilateral pheochromocytoma.

Ultrasonography. This can identify normal adrenals in 85% of
cases and can show tumors of about 2 cm diameter. It is the least
invasive, least costly and best tolerated examination and the only
one able to give sagittal and transverse images. It can also indicate whether the tumor is cystic or solid and can generally distinguish whether the mass is renal or adrenal in origin, except in the
case of voluminous tumors where the demarcation between the kidney
and the adrenal gland is no longer visible. These indications are a
guide for diagnosing the nature of the tumor; cystic forms are generally benign whereas solid forms are mostly adrenal metastases or
cortical carcinomas and in the pediatric population usually neuro-
blastomas.

Abdominal CT scan. This can be superimposed on the ultrasono-
gram. The advantages are a better definition of the organs, modifi-

cations in them, and their reciprocal relationships. The elevated analyses potential can recognise normal adrenal glands in 95% of cases. The adipose tissue surrounding the adrenals makes it easier to identify their limits and tumors of up to 1 cm diameter can be shown (Figure 2-3).

Isotope scan. This is usually carried out with radioactive cholesterol and allows a functional examination of the adrenals. Although this method can diagnose some affections such as aldosteronism or bilateral nodular hyperplasia and localize metastases of adrenal cortical carcinoma it is useful for diagnosing adrenal cortical carcinoma since the majority of these are non-functional.

Selective adrenal arteriography and venography. This may be necessary in a few cases, but it is invasive, catheterization of the relevant arteries and veins is not easy and there is the risk of complications.

PHEOCHROMOCYTOMAS

The diagnosis of a pheochromocytoma is done in two distinct stages. The first is to ascertain the existence of the tumor. Biochemical tests are used since they are least invasive and manage to diagnose with certainty almost all positive cases. Since there is a risk of false results from a single test they should be done in series.

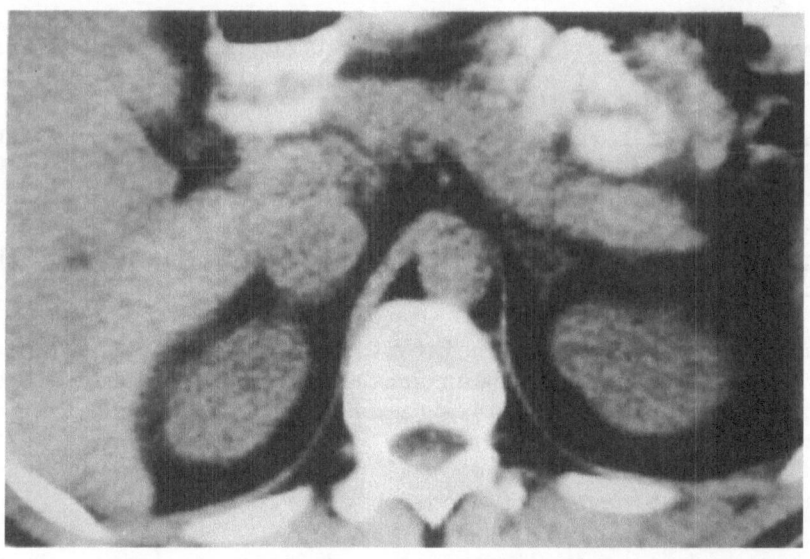

Fig. 2. Hyperplasia of right adrenal.

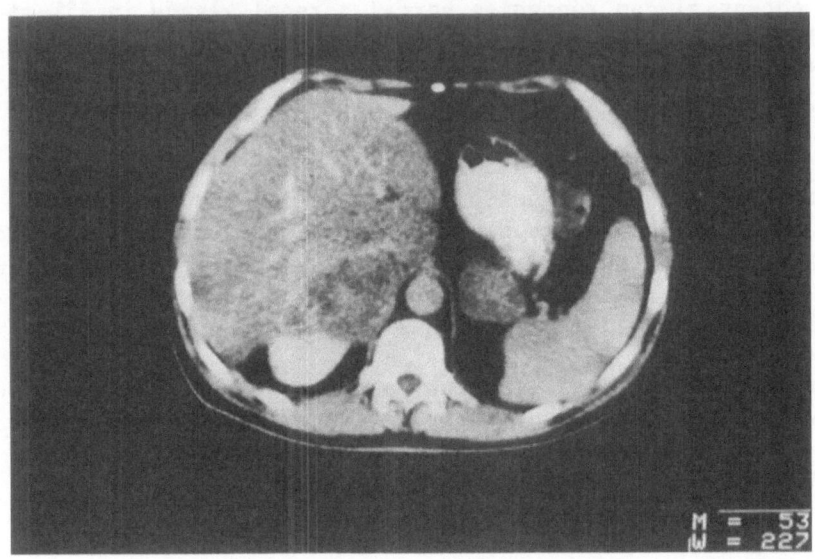

Fig. 3. Metastatic tumor of right adrenal.

     Urinary determination of vanylmandelic acid, catecholamines and
metanephrine assure together a 99% reliability. Some contrast radiology media, such as methylglucamine, modify the amount of metanephrine while intake of vanilla or physical stress modifies the
vanyl mandelic acid level. The presence of these should be ascer-
tained and eliminated before testing.

     Serial estimations of catecholamines using the radio immunoassay
(RIA) method has recently been introduced and although the statistics
are not yet abundant it appears to be very reliable.

     Pharmacodynamic tests, especially the provocatory type are now
only used in dubious cases and when the symptoms are less evident
since they are not very reliable and carry a high risk. It should be
remembered that a positive glucagon test could, because of its action
only on the adrenal cells, indicate the situation as well as the
presence of a tumor.

     Suppressive pharmacological tests, the classic phentolamine and
the more modern clonidine, which are almost 100% reliable are more
frequently used now. In dubious cases associated biochemical and
pharmacological tests are indicated. RIA determination of serial
catecholamines after histamine or glucagon stimulation will exclude,
if negative, the presence of a pheochromocytoma.

     The second stage in diagnosing is defining the site. Previously
surgery was the only possible method but now excision can be carried

out after almost 100% accurate localization by various investigative techniques in association. The most important of these are the isotope scan, arteriography, CT scan, ultrasonography and selective venous sampling.

Retroperitoneal air insufflation, necessitating the use of numerous instruments, and retrograde phlebography may cause serious hemorrhage, increase the risk of provoking hypertensive crises and are not very effective in diagnosis. These tests and the intravenous pyelogram (IVP) are not very useful.

Arteriography. Arteriography is only 80% reliable, is less accurate than ultrasonography and CT scan and is more dangerous. It is best used to indicate the exact vascularization of the tumor prior to surgery. (Figure 4-5).

Ultrasonography. Though there is not yet great experience with ultrasonography and CT scan they seem to be the most useful for

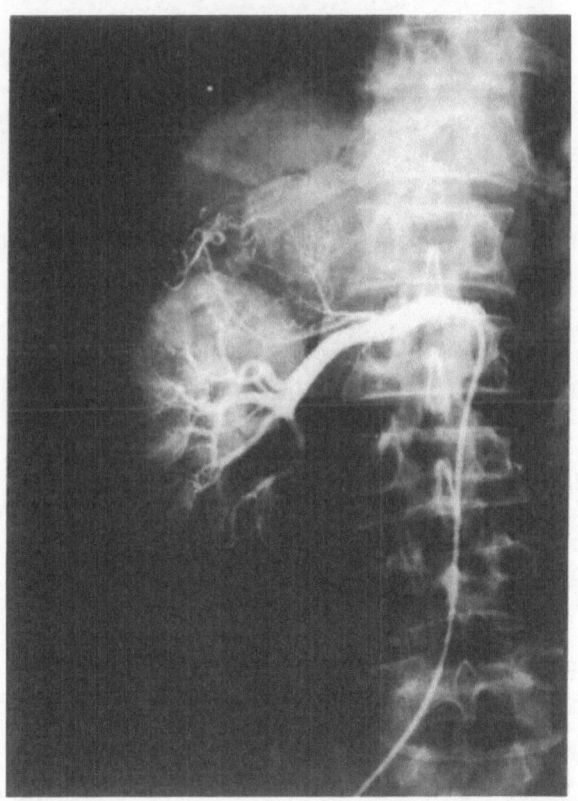

Fig. 4. Arteriography in two cases or right pheochromocytoma.

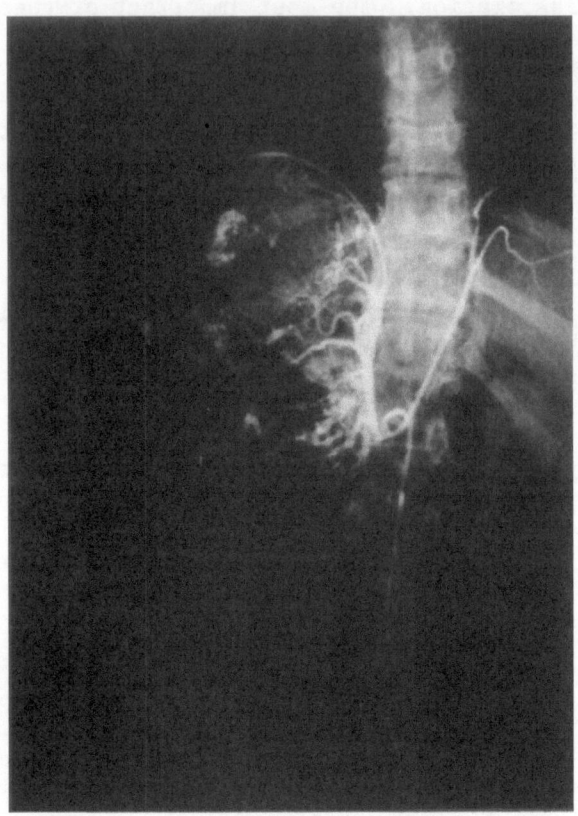

Fig. 5. Arteriography in two cases of right pheochromocytoma.

diagnosis since they can be superimposed, have high reliability and
carry very few risks. (Figure 6).

     Hypertensive crises following repeated passage of the U.S.
scannerprobe over the tumor area have been reported. CT scan has
also provoked crises but in these cases the paroxysms were triggered
by faulty administration of glucagon, given to reduce intestinal
peristalsis which disturbs the image.

     Isotope scan. The isotope scan has recently been reevaluated as
a diagnostic technique. The new isotope 131-HIBG gives very interesting results with positive findings.

     Selective sampling. At various levels of the vena cava deter-
mining the gradients of plasma catecholamines gives maximal reliability with low risk.

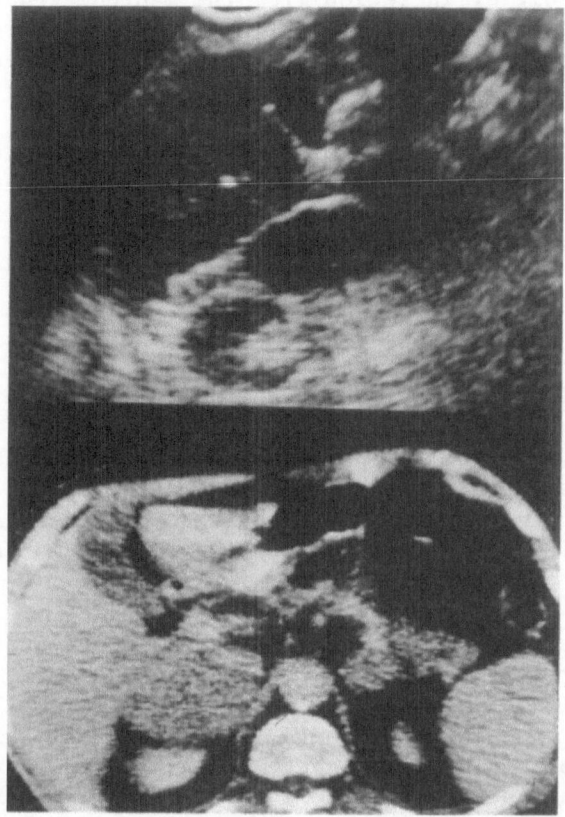

Fig. 6.   Echo and CT scan of right pheochromocytoma.

     In conclusion at present it seems opportune to combine selective
venous sampling and one of the morphological examinations, to obtain
the most complete diagnosis possible.

     Since any one method may prove physically, psychologically or
pharmacologically invasive the patient should be prepared beforehand
to make him less responsive to the secretion of catecholamine into
the system, by administering antiadrenergics. A sodium nitroprussate
infusion should be at hand to neutralise any crises.

SURGICAL THERAPY OF ADRENAL TUMORS

     Surgery is still the only radical treatment for almost all the
adrenal tumors. Their particular clinical and biological configuration, usually extremely aggressive (neuroblastoma, adrenocortical
carcinoma) or hypersecreting (pheochromocytoma) demands the most
extensive approach possible to effect both radical surgery and permit

some preliminary manouvres to avoid serious intrasurgical complications as a result of incretion from the manipulated tumor. In our
case diffusion or neoplastic implanting may happen through the lymphatic routes which reach the paraortic stations on the same side or
through the blood invading the lower right cava or left renal vein.

Problems connected with the gland's topography are more technical. Being an anatomic crossroads completely surrounded by vascular forms and the viscera of bowel and urinary tracts it is difficult
to reach surgically.

There are two groups of approaches to the adrenal: Economic and
Extended.

The economic approach is by a posterior, abdominal extraperitoneal or thoraco-abdominal incision. This approach is indicated in a
non-functioning, non tumoral pathology. The advantages are that it
is direct to the adrenal without opening the peritoneal serosa. The
disadvantage is the very limited exposure considering also the difficulty in getting an accurate diagnosis of the tumor's nature prior
to the intervention.

The extended approach is necessary for interventions on
malignant and/or functioning neoplasms (pheochromocytomas are always
potentially malignant) since both these situations call for preventive control of vascular pedicles. In the case of malignant neoplasms it is the only possible approach that avoids spreading neoplastic cells during removal of the tumor mass. With functioning
neoplasms, especially pheochromocytomas preventive control of the
vascular pedicle is of great value in preventing the hypertensive
crises which may result from the sudden massive emmission of catecholamine into the system during dissection.

The advisability of performing a lymphadenectomy (10% of cases
have lymphnodal metastases) or total excision of the surrounding
viscera infiltrated by the tumor (7-10% of cases) is the second
situation calling for the greatest possible access to the viscera and
abdominal nodes.

The third reason for choosing this approach is when there is
possible adherence (very frequent) to the large vessels, especially
the cava, necessitating extensive angioplastic intervention. Since
the incidence of bilateral and/or ectopic forms of pheochromocytoma
is fairly high (up to 20% of cases) complete control of both suprarenal regions, the abdomen, and the mediastinum, is indispensable for
radical surgery. From the data in the literature and our own experience it appears that the most common extended approach techniques are
abdominal transperitoneal and thoracoabdominal routes.

The median transperitonal approach. (Figure 7). The incision

is midline from the xiphisternum to the umbilicus with the
possibility of extending it upwards to the rib arch or downwards to
the pubis. Variants are a T or oblique costoumbilical incision.

Adrenalectomy, depending on whether one is operating on the
right or left gland presents different difficulties due to the different relationships with nearby organs.

On the right the costal arch conformation and liver volume
greatly limit the field available for freeing the retrocaval portion
of the gland, especially when there are neoplastic adhesions, and for
detaching the adrenal apex which lies behind the liver.

On the left of the stomach and pancreas are obstacles. In spite
of these difficulties the retroperitoneal approach to the adrenal
area is usually fairly easy. On the right hepatic displacement is
hardly ever necessary. We have sectioned the falciform and tri-
angular ligaments in those rare cases when a voluminous liver could
create an obstacle. The hepatic angle of the colon is easily moved
downwards and the duodenum moved to the midline by dividing the
parietal peritoneum along the right edge of the second portion.
(Figure 8). Detaching the cava is almost always necessary to free
the retrocaval face of the gland and to display the median adrenal or
principal vein which is always short.

On the left we think the best approach is through the lesser sac
after dividing the gastro-colic ligament. It should be remembered
that the lower half of the left adrenal is retropancreatic so that in
the case of tumors with possible adrenal-pancreatic adhesions it is

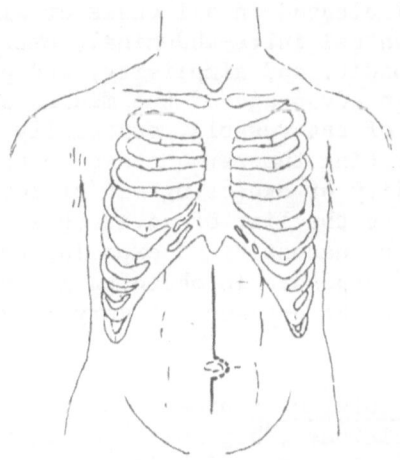

Fig. 7. The anteromedian transperitoneal approach.

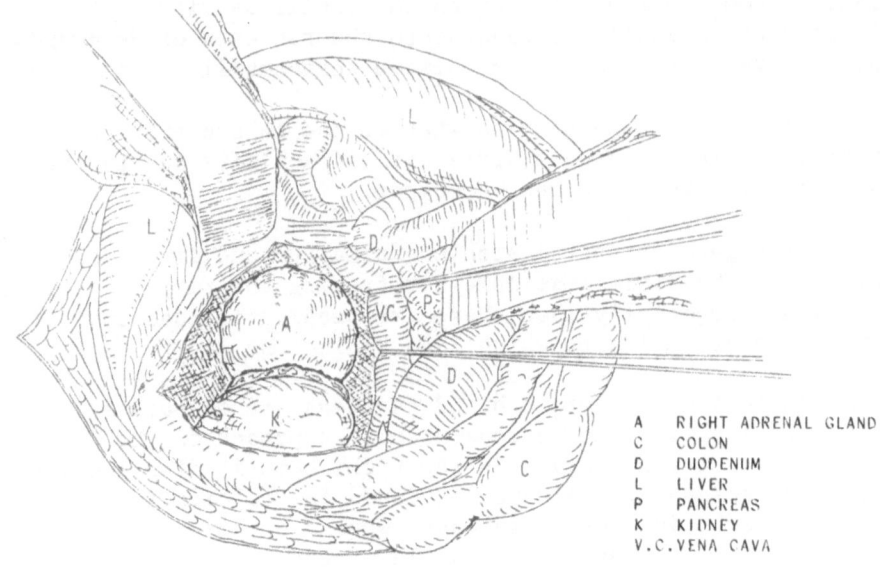

A   RIGHT ADRENAL GLAND
C   COLON
D   DUODENUM
L   LIVER
P   PANCREAS
K   KIDNEY
V.C. VENA CAVA

Fig. 8. Surgical procedure.

advisable to detach the pancreas to avoid rupturing it while exposing
the underlying adrenal. Cutting the lieno diaphragmatic ligament and
mobilizing the spleen and body and tail of the pancreas exposes the
adrenal area better and facilitates a direct approach to the pedicle
and left adrenal vein which is often double and joins veins of the
lower diaphragm or of the kidney. (Figure 9).

This approach is indicated in all cases of single or bilateral
adrenal tumors with eventual intra-abdominal, extra-adrenal exten-
sions. Besides the rapidity and simplicity, and preservation of the
pleural cavity, the main advantage of the median approach is that it
allows optimal control of the vessels, especially the lower cava, in
the presence of infiltrating tumors. Access to the entire peritoneal
cavity and the possibility of exploring and/or removing both adrenals
at the same time complete the list of advantages of this approach.
The disadvantages are the necessity of exposing the abdominal cavity
and a less than optimal exposure in obese patients with large chests
and short abdomens or in the presence of very large tumors which have
grown upwards.

In the transverse anterior approach (Figure 10) there are 2
subcostal confluent incisions ("big smile" or chevron). This approach is indicated only in cases of bilateral and or extra-adrenal
pheochromocytoma localized in the upper abdomen. It is relatively
simple and gives a good exposure of the peritoneal cavity.

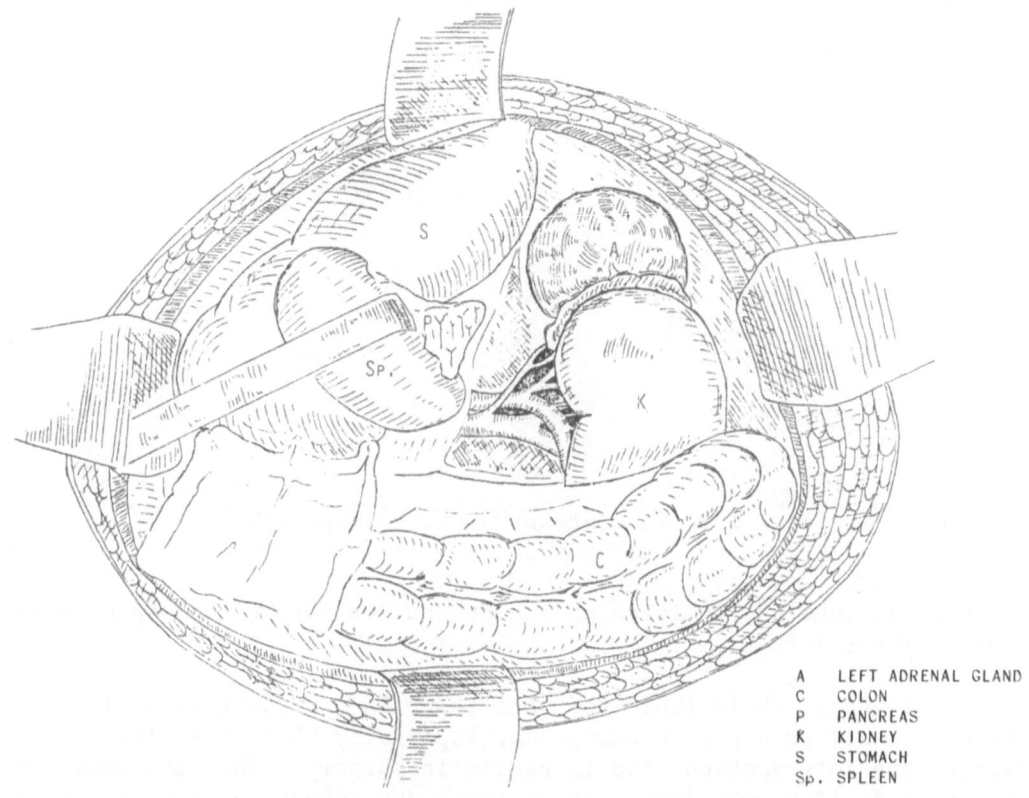

A   LEFT ADRENAL GLAND
C   COLON
P   PANCREAS
K   KIDNEY
S   STOMACH
Sp. SPLEEN

Fig. 9. Surgical procedure.

The greatest reputed disadvantage is that the operating field is
limited, especially in obese patients or when the tumor is very
large. We have however operated on a patient with bilateral pheochromocytoma where the left one weighed about 1-Kg much larger than
those usually encountered - and the right one was decidedly smaller.
Using this approach and without recourse to thoracic extension we
were able to operate, isolate and excise both tumors easily. (Figure
11). This approach is not useful when extensive lymphadenectomy is
necessary.

We consider the thoraco-abdominal-transpleural-transdiaphrag-
matic and transperitoneal route the best approach to the adrenal area
(Figure 12). It always allows an easy excision of tumors, even very
high ones, with a minimum of manipulation and permits an ample lymphadenectomy of peritoneal and mediastinal nodes. The incision we
practice most runs from the 9th intercostal space, starting from the
mid axillary line and reaching the epigastrium from where it continues downwards to the umbilicus or pubis as in a normal median or
paramedian abdominal incision. When there are large tumors growing

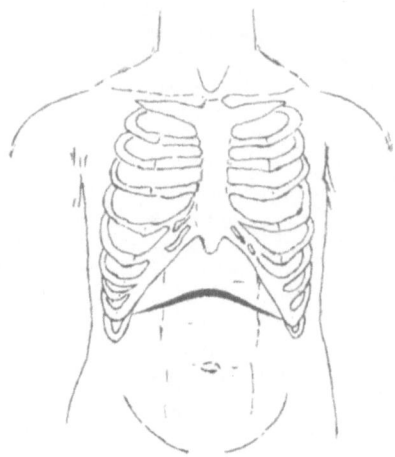

Fig. 10.   The anterotransverse approach.

upwards we make the incision at the 8th or 7th intercostal space with costal resection.

This approach is indicated when there is a unilateral malignant adrenal tumor or a pheochromocytoma, especially when it is large, for successive interventions and in pediatric surgery. The disadvantages are that it is a more difficult approach and reconstruction is rather laborious - though not difficult.

Pheochromocytoma: pre and intra surgical preparation and surgical management

For the reasons already mentioned an adequate pre-surgical medical treatment and precise intra-surgical management are necessary when dealing with pheochromocytoma. The patient must arrive at the operation with a very stable pressure and no alterations in cardiac rhythm which could lead to serious intra-surgical hazards. Beta-blockers are essential since they potentiate the hypotensive effect of alphablockers and normalize cardiac rhythm. By blocking tachy-cardiac reflexes they prevent the onset of more serious arrhythmias, for example ventricular extrasystoles.

Recently, the possibility of using a catecholamine synthesis inhibitor such as alpha-methyl-paratyrosine which blocks conversion of L-tyrosine to 3-4 dihydroxyphenylalanine roused great interest.

Pre-surgical administration of steroids is necessary when a foreseeable bilateral total adrenalectomy will probably lead to hypocortical crises in the immediate post-surgical period.

Fig. 11.   Operative findings: bilateral pheochromocytoma.

Transfusion is very important since patients with pheochromocytomas always have lower blood volumes than normal especially when
hypertension is continuous. In these cases removal of the tumor mass
is followed by a sudden drop in pressure which may cause very serious
shock due to the rapid disappearance of the arteriolar vasoconstric-
tion and acute disproportion between the vascular bed and the circulating blood volume. In serious cases transfusion therapy should
begin at least 48-72 hours prior to the intervention.

Few interventions require as much vigilance as the removal of a
pheochromocytoma and medical intra-surgical treatment is closely
linked to pre-surgical preparation.

Rapid continuous infusion of alpha and beta blockers is the best
means of counteracting a hypertensive emergency. Generally if the
pressure is maintained within acceptable levels there is no disturbance in cardiac rhythm.

To counteract serious hypotension rapid reintegrative steps
using blood and/or Plasma expanders should be taken and continued
till good perfusion pressure values are re-established.

Opinions differs regarding the surgical procedure where bilateral forms are present. Some justify total excision of both

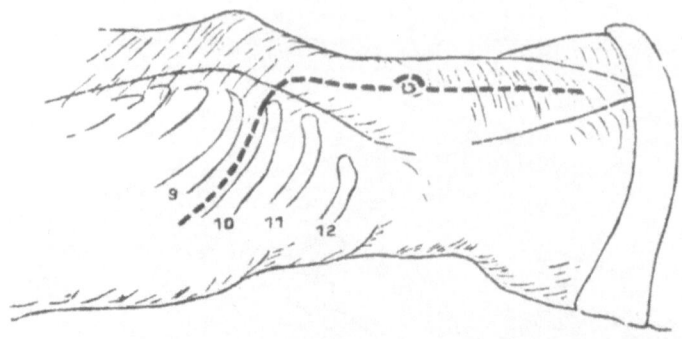

Fig. 12. The thoracophreno-abdominal approach.

glands because of the need for prophylaxis and the incapacity of the
remaining cortex to maintain a sufficient hormonal function.

However, we maintain that if one of the tumors is benign it is
worthwhile preserving the integrity of the adrenal cortex which is
carefully and accurately detached from the chromophin mass so that it
does not suffer anatomical or functional damage.

Lastly in those rare cases in which the pre-surgical diagnosis
has not localized the tumor or tumors precisely, it might be useful
to palpate the suspected area and as a result of triggering a hypertensive crisis ascertain the presence of hormone secreting tumors.

Routine contralateral biopsy and tumor section should be carried
out for all MEN, 2nd tumors and pediatric adrenal phechromocytomas
since the incidence of bilateral forms is extremely high.

Pheochromocytomas - Results of Treatment

Pre-surgical preparation, intra-surgical treatment and surgical
technique, by drastically reducing vascular and cardiac accidents
(hypertension, hypotension, arrhythmia) have revolutionized the
prognosis of these difficult tumors.

Welbourn who has used both the old and the new therapeutic
regimes reported an 18 to 25% decrease in operative mortality. The
only death in this last group was probably due to an overdose of beta
blocker.

In our series (Table 3) the sum of personal cases and those of
various medical and surgical Institutes of Perugia University, the
surgical mortality for all the 26 cases observed from 1965 is zero.

Table 3. Comparison of results from the literature with those
          obtained from treatment of 30 tumors in 26 patients at the
          University of Perugia. Of the 30 tumors, 23 were adrenal
          (right 15, left 8) and 7 extraadrenal. The 26 patients
          comprised 10 males and 16 females. The lesions were
          bilateral in one patient and multifocal in three.

	Results from Literature	Patients treated in Perugia
Death during		
Operation	2-18%[11,12,18,19]	None
Relapses	1.5-10%[12,18]	3 (11.1%)
Post Operative Blood Pressure		
Resolution of Paroxysms	100%[12]	100%
Normotension without treatment	8%	91%
Hypertension	25%[18,19]	9%
Survival up to 5 Years		
Overall	64-90%[12-19]	(from 2 months
Benign	96%[12]	to 5 years
Malignant	44%[12-18]	95%)

In the literature the incidence of relapses (by relapse we also
include subsequent manifestation of tumors in sites other than the
primary one) is from 5 to 10%. In our 26 cases there were 3 relapses
(11.1%).

     Functional results after excision of hormone secreting tumors
are not always brilliant since the disappearance of paroxysms in 100%
of cases is counterbalanced by a residual hypertension in about 25%
of the cases reported. Welbourn reported that only 8% of his
patients became normotensive in the absence of therapy. We find this
very strange as the incidence of normal pressure (diastolic pressure
values below 100) is 91% in our series.

     The 5 year survival is excellent for benign forms (up to 96%).
It is clearly less in malignant forms. In our series, with only one
malignant case, a 2 months to 5 year follow-up shows 95% total survival. Our only death was from uncontrollable hypertension resulting
from lung and lymph node metastases 2 years after an intervention for
malignant neoplasia.

Adrenal Autotransplantation

     The patient receiving bilateral adrenalectomy is obliged to have
substitution therapy for the rest of his or her life.

From 7 to 38% of these patients will develop a chromophobic tumor in the piturtary which will produce ACTH (Nelson's syndrome).

In order to avoid this, some authors have carried out an adrenal autotransplation. The graft is obtained from a portion of the adrenal gland which is implanted in a muscular pocket.

In all cases so far referred the medullary portion of the gland has always been dead, whereas the cortical secretion has allowed the patients to avoid substitution therapy in 57% of cases and to reduce the maintenance dose in all the others.

REFERENCES

1. R. Goyer, C. I. Simart, and J. Vaillant, Pseudo-kyste calcifié de la glande surrénale, J. Chir, (Paris), 88:301-310 (1964).
2. E. K. Lang, M. Nourse, D. McCallum, O. H. Mertz, and W. N. Wishard, The diagnosis of suprarenal masses by retroperitoneal air studies and arteriography, J.Urol., 94:220 (1965).
3. G. D. Foster, and A. Calif, Adrenal cysts, Arch.Surg., 92:131-143 (1966).
4. E. J. Cerise, and J. W. Hammon, Adrenal cyst, Surgery, 63:903-910 (1968).
5. J. G. Fries, and J. A. Chamberlin, Extra-adrenal pheocromocytoma: literature review and report of a cervical pheocromocytoma, Surgery, 63:268 (1968).
6. Ph.Tcherdakoff, G. Simoni, and P. Milliez, Caractères évolutifs du phéochromocytome, Nouv.Press Med., 3:861 (1974).
7. M. Melicow, One hundred cases of pheocromocytoma (107 tumors) at the Columbia-Presbyterian Medical Centre, (1926-1976) Cancer, 40:1987 (1977).
8. D. Merril, Modified Thoracoabdominal approach to the kidney and retroperitoneal tissue, J.Urol., 117:15 (1977).
9. D. G. Skinner, Considerations for managment of large retroperitoneal tumors: use of the modified thoracoabdominal approach, J.Urol., 117:605 (1977).
10. R. M. Tucker, and D. R. Labarthe, Frequency of surgical treatment for hypertension in adults at the Mayo Clinic from 1973 through 1975. Mayo Clin.Proc., 52:549 (1977).
11. W. R. Pitt, and V. F. Marshall, Radical retroperitoneal surgery: a 25-year experience, J.Urol., 37:119 (2978).
12. I. M. Modlin, J. R. Farndon, and R. B. Welbourn, Pheocromocytoma in 72 patients: clinical and diagnostic features, treatment and long term results, Br.J.Surg., 66:456 (1979).
13. E. T. Zawada, L. Dornfeld, M. Maxwell, L. Marks, and J. J. Kaufman, Pheocromocytoma and vasodepressor response to Saralasin, Ann.Int.Med., 91:494 (1979).
14. J. Chelly, J. Kieffer-Ph, D. Tournay, and J. Passelecq, Utilis-

ation de la nitroglycérine au cours de la chirurgie du phéocromocytome, Soc.Fran.Anest.Analg.Réanim., 37:325 (1980).

15. M. Echrari, M. J. McZoughlin, I. E. Rosen, E. L. St Louis, S. R. Wilson, D. J. Wise, and H. P. H. Yeung, The role of computed tomography in assessment of tumoral pathology of the adrenal glands, J.Com.Ass.Tom., 4/71 (1980).

16. G. W. Geelhoed, Cat scans and catecholamines, Surgery, 87:719 (1980).

17. J. Rubay, G. Mousset, and P. Guiot, Expérience chirurgicale des tumerus surrénaliennes, Acta.Chir.Belg., 79:415 (1980).

18. S. G. Sheps, J. A. Van Heerden, and P. F. Sheedy, Current approach to the diagnosis and management of pheochromocytoma, Minn.Med., 63:509 (1980).

19. R. B. Welbourn, Some aspects of adrenal surgery, Br.J.Surg., 67:723 (1980).

20. J. P. Yvergneaux, E. Bauwens, L. Van Outryve, W. Boels, and E. Yvergneaux, Volumineux cortico-surrénalome malin, réséque par thoracophreno-laparotomie droite, Acta Chir.Bel., 79:293 (1980).

21. E. L. Bravo, R. C. Tarazi, F. M. Fouad, D. G. Vidt, and R. W. Gifford, Clonidine-suppression test. A useful aid in the diagnosis of pheocromocytoma, New.Engl.J.Med., 305:623 (1981).

22. R. N. Brogden, R. C. Heel, T. M. Speight, and G. S. Avery, α-Methyl-p-Tyrosine: a review of its pharmacology and clinical use, Drugs, 21:81-89 (1981).

23. S. V. Moore, and J. S. Aldrete, Primary retroperitoneal sarcomas: the role of surgical treatment, Ann.J.Surg., 142:358 (1981).

24. J. C. Sisson, M. S. Frager, T. W. Valk, M. D. Gross, D. P. Swanson, D. M. Wieland, M. C. Tobes, W. H. Beierwaltes, and N. W. Thompson, Scintigraphic localization of pheocromocytoma, New.Engl.J.Med., 305:12 (1981).

25. C. Valori, and L. Corea, Il feocromocitoma ed il sistema nervoso adrenergico, Piccin Padova ed., (Italy), 1981.

26. F. Micali, and M. Porena, Le vie di accesso allargate al surrene. Atti Soc.It.Urol., Catania, 1982 (in press).

27. M. Porena, G. Vespasiani, and G. Virgili, Clinica e terapia del feocromocitoma, Atti Soc.Urol.Centro., Merid, Isole, Bari, (1982).

28. J. L. Siekavizza, M. E. Bernardino, and N. A. Samaan, Suprarenal mass and its differential diagnosis, Urology 18:625-633 (1981).

ADVANCES IN DIAGNOSIS AND TREATMENT OF

TUMORS OF RENAL PELVIS AND URETER

A. Jardin

Hopital La Pitié
Paris
France

Tumors of the renal pelvis and the ureter, like the other urothelial tumors, have not benefited from any dramatic changes in their
treatment over the last few years. However, the studies which have
been devoted to them have led to their earlier recognition and, in
the best conditions, to classification prior to treatment which may
allow a wider choice of therapy than in the past.

AETIOLOGY

The known aetiological factors remain unchanged (tobacco, phen-
acetin and benzene derivatives). Renal calculi are found in up to 8%
of cases [1,2]. Also, a few cases of tumor of the upper urinary
tract have been reported after cyclophosphamide therapy. An accurate
register of tumors of the upper tract needs to be kept in order to
evaluate the risk factors for a given population.

DIAGNOSIS

In 80% of cases diagnosis by simple means is straightforward[1].
In the remainder it can be assisted by three important tools:

(i)   explorative cytology: this examination, when positive, enables
        the diagnosis to be confirmed and can indicate the grade of
        the tumor. It is therefore essential and requires a very
        skilled cytologist if the results are to be reliable. Its
        value remains controversial, not only because of the false
        negatives, but more especially because of the false positives,
        seen particularly in patients with urate stones[3].

(ii)   computurized tomography (C.T): For the last five years, C.T.
        has been used to differentiate between a radiolucent stone and
        a tumor in the upper urinary tract. This technique is ac-
        curate, provided films are taken before and after injection of
        a contrast medium. Only tumors or stones smaller than 1 cm in
        diameter remain undetected. C.T. is also very valuable in
        determining the degree of infiltration of the wall and es-
        pecially of peri-ureteric or peri-pelvic extension of the
        tumor[4].

(iii)   Fibroscopy of the upper urinary tract: the fibroscopic examin-
        ation of the upper urinary tract is an important aid to diag-
        nosis in urology[5,6]. Prototype apparatus has been used for
        the last fifteen years in Japan, and the effectiveness and
        reliability of the Olympus model which is available to us
        suggests that it will soon be manufactured on a larger scale.
        The apparatus has a calibre of 9 charriere and is 90 cm long.
        It is easily introduced into the ureteric orifice with the aid
        of a cystoscope and is easily advanced. The positioning is
        greatly facilitated by radioscopic control. The mobility of
        the distal end facilitates penetration into the various caly-
        ces.

     The examination of the ureter and pelvicalyceal system may be
hampered by difficulties including loss of vision if the tip of the
apparatus touches the wall of the collecting system. The mobility of
the distal end avoids this difficulty. In addition the collecting
system can only be well explored when it is full of fluid. We prefer
to perform the investigation after furosemide rather than by injecting fluid into the ureter via another catheter introduced into the
ureteric orifice. This problem limits the usefulness of fibroscopy
in cases of non functioning kidney. Also, the mobility of the kidney
with respiration may also complicate the examination, as whether the
patient is awake or asleep apnoea is possible for only short periods
of time. Finally, identification of the calyceal group explored can
only be made with any precision with radioscopic control after opaci-
fying the cavities with a contrast product (by retrograde pyelography
performed at the beginning of the examination).

     We have carried out more than 100 endoscopic examinations with
this apparatus between November 1980 and May 1983 and feel that this
form of investigation will contribute to diagnosis and thus to prognosis. Several recent studies have convincingly demonstrated that
the prognosis is essentially a function of the stage and the grade:
with 100% 5 year survival for grade I tumors, 80% for grade II tumors
and 30% for grade III tumors(7,8].

     Cytology can aid the diagnosis of grade and already ureteropye-
loscopy sometimes allows tumors of the upper urinary tract to be
treated like bladder tumors, by enabling biopsies not only of the

tumor but also of other parts of the urinary epithelium. The prognostic value of such mapping has recently been emphasized[9]. Finally, the study of the presence or absence of ABO antigens on the
surface of tumors of the upper urinary tract has enabled certain
authors to make a reliable prognosis in a large percentage of
cases[10].

TREATMENT

     Over recent years, a more or less extended nephro-ureterectomy
has been recommended in the treatment of tumors of the renal pelvis
and the ureter. Some authors prefer the widest possible surgical
excision[1,2], but others[7,8] note that the survival of patients is
much more closely correlated with the state and grade of the disease
than with the type of operation performed and recommend consideration
of conservative surgery for tumors of the upper urinary tract, including resection of the relevant part of the pelvis or the ureter.
Uretero-neocystostomy after mobilization of the bladder may be appropriate for lesion of the lower ureters and auto-transplantation of
the kidney into the inguinal fossa has also been proposed[11]. This
enables cystoscopic surveillance of the remaining renal pelvis and
calyces.

     The indications for conservative surgery include low grade
tumors on cytology, well localized unilateral lesions, initially
bilateral tumors, renal failure at the time of discovery of the
tumor, and tumor in a solitary unique kidney.

     In summary, the extension of the operation has not proved its
effectiveness to the point where we should propose nephrectomy and
haemodialysis for a patient with a tumor of the upper urinary tract
in a solitary kidney. For this reason, we have treated 3 patients
with extensive pyelocalyceal tumors in a solitary kidney by heminephr-
rectomy. Only one of these 3 patients had to have the remaining half
kidney removed for recurrence (two years later) after which the
patient was treated by haemodialysis.

     Radiotherapy has also been proposed in this situation and
appears able to reduce the rate of local recurrence after surgical
excision[12,13].

     Chemotherapy has been given in some units but, to our knowledge,
no conclusive papers have been published.

CONCLUSION

     Diagnostic advances have enabled a better understanding of
tumors of the upper urinary tract. The debate between supporters of

wide excision and those of conservative surgery seems no longer
relevant as we can now consider tumors of the renal pelvis and ureter
in the same way as bladder tumors, in terms of the ease of their
diagnosis and surveillance and of what is required for their eradication.

REFERENCES

1. J. Cukier, H. Abouchaird, B. Pascal, J. P. Sueur, and E.
    Merimsky, Les tumeurs de la voie excrétrice, J.Urol.Nephrol.,
    87,2,57, (1981).
2. E. Mazeman, Les tumeurs de la voie excrétrice urinaire
    supérieure, Rapport au 66 éme Congrés de l' A.F.U., Paris
    (1972).
3. H. Rubbe, F. Hering, H. H. Dahm, and W. Lutzeyer, Value of
    exfoliative urinary cytology for differentiation between uric
    acid stone and tumor of upper urinary tract, Urology, 20:571
    (1982).
4. R. L. Baron, B. L. MacClennan, J. K. T. Lee, and T. L. Lawson,
    Computed tomography of transitional cell carcinoma of the
    renal pelvis and ureter, Radiology, 144, 1, 125 (1982).
5. I. M. Bush, B. Guinan, and J. Lanners, Ureterorenoscopy,
    Urol.clinics of North America, 131 (1982).
6. A. Jardin, Fibroscopy of the upper urinary tract, Film 16mm, Son
    optique, Congres de la S.I.U., San Francisco, (1982).
7. N. M. Heney, B. N. Nocks, J. J. Daly, P. M. Blitzer, and E. C.
    Parkhurst, Prognostic factors in carcinoma of the ureter,
    J.Urol., 125:632 (1981).
8. D. A. Wallace, D. M. Wallace, H. N. Whitfield, W. F. Hendry, and
    J. E. A. Wickham, The late results of conservative surgery
    for upper tract urothelial carcinomas, Brit.J.Urol., 53:537
    (1981).
9. P. A. Mahadevia, G. L. Karwa, and L. G. Koss, Mapping of uro-
    thelium in carcinomas of the renal pelvis and ureter. A
    report of nine cases, Cancer, 51:890 (1983).
10. M. B. Gruber, S. N. Becker, M. M. Warren, J. Rambo, and C. P.
    Davis, Specific red cell edherence test applied to tumors of
    ureter and renal pelvis, Urology, 19:361 (1982).
11. S. Petterson, H. Bryncer, L. E. Gelin, S. Johansson, and A. E.
    Nilson, Total ureterectomy, subtotal pyelectomy and calico
    vesicostomy for urothelial carcinoma of the upper urinary
    tract. Proceedings of XVIIéme Congrés de la S.I.U. Paris
    (1979).
12. T. B. Kjaer, T. M. Jorgensen, P. Frederiksen, and H. Genster,
    Transitional cell tumors of the upper urinary tract, Radical
    or conservative treatment, Scand.J.Urol Néphrol., 15:235-238
    (1981).
13. T. D. Nguyen, D. Franck, R. Bugat, J. Douchez, and P. F. Combes,
    Tumeurs primitives urétéro-pyélo-calicielles. Sem.des Hop.,
    18,58,661 (1982).

TUMORS OF THE EPIDIDYMIS

L. Denis,* R. Schoysman** and J. Pacco*

*Dept of Urology, A. Z. Middelheim
Antwerp-Belgium
** Free University of Brussels

Primary tumors of the epididymis are very rare although exten-
sions from testicular or paratesticular tumors regularly involve the
epididymis. The exact origin of some tumors involving the testicular
adnexal structures is sometimes difficult and/or impossible to establish. Clinical problems, mass diagnosis or pain in the epididymis
usually lead to the diagnosis of infection. Infection with or without subsequent obstruction is a common disorder in male infertility.
Benign and malignant tumors of the epididymis are rare. The bulk of
the true epididymal tumors is represented by benign tumors.

Epididymal Obstruction

The diagnosis of epididymal obstruction is relatively easy. One
finds a slightly enlarged caput epdidymis which is soft and rarely
painful. Sometimes however the upper part of the epididymis contains
a small hard nodule. Transscrotal ultrasound, best at 10 Mhz, provides a diagnosis in many instances (Figure 1), especially when cyst
formation or hydrocele reaction are present. Visual inspection is
indicated when ultrasound fails to provide an exact diagnosis. A
complete inspection of the cauda of the epididymis is diagnosed by
the swollen and dilated loops of the epididymal canal under the
delicate serosa. They are normally yellowish. Bluish zones or the
presence of white sclerotic tissue are signs of advanced destruction.
Dilation of the epididymal tubules is not uniform. It happens first
of all close to the obstruction and progressively spreads out higher
up. Microscopically the obstructed ducts become dilated and tortuous
with the early phagocytosis of sperm cells that collect there in
excessive numbers. As spermatogenesis continues[1] progressive
packing of the ducts occurs. Rupture of the ducts with sperm cell

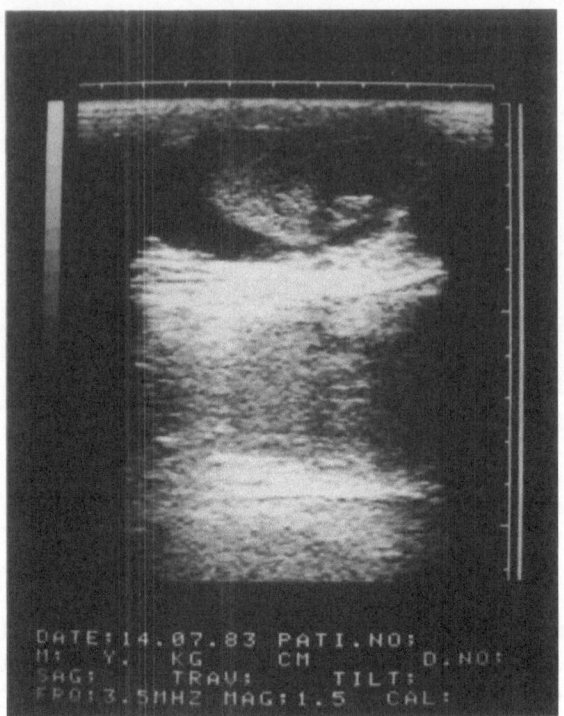

Fig. 1.   Sonogram of a cystic tumor of the epididymis with hydrocele
          reaction.

leaks may lead to sperm granuloma. This in turn may lead to a nodular formation which ultimately will be absorbed. Epididymal obstruction is almost exclusively caused by infection be it acute, subacute
or chronic. As pointed out the testes show no significant difference
with regard to the quality of spermatogenesis and prompt treatment
with antibiotics and anti-inflammatory drugs may prevent the
obstructive sequelae.

     Gonorrhoeal infections as well as tuberculosis are much less
frequent than in the past.

     Other causes of epididymal obstruction are iatrogenic trauma
e.g. hernial repair, dystrophic lesions and cysts, a common denominator in the epididymis. These cysts are either solitary or multiple. The round shape allows for easy diagnosis with an absolute
proof by ultrasound investigation. Aspiration or resection are
dictated by common sense.

     It should be remembered that epididymal cysts and/or testicular
hypoplasia has been found in 31.5 percent of 308 men exposed to
diethylstilbestrol in utero[2].

Benign Tumors

Most benign tumors of the epididymis are the so called adenomatoid tumors. A dozen synonyms or related terms are reported in the literature[3]. They are characteristic small, solid and painless tumors peculiar to the genital tract of both male and female. They can occur at any age but are most frequent between the third and the sixth decade. They are most frequently located in the cauda of the epididymis and occasionally in the testis or the cord[4]. The size varies from a couple of millimeters to a maximum of 5 cm.

Encapsulated with ovoid or spherical shape they are loosely fixed to the tissues and are easily dissected. The histogenesis was controversial but it is now accepted that these lesions are of Mullerian origin and are derived from mesenchyme, which retained the capacity to form epithelium[5]. They are mostly white on cut surface. The distinctive histology should be recognized to avoid confusion with adenocarcinoma. Treatment consists of a local excision when appropiately recognized but castration through an inguinal incision may be excused in doubt.

Among the other benign tumors, listed in Table 1, comes the leiomyoma as discrete nodules that resemble fibroma. They can be recognized after surgical excision by the microscopical characteristic appearance of spindle shaped cells. The papillary cystadenoma is usually found in the caput epdidymis and over forty cases have been reported in the literature. They can be uni- or bilateral. The microscopic appearance of papillary formations within tubules or cysts lined by cuboidal or cylindric epithelium is characteristic.

Table 1. Tumors of the Epididymis

Benign Tumors	Adenomatoid tumor
	Leiomyoma
	Papillary Cystadenoma
	Fibroma
	Lipoma
	Cholesteatoma
	Cyst
Malignant Tumors	Leiomyosarcoma
	Fibrosarcoma
	Rhabdomyosarcoma
	Epididymal carcinoma
	Teratoma
	Metastatic tumor

Table 2. Paratesticular Tumors

Benign Tumors	Adenomatoid tumor
	Papillary Cystadenoma
	Benign Cyst
	Dermoid Cyst
	Lipomatosis
	Lipoma
	Fibrolipoma
	Fibromyolipoma
	Mixed Lipoma
	Myxolipoma
	Fibroma
	Histio Cytofibroma
	Neurofibroma
	Myxofibroma
	Leiomyoma
	Myoma
	Myxoma
	Osteoma
	Lymphangioma
	Angioma
	Angiotheliomyoma
	Teratoma
Malignant Tumors	Liposarcoma
	Lipofibromyxosarcoma
	Lipoosteofibrosarcoma
	Mixed sarcoma
	Melano sarcoma
	Malign fibrous histiocytoma
	Fibrosarcoma
	Malignant neurinoma
	Mesothelioma
	Leiomyosarcoma
	Myosarcoma
	Rhabdomyosarcoma
	Myxosarcoma
	Lymphosarcoma
	Angio endothelioma
	Embryonal carcinoma
	Dysembryoma
	Reticulosarcoma
	Reticuloendotheliosarcoma
	Teratoma

There is a true relation with the von Hippel-Lindau disease[6]. This hereditary disease is diagnosed by the presence of cysts, adenoma or angioma in the brain, the retina, the abdominal organs and the epididymis. Local excision is the ideal treatment. Lipoma, fibroma and cholesteatoma though rare have been reported in the literature.

Malignant Tumors

Most malignant tumors of the epididymis are soft tissue tumors and most reported epithelial tumors are metastases from stomach or prostatic cancer[7]. They occur at all ages but rhabdomyosarcoma occurs predominantly in the first two decades of life. The tumor presents as a large intrascrotal mass and the exact origin is difficult to ascertain as in all malignant tumors of the epididymis. A complete list of reported paratesticular tumors is presented in Table 2. The classic source of misdiagnosis is epididymitis since these tumors can be painful. It is sound to explore all intrascrotal masses which give rise to a reasonable degree of suspicion of malignancy through an inguinal incision.

Rhabdomyosarcoma have a tan to red brown color in contrast to the smooth muscle tumors which are usually white. The diagnosis of fibrosarcoma depends on a degree of collagen formation. The epididymal carcinomas are extremely rare and confusion with the adenomatoid tumor may occur. Prognosis is poor for the primary as well as for the metastatic lesions. The final diagnosis depends on the histological appearance which may show substantial variation. A criterion of malignancy or an attempt to predict prognosis may depend on the mitotic activity. Most of these tumors should be classified as paratesticular tumors[8].

REFERENCES

1. R. Schoysman, Epididymal causes of male infertility, in: "Progress in Reproductive Biology" vol. 8., p.102-113 Karger, Basel (1981).
2. W. B. Gill, G. F. B. Schumacher, M. Bibbo, F. H. Straus and H. W. Schoenberg, Association of diethylstilbestrol exposure in utero with cryptorchidism, testicular hypoplasia and semen abnormalities, J.Urol., 122:36-38 (1979).
3. F. J. Dixon and R. A. Moore, Tumors of the male sex organs, Armed Forces Institute of Pathology, Washington, D. C. F32: p.127-142 (1952).
4. G. Teilum, Special Tumors of Ovary and Testis, Munksgaard, p.458-465 Copenhagen (1976).
5. N. J. Brown, Other tumors of epithelial type, in: "Pathology of the Testis", chapter 9. Pugh, Oxford: Blackwell, (1976).

6. C. J. Devine and J. F. Stecker, Urology in Practice, p.748-750
      Little, Brown and Company, Boston (1978).

7. F. Algaba, J. M. Santaularia and H. Villavecencio, Metastatic
      tumor of the epididymis and spermatic cord, Eur.Urol.,
      9:56-58 (1983).

8. G. P. Murphy and J. F. Gaeta, Campbell's Urology, 4th edn.
      p.1200-1212, W. B. Saunders Company, Philadelphia, London,
      Toronto (1979).

ADVANCES IN DIAGNOSIS AND TREATMENT

OF CANCER OF THE PENIS

G. Pizzocaro, F. Zanoni, L. Piva,
G. L. Riboldi, R. Salvioni and M. Pasi
National Tumor Institute
Tumori, Milan
Italy

INTRODUCTION

Cancer of the penis is a squamous cell carcinoma arising from the modified epitelium of the preputial sac. Sites of origin are: the glans penis, the sulcus and the inner part of the foreskin in decreasing order of frequency. Tumors outside the preputial sac are not typical penile cancers.

Accumulations of smegma and phimosis are conditions associated with cancer of the penis. The tumor could be prevented by early circumcision in phimotic children and good hygiene during life[1]. Viruses are associated with condylomata acuminata; also the association of penile carcinoma with a virus cytopathic effect has been described in our Institute[2].

Condylomata acuminata may give give to or coexist with carcinoma of the penis in 15% of cases; Buschke Löwenstein giant condyloma may be confused with verrucous carcinoma; erythroplasia of Queyrat and Bowen's disease are considered carcinomas in situ of the penis[3]. Cancer of the penis may be fungating or infiltrating, well or poorly differentiated. The natural history of the disease is characterized by local extension of the tumor, with destruction of the penis and bulky involvement of regional lymph nodes. Distant metastases are rare and occur late. Death is usually due to complications of locally advanced disease, infection and hemorrhage.

DIAGNOSIS AND STAGING

Diagnosis is established by biopsy. In phimotic patients, it may be necessary to split the foreskin to gain access to the tumor.

Distant metastases e.g., in supraclavicular nodes or the lung, are very uncommon at disease presentation, and even the inguinal nodes though frequently enlarged, are actually involved by the tumor in approximately one third of patients only[1].

Lymphangiography is not usually performed in carcinoma of the penis, even though it might be useful in improving the clinical diagnosis and in following patients after definitive treatment of the primary tumor[1,4]. However contrast medium examination of erectile tissue is useful in establishing invasion into corpora cavernosa or spongiosa, and it is particularly indicated in infiltrating tumors[5].

The TNM system (Table 1) is very appropriate for carcinoma of the penis. The main problem is the definition of regional lymph nodes. Classically, only the inguinal nodes on each side are considered as regional nodes for cancer of the penis, but Riveros and Gorostiaga[6] have reported cases with primary involvement of external iliac and obturator nodes even in patients with negative inguinal nodes. Currently, inguinal, external iliac and obturator nodes are considered as regional for cancer of the penis[1].

In our experience[7], 41 of 111 (37%) patients with cancer of the penis developed regional lymph node metastases, 25 (23%) at disease presentation, and 16 (14%) during follow-up. Occurrence of nodal metastases was seen in all category T4 tumors, in 53% in category T3, 30-35% in T1 and T2, and none in Tis.

TREATMENT

Cancer of the penis may be managed by surgery, radiotherapy, chemotherapy and combined modalities. Usually, the primary tumor is treated first, and management of regional lymph nodes follows. The problems are conservative treatment of the primary tumor, and optimal management of advanced disease (categories N3 and M1).

The surgical treatment of choice for the primary tumor is partial amputation 2 cm proximal to the neoplastic lesion[8]. Total amputation with perineal urethrostomy is mandatory in category T3 cancers infiltrating the corpora, and emasculation is the only surgical treatment of category T4. Local recurrences following radical surgery are rare (1.8), but these mutilating procedures are followed by severe psychological side effects.

Conservative surgery (circumcision) is possible only for small cancers ($\leq$ 3 cm) of the foreskin[9]. Cryosurgery for localized non infiltrating tumors of the sulcus and glans penis needs further investigation. Laser CO_2 surgery is promising.

Table 1.   TNM Pre-Treatment Clinical Classification of
Cancer of the Penis

T - Primary Tumor

Tis	Pre-invasive carcinoma (carcinoma in situ)
T0	No evidence of primary tumor
T1	Tumor 2 cm or less in its largest diameter strictly superficial or exophytic
T2	Tumor more than 2 cm but less than 5 cm in its largest dimension or tumor with minimal infiltration
T3	Tumor more than 5 cm in its largest dimension or tumor with deep extension, including the urethra
T4	Tumor infiltrating neighboring structures
TX	The minimum requirements to assess the primary tumor cannot be met.

N - Regional Lymph Nodes

N0	No evidence of regional lymph node involvement
N1	Evidence of involvement of movable unilateral regional lymph nodes
N2	Evidence of involvement of movable bilateral regional lymph nodes
N3	Evidence of involvement of fixed regional lymph nodes
NX	The minimum requirements to assess the regional lymph nodes cannot be met.

M - Distant Metastases

M0	No evidence of distant metastases
N1	Evidence of distant metastases
MX	The minimum requirements to assess the presence of distant metastases cannot be met.

In recent years, Iridium 192 interstitial radiotherapy has proved capable of curing 80% of categories T1 and T2 penile cancers[10]. Also the combination of external irradiation and bleomycin is promising[11]. In our experience, conservative management of the primary tumor was performed only in low T categories (Tis, T1 and T2) without nodal involvement (N0). We treated 32 cases conservatively; circumcision alone cured 4 of 5 patients with cancer of the foreskin and radiotherapy cured 14 of 27 (52%) tumors of the sulcus and glans penis. Nine of 14 patients who had relapse (64%) were cured by further therapy, usually radical surgery. All of the patients who did poorly developed lymph node metastases.

In the management of regional lymph nodes, surgery is superior to radiotherapy; however surgical complications (skin flap necrosis, lymph leakage, leg edema, etc.) are not uncommon[1]. As lymph node metastases are found in approximately 50% of patients with enlarged nodes and only occasionally in those who are clinically N0, it is generally accepted that one should carry out lymph node dissection only in patients with clinical N1 and N2 disease[7]. The most affected side is operated upon first, and contralateral dissection is carried out only in the presence of histologically confirmed lymph node metastases. Dissection usually extends from the groin to the external iliac and obturator nodes[6,7]. Clinical category N0 patients are usually followed at regular intervals and lymph node dissection is carried out on occurrence of enlarged regional nodes. In our experience, the 5 year survival in the 25 patients who presented with proven lymph node metastases was 36%, versus 20% in 16 patients who developed metastases during follow-up. All metastases occurred within 3 years following treatment of the primary tumor, and this is an indication of the need for a very careful follow-up during this period of time.

Chemotherapy is promising in cancer of the penis, but limited experience has been collected and no definite indication can be drawn. Bleomycin[12], methotrexate[13] and cisplatin[14] as single agents induced significant responses in 30-50% of patients with advanced disease. Combination chemotherapy is more promising. We achieved 2 complete (CR) and 3 partial (PR) responses in 5 patients with locally advanced or metastatic disease using the sequential combination of vincristine (1 mg i.v.), bleomycin (15 mg i.m. 6 and 24 hours following vincristine) and methotrexate (20 mg/m2 orally at 48 hours). This regimen was repeated weekly[15]. Responses lasted from 9 to 40+ months. New combinations including cisplatinum are being investigated, but the rarity of the disease makes evaluation of results very difficult. We believe that urologists should follow the evolution of chemotherapy for cancer of the oral cavity, because of the analogies with carcinoma of the penis.

Several challenging topics still require further thought and development including adjuvant chemotherapy in category pN1, pN2 cancer of the penis, pre-operative chemotherapy in patients with locally advanced disease (categories T4 and N3) and the use of chemotherapy in the conservative treatment of the primary tumor. Although the preliminary results are encouraging, the number of patients treated are as yet too limited to allow any conclusions to be drawn.

Summary

Advances in diagnosis and staging of carcinoma of the penis include a proper use of contrast medium examination of erectile tissue and lymphangiography.

Mutilating surgery is mandatory in T3 tumors infiltrating the corpora and in T4 lesions. There is increasing evidence that less extensive tumors can be successfully treated conservatively: Iridium 192 interstitial radiotherapy, along or in combination with external irradiation; external radiotherapy combined with bleomycin administration; and, more recently, modern combination chemotherapy followed by new conservative surgical procedures, including laser CO_2, are promising.

Lymph node dissection is mandatory in category N1 and N2 patients. The others can be followed at close intervals and operated on if nodal involvement occurs. The role of adjuvant chemotherapy in pN1, pN2 disease is under investigation.

Advanced cancer of the penis is a chemo responsive tumor, and schedules and strategies which are being developed for oral cancer, may be of relevance for this rare tumor because of its histological and biological similarity to these lesions.

REFERENCES

1. G. Pizzocaro and L Piva, Il carcinoma spinocellulare del pene, Argomenti di Oncologia, 1:371 (1980).
2. L. Piva, G. Pizzocaro, B. Stefanon, and G. De Palo, Carcinoma of the penis with virus cytopathic effect, Proc.2nd Internat. Meeting Gynec.Oncol., 225 (1982).
3. F. K. Mostofi and E. B. Price, Jr., Tumors of the male genital system, in: "Atlas of Tumor Pathology", Fasc. 8, AFIP, Washington D.C. (1973).
4. J. R. Cabanas, An approach for the treatment of penile carcinoma, Cancer, 39:456 (1977).
5. P. Edling, Contrast medium examination of erectile tissue of the penis, Urol.Internat., 18:293 (1964).
6. M. Riveros and R. Gorostiaga, Cancer of the penis, Arch.Surg., 85:377 (1962).
7. F. Zanoni and G. Pizzocaro, Linfadenectomia nel trattamento del carcinoma spinocellulare del pene, in: "I Tumori Urologici", U. Veronesi, E. Lasio, and G. Pizzocaro, eds., CEA, Milano, p.473 (1982).
8. W. J. Staubitz, L. H. Lent, and O. J. Oberkincher, Carcinoma of the penis, Cancer, 8:371 (1955).
9. R. Ekstrom and F. Edsmyr, Cancer of the penis. A clinical study of 229 cases, Acta Chir.Scand., 115:25 (1958).
10. P. Fortier, Iridium 192 interstitial radiotherapy of carcinoma of penis. Report of 87 cases, Ann.Dermat.Venerol., 106:465 (1979).
11. F. Edsmyr, Combined treatment with bleomycin in penile carcinoma, in: "Bladder Tumors and other Topics in Urological Oncology", M. Pavone, P. H. Smith, and F. Edsmyr, eds., Plenum Press, New York, p.503 (1980).

12. T. Ichikawa, Chemotherapy of penis carcinoma, in: Tumors of the Male Genital System", E. Grundmann and W. Vahlensieck, eds., Springer Verlag, Berlin and New York, p.140 (1977).

13. R. B. Sclaroff and A. Yagoda, Methotrexate in the treatment of penile carcinoma, Cancer, 45:114 (1980).

14. R. B. Sclaroff and A. Yagoda, Cis-diamminodichloroplatinum in the treatment of penile carcinoma, Cancer, 44:1563 (1979).

15. G. Pizzocaro, L. Piva, and R. Salvioni, Trattamento dei casi avanzati di carcinoma spinocellulare del pene, in: "I Tumori Urologici", U. Veronesi, E. Lasio, and G. Pizzocaro, eds., CEA, Milano, p.483 (1982).

RESULTS OF TREATMENT OF CARCINOMA OF THE PENIS

J. P. Blandy, R. T. D. Oliver,
H. F. Hope-Stone, and M. El-Demiry

The London Hospital
London, E1

Five years ago we reported the preliminary results of a method
of treatment of early carcinoma of the penis with a radioactive 192
Iridium mould. These results were so encouraging that we came to use
this as the method of choice, whenever it was possible to fit the
penis with a suitable mould, and when the patient could cooperate
fully in the technique of treatment[1]. The details of the method
have been fully described by Hope-Stone[2]. Since our early report,
we have followed up the original cases for a longer period of time
during which several other patients have been added. This longer
follow-up has made us temper our initial enthusiasm, albeit only in
certain minor particulars. These results are to be reported more
fully elsewhere and it is our intention here only to discuss some of
the more important findings.

THE SERIES

It is only possible to apply the radioactive 192 Iridium mould
technique when the tumor of the penis is not so bulky that an acrylic
cylinder cannot easily be fitted over it. Hence it is only applicable to tumors confined to the prepuce or glans penis - Stage I[3]
and only rarely when the tumor has invaded the copora cavernosa -
Stage II.

Between 1964 and 1983 59 patients were referred to the London
Hospital in Stages I and II. Of these, only 23 patients in Stage I
and 4 in Stage II were considered suitable for treatment with an
Iridium mould.

CLINICAL RESULTS

Stage I. The crude 5 year survival rate was 90.3% in this group, but it should be noted that there were 5 local treatment failures, all of them within the first 3 years of treatment. In two the disease recurred in the penis, but was easily cured by partial amputation, with prolonged survival. In three others, however, the disease recurred in the inguinal nodes: of these men only one survived after radical ablative surgery.

Stage II. Of the 4 patients in whom there was evidence of invasion of the shaft of the penis, but where it was felt that an Iridium mould could be tried as an initial therapeutic measure, three did survive for 5 years, but in only one could it be said that the disease had been cured by this form of treatment. In one patient a local recurrence required partial amputation; another underwent block dissection of the inguinal nodes for recurrence there, and the fourth died of widespread metastases without local recurrence.

Histology

As in other series, there was a close relationship between histological findings and outcome: in the entire group well differentiated tumors had a 91% crude survival whereas only 51% of undifferentiated cancers survived 5 years. Tumors that showed a "solid" pattern of growth had a 96% 5 year crude survival compared with only 73% of those with the "cord" pattern of growth[4].

SUMMARY

Although the results from the use of 192 Iridium continue to be just as good as those of partial amputation or other forms of radiotherapy, this more critical and longer-term review makes us less sanguine about the results of any treatment for what may seem at first to be localized disease. Any patient who presents with a carcinoma of the penis, no matter how well-localized, may develop inguinal and other metastases, especially when the tumor is undifferentiated. The excellent cosmetic and functional results of the Iridium mould technique must not beguile us. Carcinoma of the penis can be a lethal disease, and it demands the utmost vigilance, even when it seems to be least dangerous.

REFERENCES

1.   J. C. Salvierra, H. F. Hope-stone, A. M. I. Paris, E. A. Molland, and J. P. Blandy, Conservative treatment of carcinoma of the penis, Brit.J.Urol., 51:32-37 (1979).

2.   H. F. Hope-Stone, Carcinoma of the penis, <u>Proc.Roy.Soc.Med.</u>,
        68:777-779 (1975).

3.   S. M. Jackson, The treatment of carcinoma of the penis, <u>Brit.</u>
        <u>J.Surg.</u>, 53:33-35 (1966).

4.   I. D. D. Frew, J. D. Jefferies, and J. Swinney, Carcinoma of
        the penis, <u>Brit.J.Urol.</u>, 39:398-404 (1967).

CARCINOMA OF THE SEMINAL VESICLES

P. H. Smith

Department of Urology
St James's University Hospital
Leeds, UK

No text book of Urology or of Pathology describes tumors of the seminal vesicles as common and many make no mention of their existence. In preparing this contribution I have been unable to find an example of such a tumor in the records of the Pathology Departments of either of the two main hospitals in Leeds and have had to rely upon a computerized (Medline) and personal search of the literature for the information presented.

Primary tumors of the seminal vesicle are exceedingly rare both in animals and in man. Benign and malignant tumors of connective tissue have been reported as have malignant tumors of the epithelium. To the best of my knowledge, however, no benign epithelial tumors have yet been recognized.

The seminal vesicle is more commonly involved by secondary infiltration of tumor from primaries in adjacent organs especially the prostate, the rectum and sometimes the bladder. Metastases may however be seen in the vesicles following peritoneal seeding of growths of the gastro-intestinal tract which subsequently invade the peritoneum of the retrovesical region to involve the seminal vesicle[1].

EMBRYOLOGY, HYPERPLASIA AND CYSTS

The embryology of the seminal vesicle is described in brief in standard anatomical texts and in detail in an excellent monograph by Nilsson[2]. Both the seminal vesicle, the vas and the common ejaculatory duct arise from the Wolffian duct. The proximal part of this duct is absorbed into the urogenital sinus and thus the origin of the tissue of the vas, seminal vesicles, posterior urethra, trigone and ureter is very similar. Although the prostate gland develops in this area it is of separate origin.

Abnormalities of development of the Wolffian duct are frequently associated with absence of the ipsilateral kidney, ureter and hemitrigone, less often with absence of the vas, rarely with absence of the ipsilateral testicle and sometimes with absence or cyst formation of the seminal vesicle.

Chesterman[3] reported that hyperplasia of the seminal vesicle may occur. In one case this was associated with enuresis and in a further two with retention of urine. There appears to be no recent literature on this particular topic.

Cysts of the seminal vesicle, however, are more widely recognized and several papers on this topic have appeared since 1965 [4-15]. Lawson and Macdougall[4] reported a case and drew attention to a further eight cases previously recorded. Fuselier and Peters[11] reported one case and reviewed 19 others. Das and Amar[14] presented one patient and reviewed a further 16 reported since 1976 in which unilateral renal agenesis and cyst of a seminal vesicle were found. The symptoms are usually those of difficulty in micturition, infection of the lower urinary tract, the passage of mucinous material in the urine, or ejaculatory disturbances including pain on ejaculation or orgasm without ejaculation. Beeby[9] states that there is free communication between the cyst, the vesicle and the vas and many authors comment on the value of a seminal vesiculo-gram in diagnosis.

Such lesions are usually excised using an abdominal approach; in at least one case[10] tumor has been reported within a cyst.

ANIMAL TUMORS

These appear to be almost as rare in animals as in man. Slye et al.,[16] carried out autopsies on 19,000 mice, approximately 50% of them male, largely because of their interest in testicular tumors of which they found 28. In addition they discovered one mouse with a spindle cell sarcoma of the seminal vesicle. They also mention a transplantable tumor of the seminal vesicle in the white rat, described by Flexner and Jobling in 1910. Rowlatt[18] has reported a papillary adenocarcinoma of the seminal vesicle in a 13 month old C57 mouse who died of a reticulum cell sarcoma and who also had a hepatoma. He also quotes Guérin who in 1954 reported a second case of sarcoma in the mouse.

In 1951 Bielschowski and Hall[19] investigated the influence of 2 acetylaminofluorene (A.A.F.) as a carcinogenic agent in parabiotic albino rats derived from wistar stock (in which a normal and a castrated rat were joined by their peritoneal cavities). The normal partner was then given A.A.F. three times weekly for 17-38 doses starting three to four weeks after parabiosis was established. Liver tumors were expected and occurred in the treated parabiotic animals but not in "single rats" or in the castrated partner.

The accessory sex organs of the normal partners of the parabiotic pair were larger than normal and in five of these 20 rats treated with A.A.F. adenocarcinomas of the seminal vesicles were found - four macroscopical and one microscopical. It was assumed that since these tumors were found only in the normal partners of the parabiotic pair - animals subjected to increased adrogen secretion and then only when A.A.F. had been given, that both factors contributed. Rowlatt[18], also draws attention to the possible importance of increased androgen secretion since both the seminal vesicles in his affected males were hypertrophic.

In a study involving Rattus (mastomys) natalensis, Hollander and Higginson[19] fed Purina Chow or Purina Chow mixed with untreated corn oil, heated beef lard, or N, N'2,7-fluorenylene-bisacetamide (2,7-F.A.A.) to groups of this African rodent to investigate the influence of these agents on tumors of the glandular stomach. They obtained careful post mortems on all animals and discovered that carcino-sarcoma of the seminal vesicle had developed in eight of 150 controls (5.3%) and in nine of 63 animals fed one of the experimental diets (14.3%), four of these tumors developing in the group fed with 2,7-F.A.

The possible importance of excessive androgen secretion and the ingestion of agents such as the fluorenamines may repay further investigation.

HUMAN TUMORS

Benign Tumors

Plaut and Standard[21] reported a patient with a cystomyoma and reviewed four patients diagnosed as having a cyst and a leiomyoma, cystomyoma and fibromyoma respectively, arising from the seminal vesicles and vas deferens. Other reports of benign tumors have come from Buck and Shaw[22] who described a leiomyoma, and Damjanov and Apic[23] who reported a cystadenoma and observed that these tumors are often present without symptoms though the patient with the leiomyoma diagnosed by Buck and Shaw[22] had complained of marked frequency of micturition by day. no benign tumors of epithelial origin have been reported.

Malignant Tumors of Connective Tissue

Lazarus[24] reported on one patient with a retrovesical sarcoma and in a careful review of the literature, felt that only two of the cases could be considered with certainty as being primary sarcoma of the seminal vesicles, one reported by Zahn in 1893 and the other by Junghanns in 1930. Of these the first was found at autopsy and the second presented with cerebral metastases. Since that time further cases have been reported by Tripathi and Dick[25] who found a patient with an asymptomatic leiomyosarcoma who was alive and well one year after its removal, by Couderc et al.,[26] whose patient with a malignant myoblastoma with granular cells presented with pain radiating from the anus to the penis, by Buck and Shaw[21] whose patient with a large fibrosarcoma presented with retention, preceded by discomfort on passing urine, by Jukasz and Kiss[27] whose patient with a liposarcoma was found only at autopsy in a patient with a known carcinoma of the prostate, and by Williamson[28] whose patient with a fibrosarcoma presented with perineal pain and survived a cysto-prostatectomy.

PRIMARY ADENOCARCINOMA

Although all of us recognize the rarity of this tumor it is almost impossible to be exact about the true number reported because of the difficulty of separating those tumors which arise primarily from the prostate but affect the vesicle from those which arise within the vesicle and spread beyond its capsule.

The literature has been reviewed in detail by Lazarus[24] Dalgaard and Giersten[29] and by Kindblom and Petterson[30]. Lazarus was prepared to accept only seven of the 20 cases reported at that time as being unequivocally of seminal vesicular origin and Dalgaard and Giersten drew attention to a further 12 cases of which they accepted 11, and added one of their own. They reconsidered the cases reviewed by Lazarus and felt that a further four were true tumors of the seminal vesicle. Thus, they came to the conclusion that the number of acceptable cases was not 19 (i.e. seven of 20 reviewed by Lazarus and 12 of 13 reviewed by them) but 23, hardly of importance but certainly of interest to those of an obsessional disposition. Kindblom and Petterson[31] added a further 12 cases from the litera-ture and one of their own to make a total of 36 reported cases.

A recent Medline computer search and further review has yielded five more cases of primary adenocarcinoma (27, 30-34) to bring the total to 41 reported cases.

Several authors draw attention to the problem of defining with certainty the number of reports which can be accepted as primary carcinoma, the major problem being that most tumors which involve

prostate and seminal vesicles are actually of prostatic origin[35]
Mostofi and Price[36] also recognize this difficulty, believing that
a diagnosis of primary carcinoma should be made only when the gross
and microscopic features show that the seminal vesicle is principally
involved.

CLINICAL FEATURES

The age range and the clinical features are varied and are
similar for tumors of connective tissue and of epithelium. The ages
of the patients with carcinoma are shown in Figure 1. The symptoms,
as one might expect, include those of obstruction to the lower
urinary tract, rectal discomfort or constipation and pain in the
lower abdomen, penis, perineum or rectum. In addition bladder
irritation haematuria and haemospermia may occur.

The tumor was discovered on routine rectal examination in two
patients and presented with symptoms of metastases in four. Table 1
shows the frequency with which each symptom occurred.

DIAGNOSIS

The majority of lesions can be recognized clinically by rectal
palpation. Once suspected, further investigation can follow. The
classical technique is that of seminal vesiculography (Figure 2) but
this is likely to do little more than confirm the findings of rectal
examination. The place of angiography has also been considered by
Goldstein and Wilson[33] who reported one patient with carcinoma of
the right seminal vesicle in whom an angiogram demonstrated puddling
of contrast medium within the pelvis.

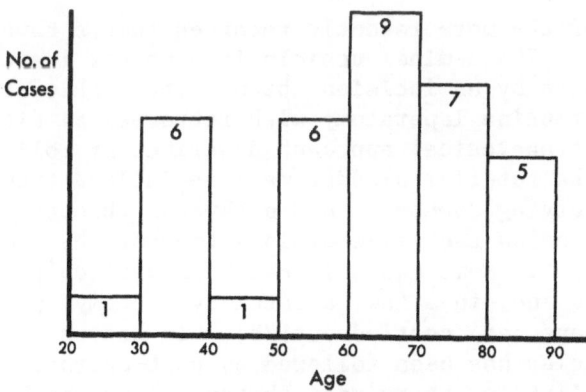

Fig. 1. Age distribution of patients with adenocarcinoma.

Table 1. Presenting symptoms in 25 of the 41 patients suffering from
          primary adenocarcinoma of the seminal vesicle. The litera-
          ture yields incomplete information on many of the reports
          in question and the general symptoms of loss of weight,
          lumbar and inguinal pain and metastases were seen more
          often in the cases reported before 1950.

Pain (lower abdomen 4, Lumbar 3, Pelvic 2, Perineal 1, Penile1, Inguinal 1, Testicular1)	11
Difficulty in Micturition	8
Haematuria	7
Frequency ± Pain on Micturition	6
Metastases	4
Nocturia	3
Retention	3
Loss of Weight	2
Routine Rectal Examination	2
Constipation	1
Watery Stool	1
Incotinence of Urine	1
Discoloured Urine	1

        Of the non-invasive techniques, transrectal ultrasonography[37]
has been shown to reveal information on the seminal vesicles in
patients about to undergo radical prostatectomy but the use of computerized tomography (CT) is now the most obvious technique[38] since
it is non-invasive and shows the vesicles more clearly than can
ultrasound (Figure 3).

TREATMENT

        The majority of the more recently recorded tumors have been
explored surgically. The seminal vesicle is somewhat inaccessible
and it may be explored by an incision through the ischiorectal fossa,
the perineum, by a routine laparatomy with retrovesical dissection[39] or by the transvesical approach described by Politano et
al.,[40] in which the anterior bladder wall is incised transversely,
the incision then curving downwards and backwards through the posterior bladder wall below the level of insertion of the ureters, i.e.
through the trigone. In some cases it has been difficult to dissect
the affected vesicle and, in a few, a total cystectomy, prostatectomy
and vesiculectomy have been carried out[28]. In others, local
removal of the vesicles has been followed by post-operative irradiation[24,32] whilst in others radiotherapy alone has been
used[29].

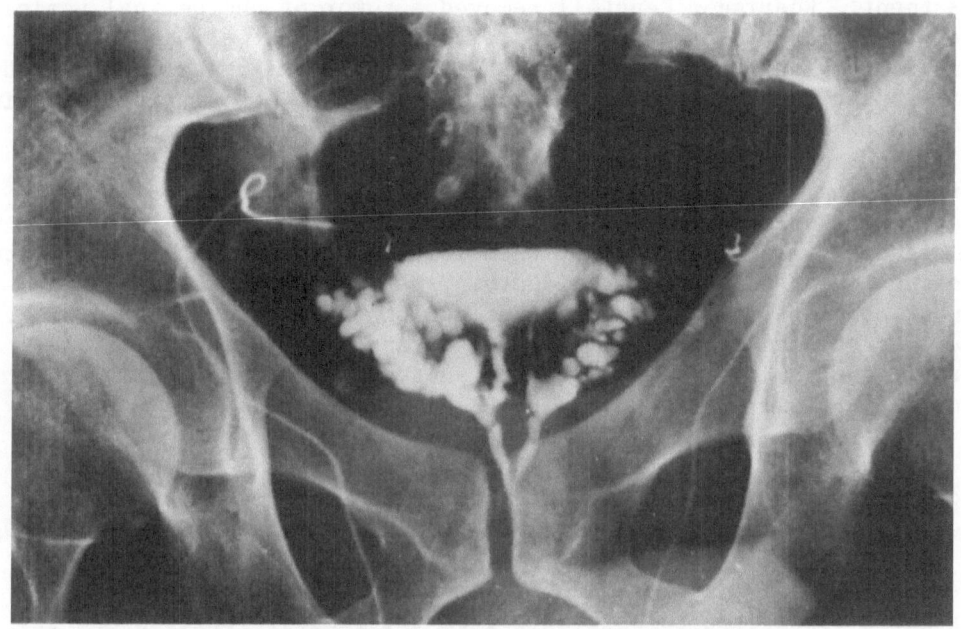

Fig. 2.   A normal seminal vesiculogram.

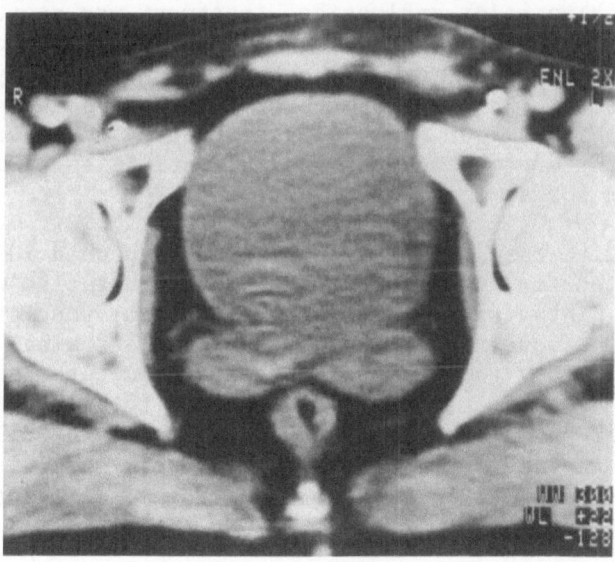

Fig. 3.   CT scan of pelvis to show seminal vesicle.

There are also at least three reports of the use of hormonal
treatment. Rodriguez Kees[41] removed a tumor of the seminal vesicle
6 cm in diameter noting that the iliac lymph nodes were positive.
Nine months later the patient had further symptoms of haematuria and
pain in the right thigh and was noted to have an osteolytic lesion in
the fourth dorsal vertebra. This was treated by radiotherapy with
calcification. Fourteen months after the original operation, bilateral neoplastic infiltration was seen in the lungs, the right
kidney was non-functioning and the inguinal nodes were palpably
enlarged. Following treatment with 100 mg Stilboestrol Diprorianate
and subsequent bilateral orchidectomy, the lung metastases showed
almost complete resolution and the inguinal nodes decreased in size.
The patient also started to gain weight and his IVP returned to
normal. He remained in good health subsequent to this, up to the
time the article was submitted for publication.

Kindblom and Petterson[30] reported a 72 year old patient in
whom a mass was demonstrated after removal of 50g of prostatic
adenoma at operation. The prostatic capsule was then incised and 50g
of greyish-brown gritty tissue was excised by blunt dissection from
the retrovesical space. Histological examination showed an adenocarcinoma considered to originate from the seminal vesicles and the
patient received 6,000 rads to the region of the prostate. Six
months later he was readmitted with increasing back pain when a bone
survey showed multiple osteolytic metastases in the ribs, vertebrae
and pelvis. Further radiotherapy (3,600 rads) to the painful areas
did not help and two months later when the patient was deteriorating
rapidly and taking major analgesics, treatment was started with
Polyestradiol Phosphate (Estradurin) 160 mg per month for the first
three months and then 80 mg per month together with Ethyniol
Oestradiol tablets (Etivex) 50 mg b.d. for 13 days and subsequently
50 mg daily. Within four weeks of this therapy the patient was free
of pain and started to resume his normal activities. His good health
was maintained for at least a further year.

In the third case, Williamson[28] reported on a 50 year old man
who was admitted for investigations of mild prostatism with some
discomfort on micturition. Following an initial unhelpful biopsy, he
was readmitted a month later having developed abdominal pain which
proved at laparotomy to be associated with a ruptured abscess of the
right seminal vesicle. This was excised sub-totally. When histology
showed adenocarcinoma, the patient was treated with Stilboestrol 1 mg
three times daily and six months later was well and gaining weight.
So far there appears to be no report on the use of cytotoxic chemotherapy in such patients.

PROGNOSIS

The prognosis in the small number of patients with benign
lesions seems to be good. The prognosis for those with carcinoma

Table 2. Results of "Curative" Therapy for Carcinoma of the Seminal
         Vesicle

Treatment	No.	Outcome
Vesiculectomy	4	Alive 2 years Died 4 months No follow-up No follow-up
Radiotherapy	1	Died 3 years of Uraemia*
Vesiculectomy & Radiotherapy	1	Well at 6 months
Cystoprostatectomy and Vesiculectomy	3	Well at 18 months Died 7 days No follow-up
Vesiculectomy and Hormone Therapy	2	Well at 21 months after operation and 14 months after start of hormone therapy Well at 6 months

* Given hormone therapy with benefit when metastases occurred.

or sarcoma is however much less satisfactory. Many of the patient
presented with advanced disease especially those in the early reports. Detailed follow-up is not available for many other patients
but some idea of survival following "curative" treatment for adenocarcinoma can be seen from Table 2. From the small amount of information available it would seem unlikely that radical surgery or
radiotherapy will prove to be the treatment of choice. Hormone
therapy should probably be used early in the course of this disease
especially in view of the diagnostic doubt which so often exists.

Acknowledgements

I should like to acknowledge my thanks to Miss A Reilly for her
help in the preparation of this paper, to Dr Philip Robinson of the
Department of Medical Imaging, St. James's University Hospital for
figure 3 and to the Department of Medical Illustration, St James's
University Hospital, for reproducing figures 1-3.

REFERENCES

1. P. A. Herbut in:"Histological Pathology," H. Kempton, ed.,
    London, p.1006 (1952).
2. S. Nilsson, The Human Seminal Vesicle - A morphogenetic and
    gross anatomic study with special regard to changes due to
    age and to prostate adenoma, Acta.Chir.Scandinav.Suppl., 296
    (1962).
3. J. T. Chesterman, Hyperplasia of seminal vesicle causing
    enuresis, Lancet 1:781 (1946).
4. L. J. Lawson, J. A. MacDougall, Multilocular cyst of the seminal
    vesicle, Brit.J.Urol., 37:440 (1965).
5. T. C. Sharma, P. S. Dorman, H. P. Dorman Bilateral Seminal
    Vesicular Cysts, J.Urol., 102:741 (1969).
6. R. T. Brooks, Cyst of the ejaculatory duct: Case report,
    J.Urol., 101:881 (1969).
7. S. K. Chatterjee and S. K. Sarkar, Retrovesical cysts in boys,
    J.Urol., 109:107 (1973).
8. A. J. Furtado, Three cases of cystic seminal vesicle associated
    with unilateral renal agenesis, Brit.J.Urol., 45:536 (1973).
9. D. I. Beeby, Seminal vesicle cyst associated with ipsilateral
    renal agenesis case report and review of literature, J.Urol.,
    112:120 (1974).
10. D. H. Bagley, N. Javadpour, F. G. Witebsky, and L. B. Thomas,
    Seminal vesicle cyst, Urology, 5:147-151 (1975).
11. H. A. Fuselier and D. H. Peters, Cyst of seminal vesicle with
    ipsilateral renal agenesis and ectopic ureter: Case report,
    J.Urol., 116:833 (1976).
12. J. B. Knudsen, B. Brun, and H. C. Emus, Familial renal agenesis
    and urogenital malformations, Seminal vesicle cyst and
    vaginal cyst with bicornuate uterus in siblings,
    Scand.J.Urol.Nephrol., 13:109 (1979).
13. P. Øgreid and K. Hatteland, Cyst of seminal vesicle associated
    with ipsilateral renal agenesis, Scand.J.Urol.Nephol., 13:113
    (1979).
14. S. Das and A. D. Amar, Ureteral ectopia into cystic seminal
    vesicle with ipsilateral renal dysgenesis and monorchia,
    J.Urol., 24:574 (1980).
15. R. E. Scully, J. J. Galdabini, and B. U. McNeely, Case records
    of the Massachusetts General Hospital, N.Eng.J.Med., 22:1246
    (1980).
16. M. Slye, H. F. Holmes, and H. G. Wells, Primary spontaneous
    tumours of the testicle and seminal vesicle in mice and other
    animals, J.Can.Res., 4:207 (1918).
17. O. Lyons, Primary carcinoma of the left seminal vesicle,
    J.Urol., 13:477 (1925).
18. C. Rowlatt, Carcinoma of the seminal vesicle in the mouse,
    Brit.J.Cancer, 20:526 (1966).
19. F. Bielschowsky and W. H. Hall, Carcinogenesis in parabiotic
    rats, Brit.J.Can., 5:106 (1951).

20. C. F. Hollander and J. Higginson, Spontaneous cancers in Praomys (Mastomys) Natalesnis, J.Nat.Can.Inst., 46:1343 (1971).

21. A. Plaut and S. Standard, Cystomyoma of seminal vesicle, Ann.Surg., 119:253 (1944).

22. A. C. Buck and R. E. Shaw, Primary tumours of the retrovesical region with special reference to messenchymal tumours of the seminal vesicles, Brit.J.Urol., 44:47 (1972).

23. I. Damjanov and R. Apić, Cystadenoma of seminal vesicles, J.Urol., 111:808 (1974).

24. J. A. Lazarus, Primary malignant tumours of the retrovesical region with special reference to malignant tumours of the seminal vesicles: Report of a case of retrovesical sarcoma, J.Urol., 55:190 (1946).

25. V. N. P. Tripathi and V. A. Dick, Primary sarcoma of the urogenital system in adults, J.Urol., 101:898 (1969).

26. M. H. Couderc and C. Chabert, Tumeur a cellules granuleuses des vesicules seminales: Myoblastome malin? Arch d'anat.Path., 19:313 (1970).

27. J. Juhasz and P. Kiss, A hitherto undescribed case of "collision" tumour: Liposarcoma, Int.Urol.Nephrol., 10:185 (1978).

28. R. C. N. Williamson, Seminal vesicle tumours, J.Roy.Soc.Med., 71:286 (1978).

29. J. B. Dalgaard and J. C. Giertsen, Primary carcinoma of the seminal vesicle, Acta.Path.Microbiol.Scand., 39:255 (1956).

30. L. G. Kindblom and G. Peterson, Primary carcinoma of the seminal vesicle, Acta.Path.Microbiol.Scand., 84:301 (1976).

31. O. Awadalla, A. C. Hunt and A. Miller, Primary carcinoma of the seminal vesicle, Brit.J.Urol., 40:574 (1968).

32. F. R. Banchieri and A. P. M. Cappa, Un cas de cancer de la vesicule seminale gauche, J.d'urol.et Nephrol., 74:890 (1968).

33. A. G. Goldstein and E. S. Wilson, Carcinoma of the seminal vesicles with particular reference to the angiographic appearances, Brit.J.Urol., 45:211 (1973).

34. J. Scheinar, F. Hruska and J. Dusek, Das samenblasen karzinom, Z.Urol.u.Nephrol., 73:415 (1980).

35. J. Zajicek, "Aspiration Biopsy Cytology Part 2," Cytology of Infradiaphragmatic organs, S.Karger (Basel), p.164 (1979).

36. F. K. Mostofi and E. B. Price, Tumours of the male genital system, Atlas of Tumor Pathology; second series fasc. 8 (Armed Forces Institute of Pathology) (1973).

37. M. I. Resnick, J. W. Willard, and W. H. Boyce, Transrectal ultrasonography in the evaluation of patients with prostatic carcinoma, J.Urol., 124:482 (1980).

38. S. O. Asbell, CT scanning in the male genital tract and in recurrent rectal carcinoma: A review, CT Vol. 5:244-255 (1981).

39. J. Boreau, Diagnosis and treatment of primary malignant tumours of the seminal vesicles, In: "Recent Results in Cancer

Research - Tumours of the Male Genital System," E. Grundman
and W. Vahlensieck, eds., Springer Verlag, Berlin,
Heidelberg, New York, p. 157-162 (1977).

40. V. A. Politano, R. W. Lankford, and R. Sysaeta, A transvesical
approach to total seminal vesiculectomy: A case report,
J.Urol., 113:385 (1975).

41. O. S. Rodriguez Kees, Clinical improvement following estrogenic
therapy in a case of primary adenocarcinoma of the seminal
vesicle, J.Urol., 91:665 (1964).

INDEX